Smart Innovation, Systems and Technologies

Volume 173

The Smart Innovation, Systems and Technologies book series encompasses the topics of knowledge, intelligence, innovation and sustainability. The aim of the series is to make available a platform for the publication of books on all aspects of single and multi-disciplinary research on these themes in order to make the latest results available in a readily-accessible form. Volumes on interdisciplinary research combining two or more of these areas is particularly sought.

The series covers systems and paradigms that employ knowledge and intelligence in a broad sense. Its scope is systems having embedded knowledge and intelligence, which may be applied to the solution of world problems in industry, the environment and the community. It also focusses on the knowledge-transfer methodologies and innovation strategies employed to make this happen effectively. The combination of intelligent systems tools and a broad range of applications introduces a need for a synergy of disciplines from science, technology, business and the humanities. The series will include conference proceedings, edited collections, monographs, handbooks, reference books, and other relevant types of book in areas of science and technology where smart systems and technologies can offer innovative solutions.

High quality content is an essential feature for all book proposals accepted for the series. It is expected that editors of all accepted volumes will ensure that contributions are subjected to an appropriate level of reviewing process and adhere to KES quality principles.

** **Indexing: The books of this series are submitted to ISI Proceedings, EI-Compendex, SCOPUS, Google Scholar and Springerlink** **

More information about this series at http://www.springer.com/series/8767

Lakhmi C. Jain · Margarita N. Favorskaya ·
Ilia S. Nikitin · Dmitry L. Reviznikov
Editors

Advances in Theory and Practice of Computational Mechanics

Proceedings of the 21st International Conference on Computational Mechanics and Modern Applied Software Systems

Springer

Editors
Lakhmi C. Jain
University of Technology Sydney
Sydney, Australia

Liverpool Hope University
Liverpool, UK

KES International
UK

Ilia S. Nikitin
Institute of Computer Aided Design
of the Russian Academy of Sciences
(ICAD RAS)
Moscow, Russia

Margarita N. Favorskaya
Department of Informatics and Computer
Techniques, Institute of Informatics
and Telecommunications
Reshetnev Siberian State University
of Science and Technology
Krasnoyarsk, Russia

Dmitry L. Reviznikov
Department of Numerical Mathematics
and Computer Programming
Moscow Aviation Institute (National
Research University)
Moscow, Russia

Federal Research Centre "Information
and Control" of the Russian Academy
of Sciences
Moscow, Russia

ISSN 2190-3018 ISSN 2190-3026 (electronic)
Smart Innovation, Systems and Technologies
ISBN 978-981-15-2599-5 ISBN 978-981-15-2600-8 (eBook)
https://doi.org/10.1007/978-981-15-2600-8

This Springer imprint is published by the registered company Springer Nature Singapore Pte Ltd.
The registered company address is: 152 Beach Road, #21-01/04 Gateway East, Singapore 189721, Singapore

Preface

The research book is a continuation of our previous books which are focused on the recent advances in computer vision methodologies and technical solutions using conventional and intelligent paradigms.

This book presents the selected papers reported at the Twenty-first International Conference on Computational Mechanics and Modern Applied Software Systems, CMMASS'2019, which was held during 24–31 May, 2019. The contributions include the modern numerical methods for solving problems of the continuum mechanics and numerical simulation of physical processes occurring in gas, fluid, and solid media. Also, methods of mathematical modeling of dynamic systems, optimization methods, and optimal control algorithms are considered. Part I "Computational Fluid Dynamics" involves Chaps. 2–6, Part II "Numerical Simulation of Multiphase Flows, Combustion and Detonation" contains Chaps. 7–12, Part III "Computational Solid Mechanics" includes Chaps. 13–19, and Part IV "Numerical Study of Dynamic Systems" consists of Chaps. 20–25.

We wish to express our gratitude to the authors and reviewers for their contributions. The assistance provided by Springer-Verlag is acknowledged.

Sydney, Australia — Lakhmi C. Jain
Krasnoyarsk, Russia — Margarita N. Favorskaya
Moscow, Russia — Ilia S. Nikitin
Moscow, Russia — Dmitry L. Reviznikov

Contents

About the Editors

Lakhmi C. Jain, Ph.D., M.E., B.E.(Hons) Fellow (Engineers Australia), is with the Faculty of Education, Science, Technology & Mathematics at the University of Canberra, Australia, and Bournemouth University, UK. Professor Jain founded the KES International for providing a professional community the opportunities for publications, knowledge exchange, cooperation, and teaming. Involving around 5,000 researchers drawn from universities and companies world-wide, KES facilitates international cooperation and generates synergy in teaching and research. KES regularly provides networking opportunities for professional community through one of the largest conferences of its kind in the area of KES. His interests focus on the artificial intelligence paradigms and their applications in complex systems, security, e-education, e-healthcare, unmanned air vehicles, and intelligent agents.

Dr. Margarita N. Favorskaya is a Professor and Head of the Department of Informatics and Computer Techniques at Reshetnev Siberian State University of Science and Technology, Russian Federation. Professor Favorskaya is a member of KES organization since 2010, the IPC member, and the Chair of invited sessions of over 30 international conferences. She serves as a reviewer in international journals (neurocomputing, knowledge engineering and soft data paradigms, pattern recognition letters, engineering applications of artificial intelligence), an Associate Editor of Intelligent Decision Technologies Journal, International Journal of Knowledge-Based and Intelligent Engineering Systems, International Journal of Reasoning-based Intelligent Systems, a Honorary Editor of the International Journal of Knowledge Engineering and Soft Data Paradigms, the Reviewer, Guest Editor, and Book Editor (Springer). She is the author/co-author of 200 publications and 20 educational manuals in computer science/engineering. She co-authored/co-edited seven books for Springer recently. She supervised nine Ph.D. candidates and presently supervising four Ph.D. students. Her main research interests are digital image and video processing, remote sensing, pattern recognition, fractal image processing, artificial intelligence, smart systems design, and information technologies.

Dr. Ilia S. Nikitin is a Professor and Director at Institute of Computer Aided Design RAS (ICAD RAS), a Professor at Moscow Aviation Institute (MAI), a member of the Russian National Committee on Theoretical and Applied Mechanics, expert RAS, expert RSF, expert Minobrnauki RF. He graduated from Moscow Institute of Physics and Technology. His scientific interests are mathematical modeling, numerical methods in continuum mechanics, moving adaptive meshes, dynamics of elastoplastic media, fatigue fracture, durability of operation, and high-frequency loading. The main scientific results are the numerical methods for solving non-stationary problems of continuum mechanics on moving and adaptive grids, methods for calculating the stress state of elements of aircraft structures and assessing the durability for various fatigue failure modes, refined models of layered and block media with different sliding conditions at the contact boundaries, the problems of propagation, transformation, and reflection of waves in such media, and models of sintering powder materials under thermomechanical and pulsed high-energy effects.

Dr. Dmitry L. Reviznikov is a Professor of the Department of Numerical Mathematics and Programming at Moscow Aviation Institute (National Research University), Russian Federation. Professor Dmitry L Reviznikov is a member of the Russian National Committee on Heat and Mass Transfer, a member of the Scientific Council of International Centre for Heat and Mass Transfer (ICHMT). He is a reviewer in international journals (International Journal of Heat and Mass Transfer, Computational Thermal Sciences, International Journal of Fluid Mechanics Research). He supervised eight Ph.D. candidates and presently supervising three Ph.D. students.

Scientific interests: mathematical modeling, computational physics, heat and mass transfer, multiphase flows, nonlinear dynamics, and data analysis. Scientific results: author of more than 100 scientific papers in Russian and international journals, 4 monographs. Fundamental results in the fields of modeling of conjugated heat and mass transfer, supersonic heterogeneous flows, thermal erosion destruction of heat-shielding coatings, anomalous diffusion, numerical methods for fractional differential equations, nonlinear wave dynamics, and interval analysis.

Chapter 1
Advances in Computational Mechanics and Numerical Simulation

Lakhmi C. Jain, Margarita N. Favorskaya, Ilia S. Nikitin and Dmitry L. Reviznikov

Abstract The chapter contains a brief description of chapters that contribute to the development and applications of computational methods and parallel algorithms in different areas of gas dynamics, aerodynamics, hydrodynamics, turbulence, solids dynamic, dynamic systems, optimal control. The first part presents the recent advances in computational fluid dynamics and aerodynamics. The second part introduces a numerical simulation of multiphase flows, combustion, and detonation. The third part is devoted to computational solid mechanics, and the fourth part provides a numerical study of dynamic systems.

L. C. Jain (✉)
University of Technology Sydney, Ultimo, Australia
e-mail: jainlakhmi@gmail.com; jainlc2002@yahoo.co.uk

Liverpool Hope University, Liverpool, UK

KES International, North Yorkshire, UK

M. N. Favorskaya
Institute of Informatics and Telecommunications, Reshetnev Siberian State University of Science and Technology, 31, Krasnoyarsky Rabochy ave., Krasnoyarsk 660037, Russian Federation
e-mail: favorskaya@sibsau.ru

I. S. Nikitin
Institute of Computer Aided Design of the RAS, 19/18, Vtoraya Brestskaya ul., Moscow 123056, Russian Federation
e-mail: i_nikitin@list.ru

D. L. Reviznikov
Moscow Aviation Institute (National Research University), 4, Volokolamskoe shosse, Moscow 125993, Russian Federation
e-mail: reviznikov@inbox.ru

Federal Research Centre "Information and Control" of the RAS, 44, Vavilova ul., Moscow 119333, Russian Federation

L. C. Jain et al. (eds.), *Advances in Theory and Practice of Computational Mechanics*, Smart Innovation, Systems and Technologies 173,
https://doi.org/10.1007/978-981-15-2600-8_1

1.1 Introduction

This book presents the selected papers reported at the Twenty-first International Conference on Computational Mechanics and Modern Applied Software Systems (CMMASS '2019), which was held during 24–31 May, 2019. The book includes the modern numerical methods for solving problems of the continuum mechanics and numerical simulation of physical processes occurring in gas, fluid, and solid media. Also, methods of mathematical modeling of dynamic systems, optimization methods, and optimal control algorithms are considered. Part I "Computational Fluid Dynamics" involves Chaps. 2, 3, 4, 5, and 6, Part II "Numerical Simulation of Multiphase Flows, Combustion, and Detonation" contains Chaps. 7, 8, 9, 10, 11, and 12, Part III "Computational Solid Mechanics" includes Chaps. 13, 14, 15, 16, 17, 18, and 19, and Part IV "Numerical Study of Dynamic Systems" consists of Chaps. 20, 21, 22, 23, 24, and 25.

1.2 Chapters in the Book

Part I presents the recent advances in computational fluid dynamics and includes five chapters. Chapter 2 reports the development of a splitting method for incompressible fluid flows aiming at modeling of various types of phenomena and processes occurring in such heterogeneous media as the atmosphere and the ocean [1, 2]. Research has led to the assumption that turbulence existing only in thin layers registered in the ocean represents the intrusion of collapsing spots mixed liquids. Four stages of splitting scheme are considered. The obtained results are well within a range with theoretical estimates, experimental data, and calculations by other authors. Chapter 3 studies the structure of accretion disks around compact astrophysical objects on the basis of numerical simulation using gas-dynamic models of the environment as a continuation of previous research findings [3, 4]. It considers two approaches for modeling the structure of the accretion disk using various mathematical models of the disk and various numerical methods for solving the modeling problem. Chapter 4 introduces a direct numerical simulation of the vortex flow formation regime in a layer of weakly compressible medium based on Navier–Stokes equations [5]. Turbulence in the two-dimensional case is studied using the plane flow of an incompressible fluid under the action of an external force, periodic in the transverse direction. Chapter 5 reports the results of numerical simulation of two-dimensional laminar flows near a regular system of cylinders. The supersonic flows around a geometrically unchanged lattice of cylinders with a variation in Mach number in the direction of increasing and decreasing in the range from 2.0 to 4.5 are visualized as the flow patterns [6, 7]. During experiments, three ranges of flow ambiguity and the corresponding hysteresis of Mach number characteristics were revealed. Chapter 6 considers a model approach for discontinuous Galerkin method implementation. Numerical results show that using discontinuous Galerkin method and applying moment limiter, slope limiter, Weighted Essentially Non-Oscillatory (WENO) limiter, or limiter based on averaging allows one to obtain a high order of accuracy on smooth solutions [8, 9]. In addition, slope limiter, WENO limiter, and averaging limiter are easy to implement and provide the generalized solutions on multidimensional unstructured grids.

Part II introduces a numerical simulation of multiphase flows, combustion, and detonation and involves six chapters. Chapter 7 conducts the numerical investigation of the grid resolution influence on the detonation initiation process in the multifocused system with the profiled end-wall [10, 11]. The study of detonation issues in multi-focused systems and clarification of the basic mechanisms accompanying the detonation process using unstructured computational grids, as well as, the study of the manifested features using the unstructured grids approach is the main goal of this chapter. Chapter 8 is dedicated to the analysis of possible flow variants with the stationary shock and detonation waves in a variable cross section channel, consisting of two consecutive Laval nozzles, with hydrogen–air and hydrogen–oxygen mixtures in a quasi-one-dimensional nonstationary formulation [12, 13]. It was obtained that a stationary detonation wave is stable in the first expanding part of the channel and unstable in narrowing parts. The authors clarify that for a hydrogen–air mixture in the investigated channel, the range of flow rates, at which a stationary detonation wave exists, can be predicted with a high degree of accuracy by the equilibrium stationary theory. Chapter 9 considers the motion of known meteor bodies in the Earth's atmosphere and their fall out on the Earth's surface [14, 15]. The mechanisms of destruction of the bodies due to thermal stresses are under consideration. The obtained results qualitatively correctly reflect the observed processes of destruction of bodies in the atmosphere. Chapter 10 presents the results of research of the neutral gas and plasma effusion into vacuum space found by computational modeling using the kinetic theory [16, 17]. The proposed physical, mathematical, and computational models are based on the computational solutions referred to Vlasov kinetic equation. Distribution functions of charged and neutral particles at various points of the region of interest, as well as, the momentums of those functions (the concentrations and velocities fields) were found through computational experiments. Chapter 11 aims to develop the condensation and evaporation in flows of two-phase gas–droplet mixture in the nozzles, jets, and external area in front of the nozzle. This study develops two ways for condensation modeling. The first one is a continual approach based on the method of moments. The second one is a kinetic approach based on a quasi-kinetic model. Using a quasi-chemical model for water vapor in nitrogen, the saturation curves in the pressure–temperature phase plane depending on the mass fraction of water vapor were obtained. Also, a qualitative agreement was obtained between the numerical and experimental pressure distributions on the plane of symmetry of the nozzle. Chapter 12 investigates the effect of various parameters of fuel injection in an oxygen–kerosene rocket thruster on the efficiency of the workflow, particularly, the droplet injection velocity components by a centrifugal nozzle in a cylindrical coordinate system and droplet size distribution parameters. The working process was modeled without and with film cooling. The main features of the numerical experiment are highlighted, as well as, the recommendations based on the obtained results are formulated.

Part III is devoted to computational solid mechanics and contains seven chapters. Chapter 13 contains the continual models of solid media with a discrete set of slip planes (layered, block media) and with nonlinear type slip conditions at the contact boundaries of structural elements [18, 19]. The developed model can be used for the numerical simulation of the seismic survey process in complex geological fractured media. Numerical simulations of the dynamic scattering process for subsurface layered and block objects in elastic 2D and 3D media were carried out using high-performance computing systems. Chapter 14 explores the explicit and implicit non-matrix finite element algorithms for calculating contact interactions between elastic–plastic bodies [20, 21]. The algorithms of Lagrange multiplier methods for explicit schemes and algorithms of penalty functions for implicit schemes are considered in detail. The effectiveness of the algorithms is illustrated by two nontrivial examples: the impact of two plates at an angle and axisymmetric welding of two dissimilar tube samples under the action of a detonation wave. Chapter 15 examines the inverse seismic problem for oil and gas exploration using different approaches [22, 23]. First, the functional of minimization based on synthetic responses from layered and fractured media is constructed, and all parameters of the model are estimated by the appropriate minimization procedure. Second, machine learning techniques are used to reconstruct the fractured structure of the geological medium. Third, the classic migration problem using adjoint operators and the grid-characteristic method on structured meshes is solved. Chapter 16 discusses the features of the scattering of plane P-waves on gas-filled fractures located along the motion of the incident wave front [24, 25]. This problem has practical mining in the areas of nondestructive testing and seismic exploration, primarily in the area of railway nondestructive testing. The analytical formulas for the scattering of a plane P-wave on a gas-filled fracture located along the motion of the front of this wave were derived. Scattered S-waves have been identified and studied. Chapter 17 focuses on the elastic behavior of modeled structures using a novel multiscale method for materials modeling, which requires the information only from the atomic level (atomic structure and potential of atomic interaction). Modeled structures are virtually divided into tetrahedral elements. Each element contains a small but representative sample of atomic structure [26, 27]. The whole system evolution is governed by equations of motion for every element vertex. Computational experiments show a good correspondence of the results obtained from the proposed model with classical molecular dynamics results, which can be considered as the exact solution. Chapter 18 develops the analytical calculation model to determine the repair joint parameters of the aircraft structural elements. The model utilizes the inclusion methodology for heterogeneous materials in the joint and Hart-Smith model of adhesive layer. This study covers the variation of the longitudinal and transversal elastic moduli of the carbon fiber plastic specimens at different temperatures and variations of Poisson ratio under cyclic load at the same values of temperature. The proposed analytical technique can be tuned for specific structural and repair materials and solutions. Chapter 19 covers a parametric identification of Tersoff potential for one-component and two-component materials based on the molecular-dynamic modeling [28]. Each potential has a certain set of parameters, the values of which are unique for each material. The problem of parametric identification is multi-extremal. Therefore, a comparison of Monte Carlo and simulated annealing methods for global minimization and Hooke–Jeeves and Radial Granular Search methods for local minimization was implemented. Also, a software tool for parametric identification of certain materials that used the parallel calculations is discussed.

Part IV provides a numerical study of dynamic systems and includes six chapters. Chapter 20 develops the algorithms and software of three metaheuristic multi-agent methods: fish school search, krill herd, and imperialist competitive algorithm. Using multi-agent approach, one can optimize not only multi-extremal functions of many variables, but also find a solution for optimal open-loop control problems in aviation and space technology [29, 30]. On the basis of krill herd and imperialist competitive algorithm, a hybrid extremum search algorithm is formulated. Also, an algorithm for finding open-loop control for a single class of dynamic systems based on the use of the described multi-agent algorithms is suggested. Software that allows to find the optimal open-loop control, criterion value, and coordinates of switching points of the control law on the basis of the suggested algorithms was formed. It is shown that the numerical solution is closed to the optimal one. Chapter 21 discusses the use of the spectral form of mathematical description for the statistical analysis of stochastic dynamical systems [31, 32]: diffusions and jump diffusions, i.e., for solving Fokker–Planck–Kolmogorov equation and Kolmogorov–Feller equation for the probability density of the state vector for these dynamical systems. A detailed description of the proposed methods was supplemented by step-by-step algorithms for solving analysis problems and numerical experiments. Dryden wind turbulence model and its jump diffusion modification are used for testing the spectral method. Chapter 22 analyzes the dynamic effects of the Earth's pole motion in the celestial mechanical problem statement as the "deformable Earth–Moon problem in the gravitational field of the Sun" [33, 34]. The aim of this research is to study the effect of lunar–solar long-period disturbances on the Earth's pole motion, in other words, to find the influence of the perturbations from the Earth–Moon system spatial motion on the Earth's pole oscillatory process. A mathematical description of the Earth's pole motion model and gravitational–tidal lunar–solar disturbances, as well as, the gravitational–tidal mechanism of the Earth pole motion with a frequency close to the frequency of the lunar orbit precession are discussed. Chapter 23 discusses the application of modified metaheuristic global optimization algorithm "fireworks" in order to solve the problems of multiobjective optimization [35, 36]. A solution is a set of Pareto optimal possible solutions. Searching of control includes two stages. At the first stage, the optimization problem is solved for each of the objectives with penalties. The values of the penalties are selected to satisfy the terminal constraints. At the second stage, the penalties found are used to solve a multiobjective optimization problem. Modification of the one-objective optimization "fireworks" algorithm and its application to find programmed control which stabilizes a satellite is proposed. Chapter 24 contributes in solving the optimal filtering problem for nonlinear continuous-time stochastic observation systems [37]. Particle filters are proposed on the basis of Duncan–Mortensen–Zakai equation, as well as, on the basis of the robust Duncan–Mortensen–Zakai equation. To find the mode of the conditional distribution approximately, Edgeworth series is used for the conditional probability density expansion that allows to reduce significantly a computation time in contrast to find the mode by estimating the conditional probability density, for example, by the histogram or kernel estimations. Chapter 25 describes fractal programming as a programming paradigm based on the concept of "elastic objects" [38, 39]. The concept supposes that elastic objects can be transformed (unfolded and folded) dynamically at runtime using strategy planning model and production rules. These rules are keeping the object structure self-similar that defines a fractal property. This provides a new type of abstraction, encapsulation, inheritance, modularity, and concurrency of objects in fractal-oriented programming.

1.3 Conclusions

The book presents the research work of major experts in the field of numerical methods and mathematical modeling the dynamics of continuous media: gases, liquids, deformable solids, as well as, dynamic systems and optimal control. Using computational methods of continuum mechanics, such diverse gas and hydrodynamics processes have been studied as inhomogeneous flows in the ocean and atmosphere, the formation of accretion disks near astrophysical objects, laminar, and turbulent flows near bodies systems, shock, and detonation waves in channels of variable cross section. In the dynamics of solids, numerical methods have been developed to study the processes of wave propagation and scattering in structured media (seismic survey), the motion and destruction of meteoroids in the atmosphere, and the contact interaction of inelastic bodies. To solve most of the problems, researchers reported the use of parallel algorithms for multiprocessor high-performance computing systems (supercomputers). The book will be useful to scientists, researchers, undergraduate, and postgraduate students specializing in the field of computational methods, parallel algorithms, gas dynamics, aerodynamics, hydrodynamics, turbulence, multiphase flows, combustion and detonation, solids dynamic, dynamic systems, and optimal control.

References

1. Gushchin, V.A.: Family of quasi-monotonic finite-difference schemes of the second order of approximation. Math. Model. Comput. Simul. **8**(5), 487–496 (2016)
2. Gushchin, V.A., Matyushin, P.V.: Simulation and study of stratified flows around finite bodies. Comput. Math. Math. Phys. **56**(6), 1034–1047 (2016)
3. Babakov, A.V., Popov, M.V., Chechetkin, V.M.: Mathematical simulation of a massive star evolution based on a gasdynamical model. Math. Model. Comput. Simul. **10**(3), 357–362 (2018)

4. Babakov, A.V., Lugovsky, A.Y., Chechetkin, V.M.: Mathematical modeling of the evolution of compact astrophysical gas objects. In: Petrov, I.B., Favorskaya, A.V., Favorskaya, M.N., Simakov, S.S., Jain, L.C. (eds.) Proceedings of the Conference on 50 Years of the Development of Grid-characteristic Method, SIST, vol. 133, pp. 210–227. Springer (2019)
5. Fortova, S.V., Oparina, E.I., Belotserkovskaya, M.S.: Numerical simulation of the Kolmogorov flow under the influence of the periodic field of the external force. J. Phys. Conf. Ser. **1128**, 012089.1–012089.5 (2018)
6. Maksimov, F.A., Churakov, D.A., Shevelev, Y.D.: Development of mathematical models and numerical methods for aerodynamic design on multiprocessor computers. Comput. Math. Math. Phys. **51**(2), 284–307 (2011)
7. Guvernyuk, S.V., Maksimov, F.A.: Supersonic flow past a flat lattice of cylindrical rods. Comput. Math. Math. Phys. **56**(6), 1012–1019 (2016)
8. Ladonkina, M., Nekliudova, O., Ostapenko, V., Tishkin, V.: On the accuracy of the discontinuous Galerkin method in the calculation of shock waves. Comput. Math. Math. Phys. **58**(8), 1344–1353 (2018)
9. Ladonkina, M., Nekliudova, O., Tishkin, V.: Construction of the limiter based on averaging of solutions for discontinuous Galerkin method. Math. Model. **30**(5), 99–116 (2018)
10. Lopato, A.I., Utkin, P.S.: Numerical study of detonation wave propagation in the variable cross-section channel using unstructured computational grids. J. Combustion **3635797**, 1–8 (2018)
11. Lopato, A.I., Utkin, P.S.: The usage of grid-characteristic method for the simulation of flows with detonation waves. In: Petrov, I.B., Favorskaya, A.V., Favorskaya, M.N., Simakov, S.S., Jain, L.C. (eds.) Proceedings of the Conference on 50 Years of the Development of Grid-characteristic Method, SIST, vol. 133, pp. 281–290. Springer (2019)
12. Gidaspov, V.Y., Severina, N.S.: Numerical simulation of the detonation of a propane-air mixture, taking irreversible chemical reactions into account. High Temp. **55**(5), 777–781 (2017)
13. Gidaspov, V.Y.: Numerical simulation of chemically non-equilibrium flow in the nozzle of the liquid-propellant rocket engine. Aerospace MAI J. **20**(2), 90–97 (2013)
14. Syzranova, N.G., Andrushchenko, V.A.: Simulation of the motion and destruction of bolides in the Earth's atmosphere. High Temp. **54**(3), 308–315 (2016)
15. Andrushchenko, V.A., Maksimov, F.A., Syzranova, N.G.: Simulation of flight and destruction of the Benešov bolid. Comput. Res. Model. **10**(5), 605–618 (2018)
16. Kotelnikov, M.V.: The distribution functions of charged particles in the vicinity of a cylindrical body in a flow of collisionless plasma in magnetic field. High Temp. **46**(6), 757–762 (2008)
17. Kotelnikov, V.A., Kotelnikov, M.V.: Current–voltage characteristics of a flat probe in a rarified plasma flow. High Temp. **54**(1), 20–25 (2016)
18. Nikitin, I.S., Burago, N.G., Nikitin, A.D.: Continuum model of the layered medium with slippage and nonlinear conditions at the interlayer boundaries. Solid State Phenom. **258**, 137–140 (2017)
19. Golubev, V.I.: The usage of grid-characteristic method in seismic migration problems. In: Petrov, I.B., Favorskaya, A.V., Favorskaya, M.N., Simakov, S.S., Jain, L.C. (eds) Proceedings of the Conference on 50 Years of the Development of Grid-characteristic Method, SIST, vol. 133, pp. 143–155. Springer (2019)
20. Burago, N.G., Nikitin, I.S.: Matrix-free conjugate gradient implementation of implicit schemes. Comput. Math. Math. Phys. **58**(8), 1247–1258 (2018)
21. Burago, N.G., Nikitin, I.S., Nikitin, A.D., Stratula, B.A.: Algorithms for calculation damage processes. Frattura ed Integrità Strutturale **13**(49), 212–224 (2019)
22. Golubev, V.I., Voinov, O.Y., Petrov, I.B.: Seismic imaging of fractured elastic media on the basis of the grid-characteristic method. Comput. Math. Math. Phys. **58**(8), 1309–1315 (2018)
23. Muratov, M.V., Petrov, I.B.: Application of fractures mathematical models in exploration seismology problems modeling. In: Petrov, I.B., Favorskaya, A.V., Favorskaya, M.N., Simakov, S.S., Jain, L.C. (eds.) Proceedings of the Conference on 50 Years of the Development of Grid-characteristic Method, SIST, vol. 133, pp. 120–131. Springer (2019)

24. Favorskaya, A.V., Zhdanov, M.S., Khokhlov, N.I., Petrov, I.B.: Modeling the wave phenomena in acoustic and elastic media with sharp variations of physical properties using the grid-characteristic method. Geophys. Prospect. **66**(8), 1485–1502 (2018)
25. Favorskaya, A.V.: The use of multiple waves to obtain information on an underlying geological structure. Procedia Comput. Sci. **126**, 1110–1119 (2018)
26. Abgaryan, K.K., Zhuravlev, A.A., Zagordan, N.L., Reviznikov, D.L.: Discrete-element simulation of a spherical projectile penetration into a massive obstacle. Comput. Res. Model. **7**(1), 71–79 (2015)
27. Abgaryan, K.K., Eliseev, S.V., Zhuravlev, A.A., Reviznikov, D.L.: High-speed penetration. Discrete-element simulation and experiments. Comput. Res. Model. **9**(6), 937–944 (2017)
28. Abgaryan, K.K., Posypkin, M.A.: Optimization methods as applied to parametric identification of interatomic potentials. Comput. Math. Math. Phys. **54**(12), 1929–1935 (2014)
29. Panteleev, A.V., Metlitskaya, D.V.: Using the method of artificial immune systems to seek the suboptimal program control of deterministic systems. Autom. Remote Control **75**(11), 1922–1935 (2014)
30. Karane, M.M.C.: Comparative analysis of multi-agent methods for constrained global optimization. In: IV International Conference on Information Technologies in Engineering Education, pp. 128–133 (2018)
31. Rybakov, K.A., Sotskova, I.L.: Spectral method for analysis of switching diffusions. IEEE Trans. Autom. Control **52**(7), 1320–1325 (2007)
32. Baghdasaryan, G., Mikilyan, M.: Effects of magnetoelastic interactions in conductive plates and shells. Springer, Cham (2016)
33. Markov, YuG, Perepelkin, V.V., Filippova, A.S.: Analysis of the perturbed Chandler wobble of the Earth pole. Dokl. Phys. **62**(6), 318–322 (2017)
34. Perepelkin, V.V., Rykhlova, L.V., Filippova, A.S.: Long-period variations in oscillations of the Earth's pole due to lunar perturbations. Astron. Rep. **63**(3), 238–247 (2019)
35. Panteleev, A., Kryuchkov, A.: Metaheuristic optimization methods for parameters estimation of dynamic systems. Civ. Aviat. High Technol. **20**(2), 37–45 (2017)
36. Panteleev, A., Pis'mennaya, V.: Application of a memetic algorithm for the optimal control of bunches of trajectories of nonlinear deterministic systems with incomplete feedback. J. Comput. Syst. Sci. Int. 57(1), 25–36 (2018)
37. Rybakov, K.A.: Solving the nonlinear problems of estimation for navigation data processing using continuous particle filter. Gyroscopy Navig. **10**(1), 27–34 (2019)
38. Semenov, A.S.: Pattern-type reachability analysis of distributed systems based on fractal Petri nets. In: 5th International Conference on Control, Decision and Information Technologies. Thessaloniki, Greece, pp. 346–351 (2018)
39. Semenov, A.S.: Pattern recognition technique for synthesis fractal Petri nets. In: 6th International Conference on Control, Decision and Information Technologies. Paris, France, pp. 1798–1803 (2019)

Part I
Computational Fluid Dynamics

Chapter 2
The Splitting Scheme for Mathematical Modeling of the Mixed Region Dynamics in a Stratified Fluid

Valentin A. Gushchin and Irina A. Smirnova

Abstract Study of wave motions' fluid is one of the most important and complex problems of modern hydrodynamics. A mathematical model for dynamics of incompressible uniform viscous liquid spots in the stratified medium is considered. This model is described by Navier–Stokes equations in Boussinesq approximation. Stratification component of the medium is saltiness. Bearing in mind that in such flows there are areas with large gradients of hydrodynamic parameters, required methods should possess such properties as a high order of accuracy, minimum scheme dissipation and dispersion, as well as monotony. To solve the task, the authors are developing a method of splitting by physical factors called as Splitting on physical factors Method for Incompressible fluid Flows (SMIF) possessing by the above-mentioned properties. Four stages of splitting scheme are considered. This chapter provides a brief description of SMIF method. The test calculations and comparison with some theoretical and experimental data respect to the calculations of other authors are demonstrated.

2.1 Introduction

Study of various types of phenomena and processes occurring in such heterogeneous medium as the atmosphere and the ocean is both an academic and practical interest. The heterogeneity of these media is linked to the effects of buoyancy that is the presence of gravity. It is known that density of sea and ocean medium depends on temperature, pressure, and salinity. A number of mathematical models describing the dynamics of stratified fluids have been suggested [1–4]. Fine structure of hydrophysical fields observed in the ocean is an alternation of deep sites with low and

V. A. Gushchin (✉) · I. A. Smirnova
Institute for Computer Aided Design of the RAS, 19/18, Vtoraya Brestskaya ul., Moscow 123056, Russian Federation
e-mail: gushchin47@mail.ru

I. A. Smirnova
e-mail: o-ira@yandex.ru

L. C. Jain et al. (eds.), *Advances in Theory and Practice of Computational Mechanics*, Smart Innovation, Systems and Technologies 173,
https://doi.org/10.1007/978-981-15-2600-8_2

high vertical gradients of varying characteristics. Almost uniform in its properties, layers, vertical dimensions that vary from tens of centimeters to tens of meters, are separated by streaks. In these streaks, vertical gradients of physical properties can significantly exceed their averages. Time scales of such irregularities ranges from several hours to several days and more. The first systematic analysis of the physical mechanisms, forming fine structure, was taken in [3].

The causes of the initial formation of the mixed areas in stratified fluids are the following: shear instability benthic stratified large-scale currents and tidal waves, overturn of surface and internal waves, and convection in layers with unstable density stratification. Next, a shear velocity, under the action of which the initially mixed areas of turbulent energy balance can be positive and begins to increase the turbulence, starts playing a role [5]. In the ocean, turbulence exists only in thin layers—"turbulent pancakes." Field studies in the ocean have shown that turbulence has a pronounced "island," intermittent nature [4]. Research has led to the assumption that "pancake" structures registered in the ocean represent the intrusion of collapsing spots mixed liquids, mainly in viscous phase of its evolution. Collapse process of mixed areas, where turbulence lives or degenerates, is one of the fundamental processes responsible for the formation of the fine structure of the ocean waters [3–6]. It is known [5] that the arising and development of turbulence in fluid density stratified steadily is inseparable from the dynamics of internal waves and is as follows. Under the influence of outside forces, the large-scale internal waves originate in the stratified fluids. As a result, their nonlinear interactions and subsequent overturning or buckling occur field mixed liquid-stains (sometimes referred to as mixing zones). These spots of mixed turbulent fluid are evolving, gradually flattening (collapse of turbulent spots) that, in turn, leads to the formation of new spots, etc.

It is naturally consider as the three main stages of its development in the process of evolution [7]:

- Stage I. Initial stage: a force is acting on a particle of the fluid inside the spot, vastly superior force of resistance; an intensive generation of internal waves by spot takes place.
- Stage II. Intermediate fixed stage: a driving force is balanced by mainly resistance forms and wave resistance caused by radiation of internal waves; the size of the spots is increasing depending on the time of happening almost linear, i.e., acceleration curve is negligibly small.
- Stage III. Final viscous stage: a driving force is balanced by mostly viscous resistance; the horizontal size of the spots varies slightly.

Stages I and II are short and were estimated in [7–9] completed through a period of time, approximately equal to $4T_b$, where $T_b = 2\pi/N$ is the period and N is Brent–Viasel frequency. The duration of Stage I does not exceed $T_b/2$. Basically, the same observed spots of turbulence are at the final Stage III of evolution. Further, as a result of the diffusion the stain is mixed with the ambient fluid and disappears.

In this chapter, we adapt a mathematical model [10] previously used in tasks of stratified fluid flows around a sphere and circular cylinder [11, 12] to the task on collapse spots that we solved earlier excluding diffusion of stratification component

[13, 14]. Bearing in mind that in such flows there are areas with large gradients of hydrodynamic parameters, the required methods should possess such properties as a high order of accuracy, minimum scheme dissipation and dispersion, as well as monotony. This is essentially impotent for flows with internal and surface waves, where arising of "numerical" waves is unacceptable. To solve the task, we develop SMIF method possessing by the abovementioned properties [15]. Four stages of splitting scheme are considered.

The chapter is structured as follows. Section 2.2 provides a foundation of the problem, mathematical model, and initial and boundary conditions. Four stages of splitting scheme are considered in Sect. 2.3. Short review of finite-difference scheme is described in Sect. 2.4. The results and comparison with theoretical estimates, experimental data, and calculations of other authors are presented in Sect. 2.5. Section 2.6 gives the conclusion.

2.2 Mathematical Model. Foundation of the Problem

Consider 2D unsteady flow that occurs, when collapse area of homogeneous fluid *A* is surrounded by sustainably and continuously stratified by density (for clarity, use a linear low) incompressible fluid (Fig. 2.1).

The course develops in a uniform gravity field with the acceleration of free fall *g*. Undisturbed linear density distribution [10] is defined as:

$$\rho(x, y) = \rho_0\Big(1 - \frac{y}{\Lambda} + s(x, y)\Big),$$

where Λ is the stratification scale, $\Lambda = \left|\frac{1}{\rho_0}\left(\frac{\partial\rho}{\partial y}\right)\right|^{-1}$, N is the buoyancy frequency, $N = \sqrt{g/\Lambda}$, T_b is the buoyancy period, $T_b = 2\pi/N$, $C = \Lambda/R_0$ is the scale ratio, R_0 is the radius of spot, s is the perturbation of salinity (stratification component), includes salt ratio of compression.

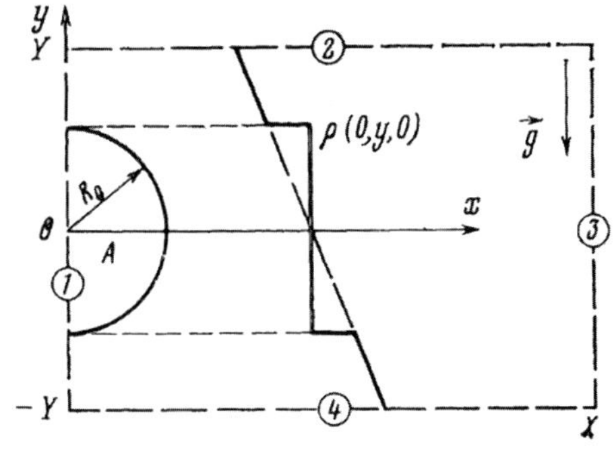

Fig. 2.1 Initial and boundary conditions

Navier–Stokes equations in Boussinesq approximation describing the flow of this type can be written as

$$\frac{\partial \mathbf{v}}{\partial t} + (\mathbf{v} \cdot \nabla)\mathbf{v} = -\nabla p + \frac{1}{Re}\Delta \mathbf{v} + \frac{1}{Fr} s \frac{\mathbf{g}}{g},$$
$$\nabla \cdot \mathbf{v} = 0,$$
$$\frac{\partial s}{\partial t} + (\mathbf{v} \cdot \nabla)s = \frac{1}{Sc \cdot Re}\Delta s + \frac{v}{C},$$

where $\mathbf{v}$ is the velocity vector with components u, v along the x- and y- axes of Cartesian coordinate system selected as indicated in Fig. 2.1, respectively, ρ is the density, p is the pressure minus hydrostatic one, s is the perturbation of salinity, Re is the Reynolds number, $Re = \rho_0 R_0^2 N/\mu$, Fr is the Froude number, $Fr = R_0 N^2/g$, Sc is the Schmidt number $= \mu/\rho_0 k_s$, k_s is the diffusion coefficient of salts, μ is dynamic viscosity coefficient, $\mathbf{g} = (0, -g)$, g is acceleration of free fall, ρ_0 is the density at the level $y = 0$, $C = \Lambda/R_0$ is the scale ratio.

We assume that the initial time $t = 0$ the system on the plane $\boldsymbol{R}^2$ is at rest, i.e.,

$$u = v = 0, (x, y) \in \boldsymbol{R}^2,$$

the density of fluid at the spot A is

$$\rho = 1, \quad (x, y) \in A,$$

and outside of spot, i.e., in the area of $\boldsymbol{R}^2 \backslash A$, is

$$\rho = 1 - \frac{y}{C} + s, (x, y) \in \boldsymbol{R}^2 \backslash A,$$

the peturbation of salinty is

$$s = \begin{cases} y/C & (x, y) \in A, \\ 0 & (x, y) \in R^2 \backslash A. \end{cases}$$

As an initial approximation necessary to solve the equation for pressure distribution, the following system is selected:

$$p = \begin{cases} -y/Fr, & (x, y) \in A, \\ -\left(y - y^2/2C\right)/Fr, & (x, y) \in R^2 \backslash A. \end{cases}$$

As the pressure in the case of an incompressible fluid shall be determined with an accuracy of up to an arbitrary constant (without limiting a generality), we can select it to zero, at level $y = 0$.

Effect of symmetry tasks concerning the plane $x = 0$, naturally seeks a solution in only one half-plane, for example, if $x \geq 0$. Solution will search in the rectangular area $\{x, y: 0 \leq x \leq X, -Y \leq y \leq Y\}$. In the left boundary (line 1 in Fig. 2.1), this area is the conditions of symmetry:

$$u = 0, \quad \frac{\partial v}{\partial x} = \frac{\partial p}{\partial x} = \frac{\partial \rho}{\partial x} = \frac{\partial s}{\partial x} = 0.$$

The top (line 2), bottom (line 4), and right (line 3) borders should be chosen far enough away from the source of disturbance (from spots) so that setting any boundary conditions at these borders, which are necessary for the solution of the problem, is not providing a significant influence on the flow.

To solve the task, we use one of the latest versions of SMIF method. Finite-difference scheme of this method possesses by properties such as a second-order approximation for the spatial variable, minimum scheme viscosity and dispersion, functioning in a wide range of Reynolds and Froude numbers, and more importantly, when solving such problems, the monotony [15].

2.3 Splitting Scheme

Let some time be $t_n = n \cdot \tau$, where τ is the time step, n is the number of steps with known velocity $\mathbf{v}$, pressure p, and perturbation of salinity s. Then the unknown scheme functions at time $t_{n+1} = (n + 1) \cdot \tau$ can be represented as follows:

$$\frac{\tilde{\mathbf{v}} - \mathbf{v}}{\tau} = -(\mathbf{v}^n \cdot \nabla)\mathbf{v}^n + \frac{1}{Re}\Delta\mathbf{v}^n + \frac{1}{Fr}s^n\frac{\mathbf{g}}{g},$$

$$\tau\,\Delta p = \nabla \cdot \tilde{\mathbf{v}},$$

$$\frac{\mathbf{v}^{n+1} - \tilde{\mathbf{v}}}{\tau} = -\nabla p,$$

$$\frac{s^{n+1} - s^n}{\tau} = -(\mathbf{v}^{n+1} \cdot \nabla)s^n + \frac{1}{Sc \cdot Re}\Delta s^n + \frac{v^{n+1}}{C}.$$

At Stage I, it is expected that the transfer of momentum (the momentum of a unit of mass) is performed only by the convention, diffusion, and buoyancy forces. At Stage II, using the found interim velocity $\tilde{\mathbf{v}}$, a pressure field is calculated using Poisson equation. Here, we take into account that $\nabla \cdot \mathbf{v}^{n+1} = 0$. At Stage III, it is anticipated that the transfer is carried out only at the expense of the pressure gradient (convection and diffusion are not available). At Stage IV, using the found velocity field $\mathbf{v}^{n+1}$, a perturbation of salinity is calculated.

Fig. 2.2 Grid stencil

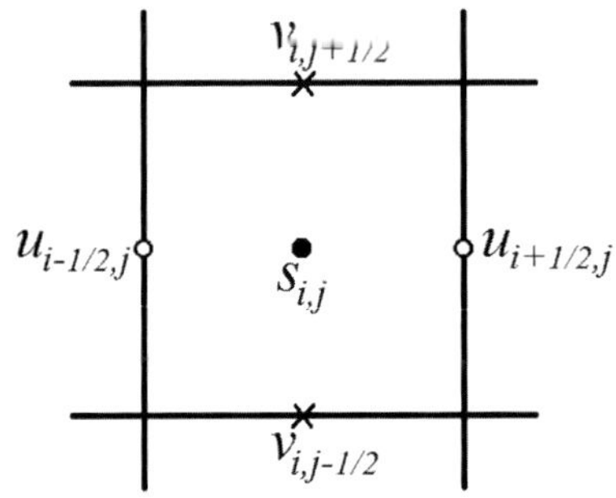

2.4 Finite-Difference Scheme

The study area is covered by a uniform x and y grid cells:

$$\Omega = \begin{cases} x_{i+1/2} = i \cdot \delta x, & \delta x > 0, \quad i = 0, 1, \ldots, L; \quad L \cdot \delta x = X, \\ y_{j+1/2} = j \cdot \delta y, & \delta y > 0, \quad j = 0, 1, \ldots, M; \quad M \cdot \delta y = 2Y, \end{cases}$$

where δx, δy are the dimensions grid steps, L and M are the numbers of grid cells in the directions of x and y, respectively (the point with coordinates (i, j) is at the center of the cell). Here, as in the original method of splitting, we use the "Chess" grid, i.e., the grid coordinates functions are separated in space, as shown in Fig. 2.2.

This enables to interpret visually each cell, such as volume, which is characterized by calculations in its central pressure $p_{i,j}$, density $\rho_{i,j}$ (possibly, temperature, energy, etc.), as well as, divergence of $D_{i,j}$ (D in according to the sign determines whether source or drain in this volume). Knowledge of the normal components of the velocity vector on the sides of the cell is able to directly calculate the flow of momentum through this side. The finite-difference scheme for 1D case is presented in [15]. For the solution of Poisson equation for pressure the successive over the relaxation method is used.

2.5 Results

Calculations were carried out in the field with dimensions $X = 10$, $Y = 5$, $R_0 = 1$ with the following coefficients and parameters: $\mu/\rho_0 = 0.01$ cm^2/s^{-1}, $k_s = 1.41\ 10^{-5}$ cm^2/s^{-1}, $N = 1$ s^{-1}, $T_b = 2\pi$ s, $\Lambda = 10$ cm, $C = 10$, Re $= 100$, $Fr = 0.1$, $Sc = 709.2$ that is close to the laboratory experimental conditions. As boundary conditions at the top, the resting states of bottom and right borders of the computational domain are chosen, i.e., $u = v = s = 0$. The computational domain is covered with a uniform grid with the steps in both directions $\delta x = \delta y = 0.1$. With a view to program verification, the calculations in the absence of stain and on different grids were performed. The results confirmed the execution of conservation laws with the required accuracy.

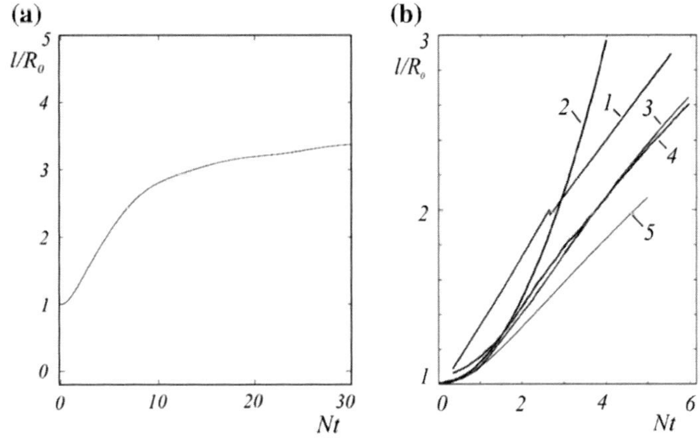

Fig. 2.3 Time dependence of horizontal spot size: **a** received by our method, **b** comparison with other authors

Time dependence of horizontal spot size l at the level of $y = 0$ is shown in Fig. 2.3a. The comparison with analytical estimations [7]—curve 2, experimental data [5]—curve 1, and calculation of other authors [4, 6]—curves 3, 4, is shown in Fig. 2.3b. Here, curve 5 is our numerical results. Time dependence of vertical spot size h at the level of $x = 0$ is shown in Fig. 2.4. Here, thick lines are our results obtained with physical model without diffusion [13] and thin lines are our present results, where diffusion of stratification component (perturbation of salinity) is taken into account. It should be noted that, as in [13], changes in both horizontal and vertical sizes of spots have non-monotonous character.

The isoclines of stream function for $t = 2, 4, 6, 8$ are shown in Fig. 2.5. Figure 2.5b shows the emergence of a second vortex pair, Fig. 2.5c depicts a clear visibility of two developed vortex pairs, and by the time $t = 8$ a third vortex pair arises (Fig. 2.5d).

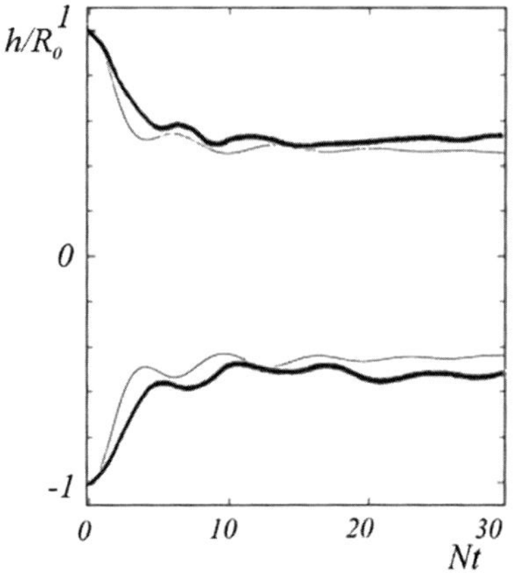

Fig. 2.4 Vertical spot size. Thick lines are taken from [13], thin lines are our present results

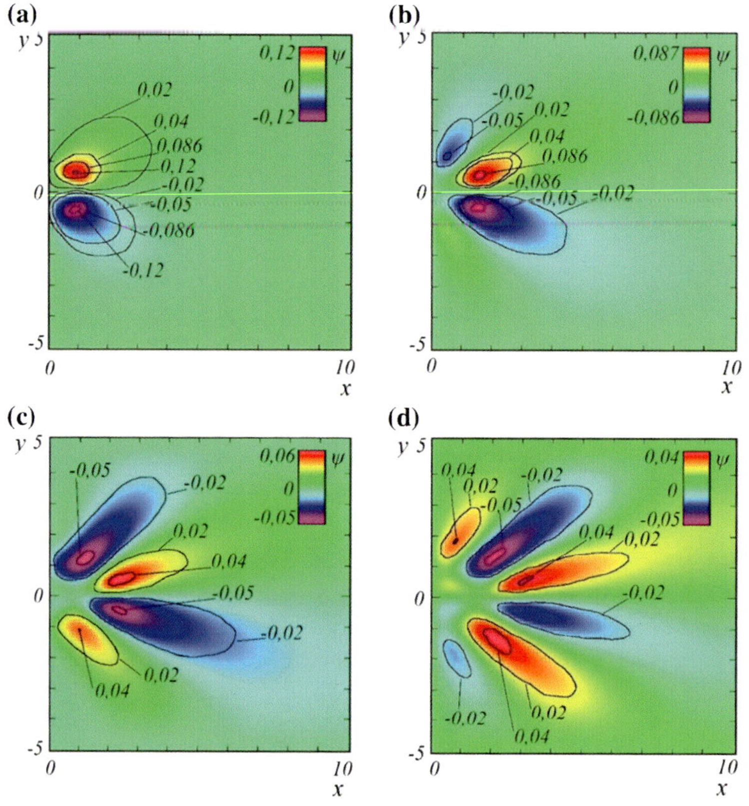

Fig. 2.5 The isoclines of stream function: **a** $t = 2$, **b** $t = 4$, **c** $t = 6$, **d** $t = 8$

It should be noted that the vortices located in the upper half-plane have a greater intensity than those vortices from the lower half-plane.

The isoclines of perturbation of salinity for $t = 2, 4, 6, 8$ are shown in Fig. 2.6, while the isoclines of perturbation of salinity for $t = 20$ and $t = 30$ are shown in Fig. 2.7.

The isoclines of stream function for $t = 20$ and 30 are shown in Fig. 2.8.

From Figs. 2.6, 2.7, and 2.8, it is possible to estimate a length of internal waves. Simultaneously, Figs. 2.7, 2.8 show us that for correct calculations, outer boundaries X and Y should be taken larger for time moments more than $t = 20$.

2.6 Conclusions

The mathematical model of the dynamics of the spots in the stratified fluid is suggested. The model is based on Navier–Stokes equations in Boussinesq approximation

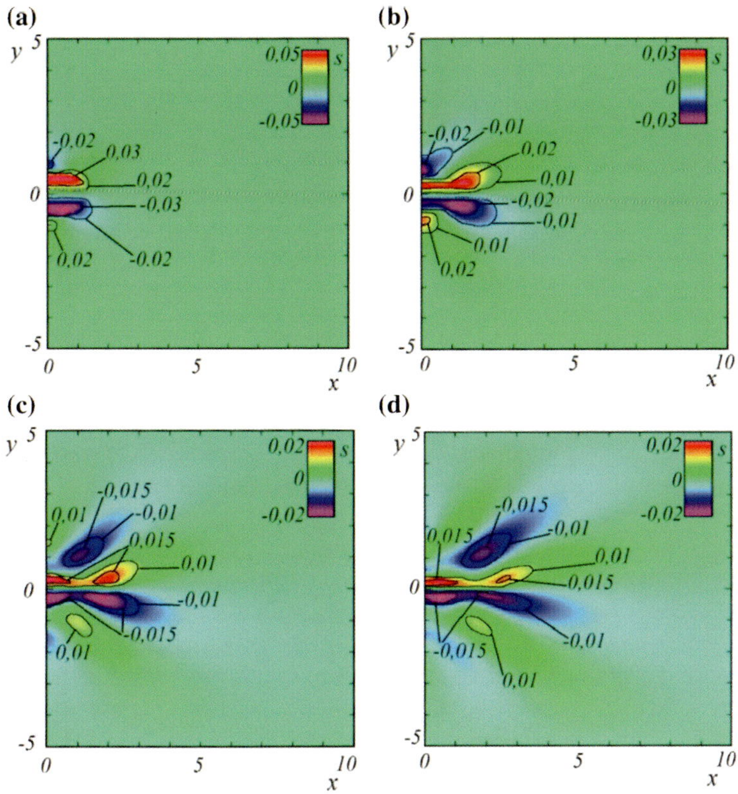

Fig. 2.6 The isoclines of perturbation of salinity: **a** $t = 2$, **b** $t = 4$, **c** $t = 6$, **d** $t = 8$

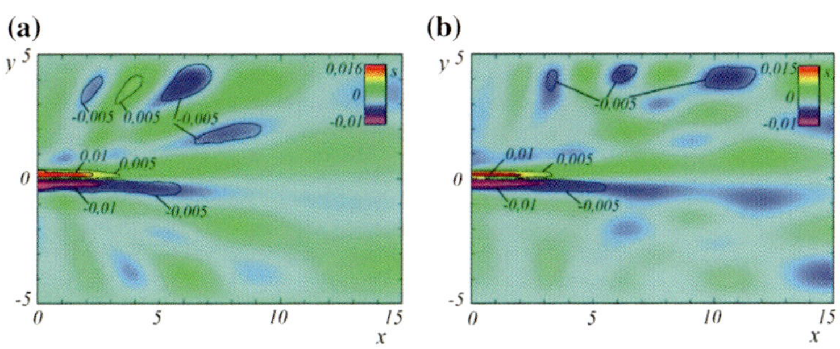

Fig. 2.7 The isoclines of perturbation of salinity: **a** $t = 20$, **b** $t = 30$

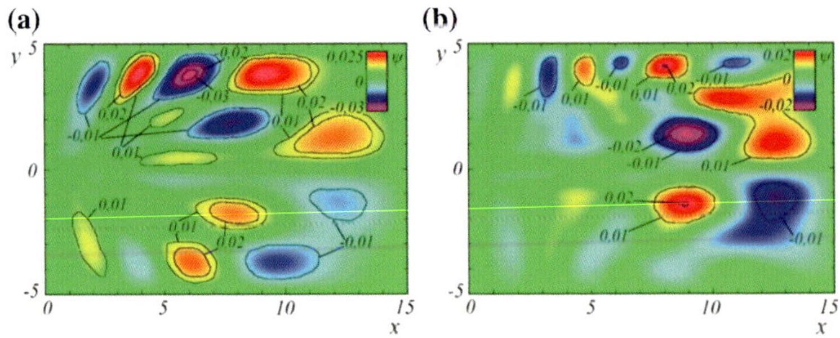

Fig. 2.8 The isoclines of stream function: **a** $t = 20$, **b** $t = 30$

taking into account the diffusion of stratification component (salt). To solve the task, four-stage scheme of splitting was considered. The code is tested on the example of calculation of a problem in the absence of spot and on different grids. The results confirmed the execution of conservation laws with the required accuracy.

As the results of the calculations, the horizontal and vertical sizes of spots are changing depending on the time. The obtained results are well coinciding with theoretical estimates, experimental data, and calculations of other authors. Previously detected not monotonous changes of the linear sizes of the spots are confirmed. The isoclines of stream function and perturbation of salinity are presented for different moments of time. It is anticipated that the proposed model will receive better results, when dealing with similar tasks.

References

1. Turner, J.S.: Buoyancy Effects in Fluids. Cambridge University Press (1973)
2. Scorer, R.S.: Environmental Aerodynamics. Ellis Horwood Publishers (1978)
3. Fedorov, K.N.: Fine Structure of the Thermohaline Ocean Waters. Gidrometeoizdat, Leningrad (1976) (in Russian)
4. Monin, A.S., Ozmidov, R.B.: Ocean Turbulence. Gidrometeoizdat, Leningrad (1981) (in Russian)
5. Barenblatt, G.I.: Similarity, Self-Similarity and Intermediate Asymptotics. Gidrometeoizdat, Leningrad (1982). (in Russian)
6. Zatsepin, A.G., Fedorov, K.N., Voropaev, S.I., Pavlov, A.M.: Experimental research of collaps mixed spot in stratified fluid. Izv. of the Academy of Sciences of the USSR. Phys. Atmos. Ocean **14**(2), 234–237 (1978) (in Russian)
7. Wu, J.: Mixed region collapse with internal wave generation in a density-stratified medium. J. Fluid Mech. **35**(3), 531–544 (1969)
8. Wessel, W.R.: Numerical study of the collapse of a perturbation in an infinite density stratified fluid. Phys. Fluids **12**(12), 170–176 (1969)
9. Kao, T.W.: Principal stage of wake collapse in a stratified fluid: two-dimensional theory. Phys. Fluids **19**(8), 1071–1074 (1976)
10. Chashechkin, Y.D., Zagumennyi, Y.V., Dimitrieva, N.F.: Dynamics of formation and fine structure of flow pattern around obstacles in laboratory and computational experiment. In: Voevodin,

V., Sobolev, S. (eds.) RuSCDays 2016, CCIS, vol. 687, pp. 41–56. Springer International Publishing AG (2016)
11. Gushchin, V.A., Mitkin, V.V., Rozhdestvenskaya, T.I., Chashechkin, Y.D.: Numerical and experimental study of the fine structure of a stratified fluid flow over a circular cylinder. J. Appl. Mech. Tech. Phys. **48**(1), 34–43 (2007)
12. Gushchin, V.A., Matyushin, P.V.: Simulation and study of stratified flows around finite bodies. Comput. Math. Math. Phys. **56**(6), 1034–1047 (2016)
13. Gushchin, V.A.: The splitting method for problems of the dynamics of an inhomogeneous viscous incompressible fluid. U.S.S.R. Comput. Math. Math. Phys. **21**(4), 190–204 (1981)
14. Gushchin, V.A., Kopysov, A.N.: The dynamics of a spherical mixing zone in a stratified fluid and its acoustic radiation. U.S.S.R. Comput. Math. Math. Phys. **31**(6), 51–60 (1991)
15. Gushchin, V.A.: Family of quasi-monotonic finite-difference schemes of the second order of approximation. Math. Models Comput Simul **8**(5), 487–496 (2016)

Chapter 3
Modeling of Some Astrophysical Problems on Supercomputers Using Gas-Dynamic Model

Alexander V. Babakov, Alexey Y. Lugovsky and Valery M. Chechetkin

Abstract In the current study, the vortex structures that occur in accretion disks are investigated using mathematical modeling methods. The simulation of the processes of formation of large-scale vortex structures in stellar accretion disks is carried out by two methods with different numerical schemes. The first numerical technique is based on conservative difference scheme with "upwind" approximation for fluxes. The second numerical technique is based on an explicit, conservative, monotone in the linear approximation Godunov-type Roe–Einfeldt–Osher scheme, which approximates, with order no higher than the third, the conservation laws in the form of Euler equations. Visualized pictures of the vortex structure are given by both methods for accretion disks. The qualitative similarity of the obtained results is discussed. Evolutionary calculations are carried out on the basis of parallel algorithms implemented on the supercomputing complex of the cluster architecture.

3.1 Introduction

This chapter of the studying the structure of accretion disks around compact astrophysical objects on the basis of numerical simulation using gas-dynamic models of the environment is a continuation of researches [1–4]. In these researches, the calculations of the evolution of a massive third-generation star, which is the predecessor of

A. V. Babakov (✉)
Institute for Computer Aided Design of the RAS, 19/18, Vtoraya Brestskaya ul., Moscow 123056, Russian Federation
e-mail: avbabakov@mail.ru

A. Y. Lugovsky · V. M. Chechetkin
Keldysh Institute of Applied Mathematics of the RAS, 4, Miusskaya pl., Moscow 125047, Russian Federation
e-mail: alex_lugovsky@mail.ru

V. M. Chechetkin
e-mail: chechetv@gmail.com

L. C. Jain et al. (eds.), *Advances in Theory and Practice of Computational Mechanics*, Smart Innovation, Systems and Technologies 173,
https://doi.org/10.1007/978-981-15-2600-8_3

a supernova, are carried out and the simulation of the development of hydrodynamic instability in accretion disks with the formation of large vortexes is carried out also.

One of the main issues in the study of the structure of accretion disks is the physics of redistribution and transfer of angular momentum [5]. Currently, within the framework of the hydrodynamic approach, there are at least two mechanisms that influence this process. The first mechanism is associated with the viscosity and development of turbulence [6], while the second mechanism is associated with the development of shear instability and with the formation of a large-scale structure of turbulence [7–9]. Since the characteristic scales of turbulence in them are different, there are also different times of transition of kinetic energy into thermal energy, and different temperatures of accretion disks are corresponded to them. This leads to different flow structures and times of evolution of accretion disks, as well as, to the different rates of angular momentum transfer from the accretion disk [8, 9]. This research is devoted to the second mechanism and explores the processes of formation of large vortexes in the flow of a disk.

It should be noted that the formation of a large-scale vortex flow structure in accretion disks is observed not only in the framework of the hydrodynamic approach. For example, it is shown in [10] that in a gas-dust cloud rotating around a protostar in the presence of a magnetic field, the large vortex structures leading to the transfer of angular momentum are appeared. The mechanism of the occurrence of such structures in the accretion gas-dust disk is the magneto-rotational instability. The occurrence of a large-scale flow structure in the disks of spiral galaxies is demonstrated in [11].

In this chapter, two approaches with different formulations and difference schemes are used to study the formation of large vortices in the flow of the disk, allowing us to study this issue from different angles.

This chapter is organized as follows. Section 3.2 shortly presents the mathematical models of studied problems. The results of the numerical modeling of the vortex structures in accretion disks are introduced in Sect. 3.3. Lastly, Sect. 3.4 presents the conclusions.

3.2 Mathematical Models of the Evolution of the Stellar Accretion Disk

The chapter considers two approaches to modeling the structure of the accretion disk using various mathematical models of the disk and various numerical methods for solving the modeling problem.

The first approach to simulate a nonstationary motion of matter in fast-rotating accretion disks uses a conservative numerical flux method [12, 13]. It is based on a finite-difference representation of the conservation laws of density, momentum components, and total energy for each finite volume that occurs when the integration region is discretized. In this case, for the approximation of the vectors of the flux densities of the indicated characteristics, "upwind" approximations are used.

Previously, based on the algorithms of the method, a complex of parallel programs for computing systems with cluster architecture was developed [14]. Parallel algorithms of the complex are based on the standardized message transfer system called as Message Passing Interface (MPI). During modeling spatial-nonstationary problems, paralleling in space in two or three directions is implemented, depending on the specifics of the task, the size of the integration domain, and the parameters of the computational mesh. During modeling fast-rotating gas objects, for which a rotation is a substantially dominant motion, the modified finite-difference technique of the flux method is used. It allows to preserve the components of the angular momentum, for which conservation laws are not included in the basic system of equations based on the laws of conservation of mass, components of momentum, and total energy (when using the finite-difference analogue of conservation laws and time integration for large times, the conservation of components of the angular moment in the area of integration may be disturbed). To preserve these components, the modified finite-difference approximations of fluxes on the surface of a finite volume are used. This make it possible, when integrating even for long times, to achieve the conservation of the component of the angular momentum associated with the rotation with an accuracy of 0.5% during the entire integration time. Moreover, for the other two components, the angular momentum errors do not exceed 10^{-12}.

The second approach also uses a conservative numerical method based on the finite-difference representation of the density, components of the momentum, and total energy conservation laws for each finite volume in a flux form using the integro-interpolation method. To obtain the fluxes at the cell boundaries, Godunov-type Roe–Einfeldt–Osher method [15–18] based on an approximate solution of Riemann problem is used. As is known, a linear monotone difference scheme cannot have an order higher than the first. In this scheme, to circumvent this limitation, the nonlinear limiters of anti-diffusion fluxes are used to increase the order of approximation to the third, while preserving the monotony of the scheme in the linear approximation. The system of Euler equations for gas dynamics is a nonlinear hyperbolic system, one of the important features of which is a possibility of the appearance of discontinuous solutions from smooth initial data. For problems of gas dynamics and especially for astrophysical problems, flows are typical, in which shock waves and contact discontinuities arise [11]. The accuracy of the numerical solution depends strongly on the ability of the difference scheme to resolve gas-dynamic features. The used Godunov-type Roe–Einfeldt–Osher scheme is conservative by recording the scheme in the flux form and has a low numerical viscosity, but at the same time retains a monotony of the solution. The efficiency of the schemes was confirmed in [8, 9, 19], in the same works it was numerically shown that the scheme retains the angular momentum including the rotating component, which is especially important for fast-rotating accretion disks. The built software package, when implementing a parallel algorithm, uses a method of decomposition of the domain of calculation into subdomains and the communication library MPI [4].

Below, on the basis of these program complexes, methods of mathematical modeling are used to simulate the nonstationary motions of matter in accretion disks with fast rotation. A gas-dynamic model of a perfect, non-viscous gas is used.

3.3 Numerical Simulation of the Evolution of the Stellar Accretion Disk

Below we consider two model problems on the structure of an accretion disk that is rapidly rotating around the gravitating center. Simulation of the evolution of the external area of stellar accretion disk is represented in Sect. 3.3.1, while numerical results of the formation of vortex structures in stellar accretion disks are discussed in Sect. 3.3.2. The tasks differ in the geometric and mass parameters of the accretion disk and the gravitating center.

3.3.1 *Simulation of the Evolution of the External Area of Stellar Accretion Disk*

The results of modeling the structure of the outer region of the accretion disk, rapidly rotating around a neutron star of mass $M = 2.7846 * 10^{33}$ g (which corresponds to $1.4M_{\odot}$) and radius $r_0 = 1.0 * 10^6$ cm, are considered. Hereinafter, the subscript $\odot$ indicates the value of the solar parameter. The results of modeling the nonstationary behavior of the outer part of the accretion disk of size $R_0 = 10r_0$ and mass $1.088 * 10^{31}$ g are presented. The ratio of specific heats of gas γ in the disk is taken to be 5/3. Simulation is carried out on the basis of parallel flux method algorithms [10, 11].

The mass of the central gravitating region (neutron star) exceeds the mass of matter in the accretion disk by 257 times. Given this mass ratio, self-gravity inside the accretion disk is neglected.

To divide the region of integration into finite volumes, a cylindrical coordinate system (r, φ, z) is introduced. Finite volumes Ω_{m} are formed by splitting in constant steps along the coordinate z, the radial r, and angular φ coordinates. The integration is carried out in the cylindrical region Ω bounded by the inner surface of radius r_0 (the surface of a neutron star), the outer surface of radius R_1, and the lateral planes $z = -z_0$ and $z = z_0$: $\Omega = (r_0 \leq r \leq R_1) * (0 \leq \varphi \leq 2\pi) * (-z_0 \leq z \leq z_0)$.

The boundary conditions at the outer boundary of the integration region $r = R_1$ are determined by the zero derivative of the normal to the external integration region with a positive value of the normal velocity component and the prohibition of convective transfer into the integration region and zero gas dynamic variables with a negative normal velocity component. On the inner boundary of the integration region $r = r_0$, the conditions on a solid surface (impermeability condition) are specified. Along with a cylindrical coordinate system intended for building a computational grid, Cartesian coordinate system is introduced with the center coinciding with the center of the cylindrical coordinate system. The equations of motion are written in Cartesian coordinate system for Cartesian components of the velocities, and the approximation of Cartesian vectors of flux densities is carried out by reference points of finite volumes Ω_{m}. The following geometrical parameters of the integration domain are used in the calculation: $R_1 = 1.5R_0$, $z_0 = 0.5R_0$. The calculations apply the computational grids containing up to 30 million finite volumes. The simulation is carried out on computing complexes of cluster architecture using up to 2000 multicore processors.

The initial fields of pressure p and density ρ are set in such a way that at zero speeds of motion equilibrium in the region of the accretion disk located in the central gravitational field (neutron star) is realized. The polytropic equation of state of the gas environment $p = K\rho^{\gamma}$ is used. The entropy in the initial field is constant throughout the integration domain, which means hydrodynamic equilibrium, which, however, does not prevent the formation of flows.

Further, dimensionless variables will be used: linear dimensions will be assigned to R_0, pressure $\bar{p} = p/p_0$, density $\bar{\rho} = \rho/\rho_0$, Cartesian velocity components $\bar{v}_k = v_k/a_0$, temperature $\bar{T} = T/T_0$, and time $\bar{t} = t/t_0$, where p_0, ρ_0, and T_0 are the pressure, density, and temperature on the surface of the accretion disk ($r = r_0$) at the initial time ($t = 0$), respectively, a_0 is the characteristic speed, $t_0 = r_0/a_0$, $a_0^2 = p_0/\rho_0$.

Figure 3.1 shows the fields of the density (Fig. 3.1a) and angular velocity (Fig. 3.1b) in the $\varphi = 0$ plane, as well as, their profiles in the radial direction (Fig. 3.1c, d) at the initial time $t = 0$ (the density is in the logarithmic scale).

In the evolutionary calculation in the region, in which the density takes a value less than the specified minimum $\rho_{\min}$, the convective transfer velocities, pressure, and temperature are taken equal to 0. In the calculation $\rho_{\min} = 10^{-8}$.

The differences between the discrete assignment of gas-dynamic variables of the initial field in the accretion disk and the gravitational field from the analytical representation are a type of small perturbations. In the evolutionary calculation (especially under conditions of a large gravitational mass of the star), these perturbations can lead to a nonstationary numerical solution.

In the numerical integration over a sufficiently large time interval ($t < 30$), the main parameters of the environment do not undergo noticeable changes. However, with further integration over time, the numerical solution at the outer boundary of the accretion disk in the region of rarefaction acquires a nonstationary character with the formation of characteristic structures. We note here that a similar behavior of a numerical solution was observed previously [20] when simulating subsonic flow around a circular cylinder, where also initially implemented a stationary, but unstable in a hydrodynamic sense, numerical solution is realized. This solution later is integrated without introducing external disturbances and passed in the near wake into the nonstationary, but stable mode of the vortex track—an analogue of Karman vortex street. Thus, in Fig. 3.2 for the different points in time, the isosurfaces of the density field of level $1 * 10^{-4}$ are shown in plane $z = 0$.

The accretion disk loses its axial symmetry. Proof of this in the form of density fields is presented in Fig. 3.3 for different times (in a logarithmic scale). The pictures are shown in the plane $\varphi = 0$ and $\varphi = \pi$, passing through the axis of rotation of the accretion disk.

These pictures give an idea of the spatially nonstationary structure of the outer boundary of the accretion disk.

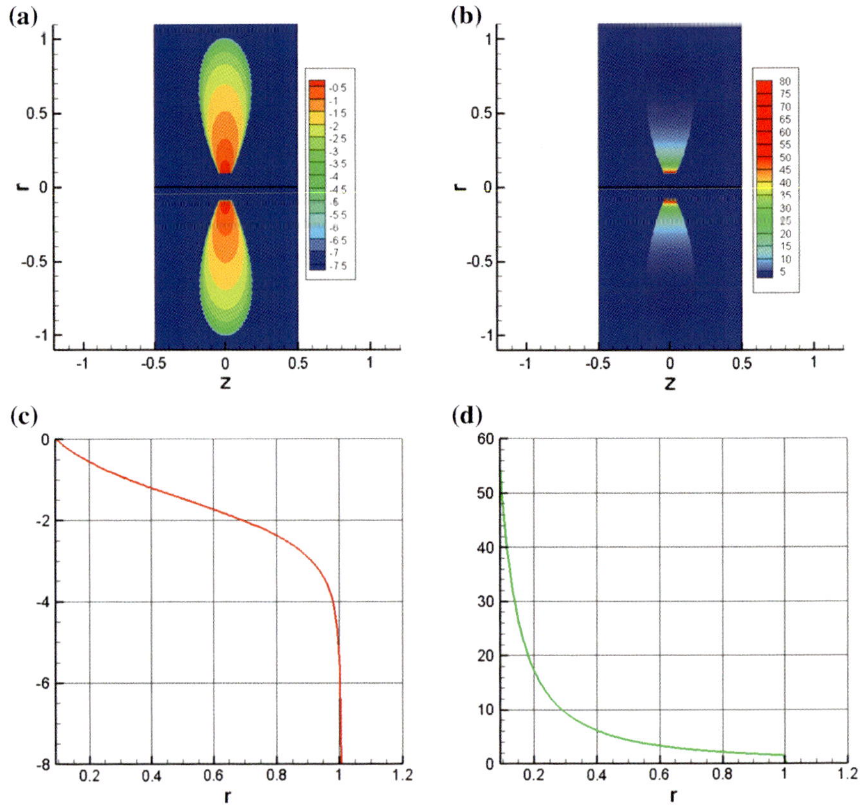

Fig. 3.1 Visualization of initial fields and profiles in the equatorial plane at the initial moment of time (the density is in the logarithmic scale): **a** initial field of the density, **b** initial field of angular velocity, **c** profile of the density, and **d** profile of the angular velocity

3.3.2 *Numerical Results of the Formation of Vortex Structures in Stellar Accretion Disks*

The task of modeling the structure of an accretion disk fast rotating around a central gravitating compact object with mass $M = 2 * 10^{33}$ g (which corresponds to $M_{\odot}$) is considered. Specific radius of the accretion disk is $R_0 = 10^{11}$ cm. The ratio of specific heats γ is equal to 5/3. Self-gravitation of the matter of a disk is not taken into consideration, since it is consumed that the mass of the central gravitating body is almost two orders of magnitude greater than the mass of the matter of the disk.

The integration domain is divided into finite volumes Ω_m by entering a uniform grid in cylindrical coordinates (r, φ, z), where r is the cylindrical radius, φ is the polar angle, and z is the height. Integration is carried out in a cylindrical area Ω bounded by an inner radius r_0, by an outer radius R_1 and by side planes $z = -z_0$ and $z = z_0$: $\Omega = (r_0 \leq r \leq R_1) * (0 \leq \varphi \leq 2\pi) * (-z_0 \leq z \leq z_0)$.

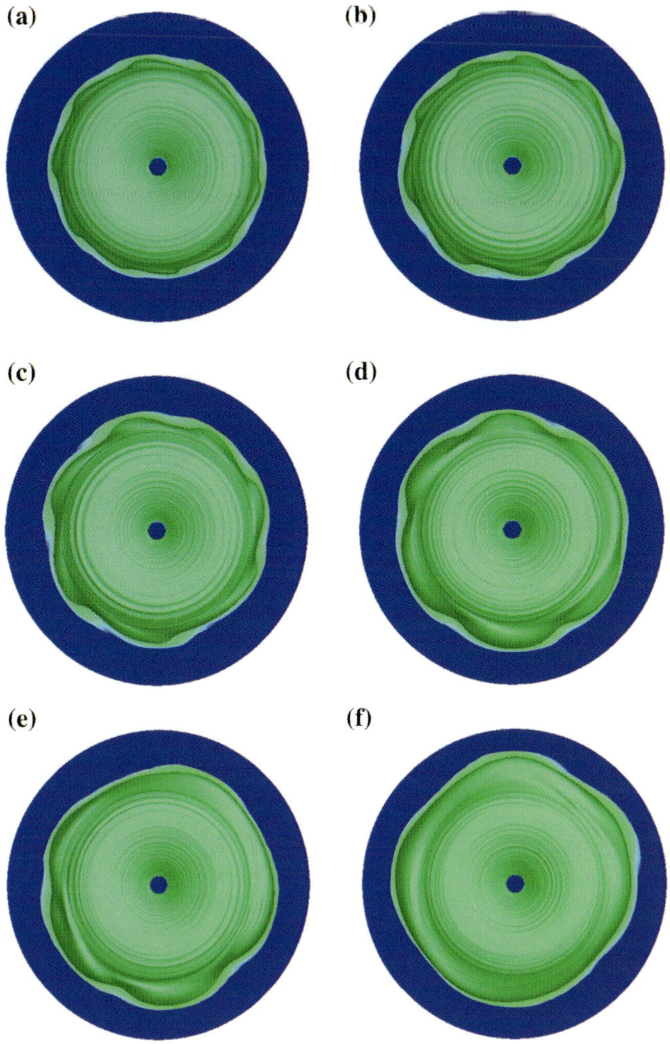

Fig. 3.2 Isosurfaces of the density in plane $z = 0$ of the accretion disk at different points in time: **a** $t = 50$, **b** $t = 70$, **c** $t = 100$, **d** $t = 150$, **e** $t = 170$, and **f** $t = 200$

On the boundaries $r = r_0$, $r = R1$, $z = -z_0$, $z = z_0$ of the computational domain, we set "free" boundary conditions, which are determined by the zero of the normal derivative. In the calculations below geometrical parameters $r_0 = 0.15R_0$, $R_1 = 1.8R_0$, $z_0 = 0.2R_0$ and the region $\Omega = (0.15 \leq r \leq 1.8) \times (0 \leq \varphi \leq 2\pi) \times (-0.2 \leq z \leq 0.2)$ are used. Computational meshes containing up to several million finite volumes are used in calculations. The modeling is carried out by cluster architecture computer systems with up to 256 processors.

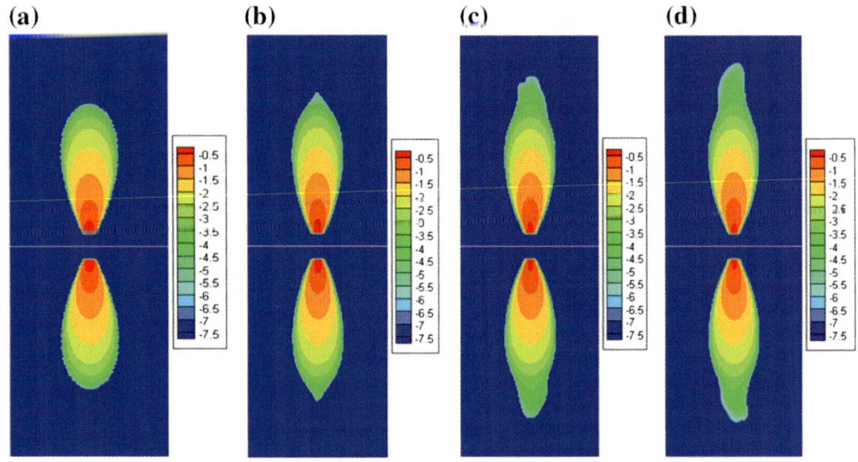

Fig. 3.3 The density fields in the plane $\varphi = 0$ and $\varphi = \pi$ of the accretion disk at different points in time: **a** $t = 0$, **b** $t = 50$, **c** $t = 200$, and **d** $t = 300$ (in the logarithmic scale)

The initial fields of the pressure p, the density ρ, and the velocity $\bar{v}$ are selected by the equilibrium state obtained in work [21] as an analytical solution for the equilibrium gas configuration near the gravitating center.

Further dimensionless variables are used. As scale factors, we choose R_0, M, and G, where G is the gravitational constant. The dimensionless variables, which are marked with a prime, are introduced according to the formulas:

$$r = R_0 r', z = R_0 z', \bar{\mathbf{n}} = \mathrm{v}_0 \bar{\mathbf{n}}', t = t_0 t', p = p_0 p', \rho = \rho_0 \rho', e = e_0 e'.$$

The multipliers v_0, t_0, p_0, ρ_0, e_0 are given by

$$\mathrm{v}_0^2 = \frac{GM}{R_0}, \quad e_0 = \frac{GM}{R_0}, \quad t_0^2 = \frac{R_0^3}{GM}, \quad \rho_0 = \frac{M}{R_0^3}, \quad p_0 = \frac{GM^2}{R_0^4}.$$

In Fig. 3.4, the fields of the density (Fig. 3.4a) and the angular velocity (Fig. 3.4b) in the plane $\varphi = 0$ are shown, as well as their profiles in the radial direction (Fig. 3.4c, d) at the initial time $t = 0$.

In the evolutionary calculation in the area, in which the density takes a value less than the specified minimum $\rho_{\min}$ convective transfer velocities, the pressure and the temperature are taken equal to 0. In the calculation, the minimum density is $\rho_{\min} = 10^{-6}$. Using the well-known technique of introducing small perturbations in a small region into the initial state of the accretion disk [4], let us follow the evolution of the flow in the accretion disk. Note that in contrast to [4], perturbations are injected to density and closer to the outer edge of the disk.

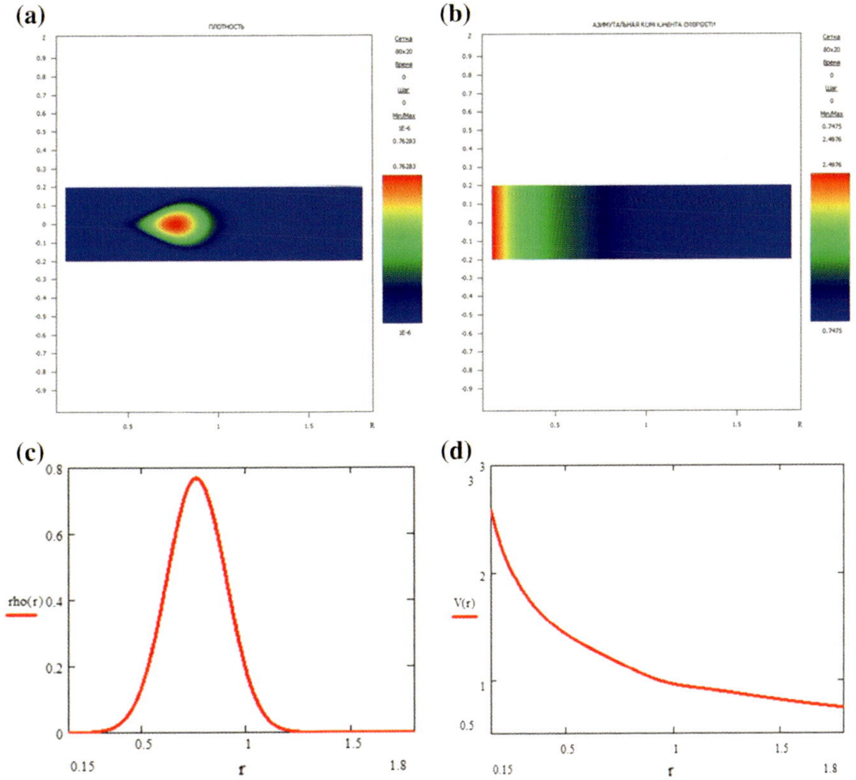

Fig. 3.4 The initial fields and profiles in the equatorial plane at the initial moment of time: **a** initial fields of the density, **b** initial fields of the angular velocity, **c** profiles of the density, and **d** profiles of the angular velocity

In numerical integration, the main parameters of the environment undergo fairly rapid changes, and the numerical solution in the accretion disk acquires a nonstationary character with the formation of characteristic vortex structures. Thus, in Fig. 3.5 for different points in time, the density field patterns in the plane $z = 0$ are shown.

The accretion disk loses its axial symmetry. Proof of this is the density fields presented in Fig. 3.6 for different times. The pictures are shown in plane $\varphi = 0$ passing through the axis of rotation of the accretion disk.

These pictures reflect the vortex structure of the flow of the accretion disk.

3.4 Conclusions

The important result of the completed numerical studies is the occurrence of vortex motion in stellar accretion disks obtained by different numerical methods. In this

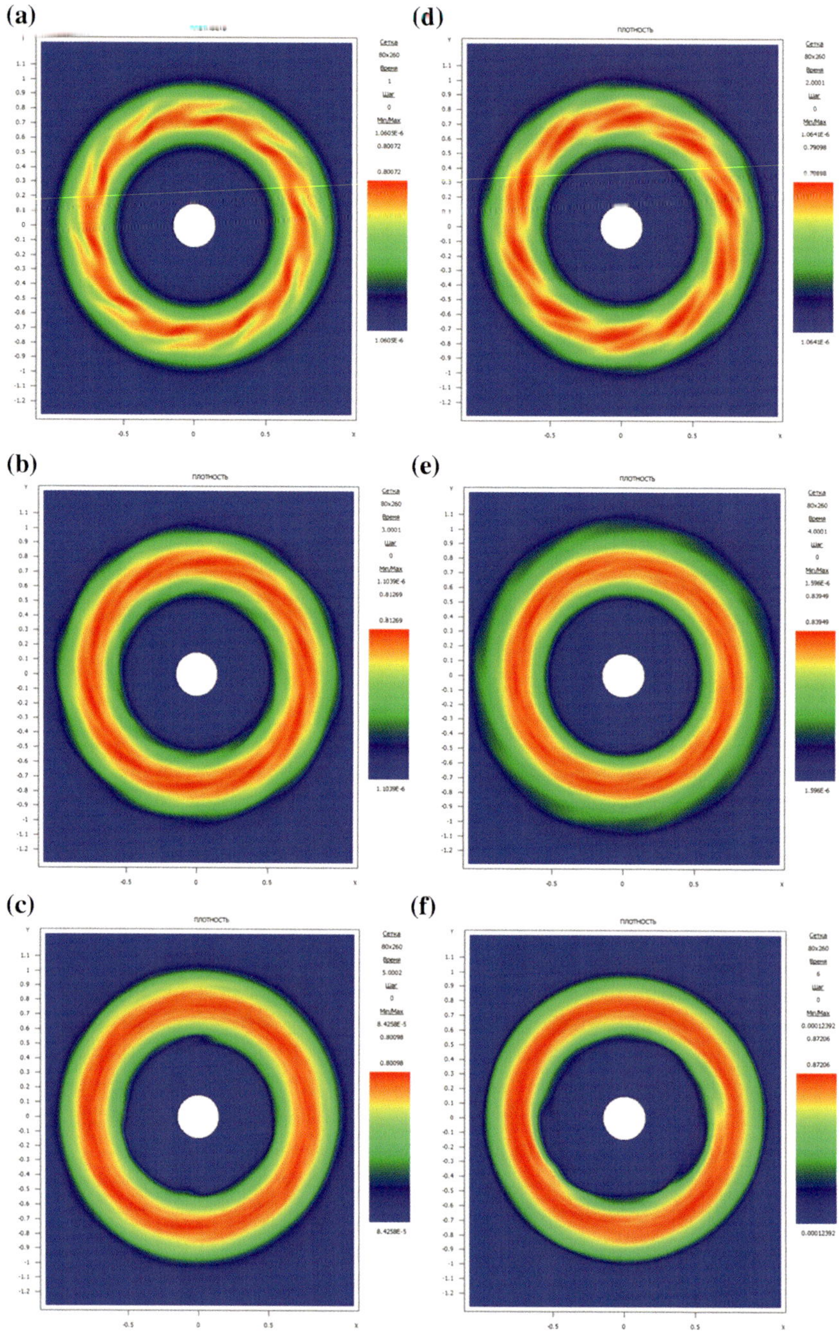

Fig. 3.5 Density patterns in the plane $z = 0$ of the accretion disk at various points in time: **a** $t = 1$, **b** $t = 2$, **c** $t = 3$, **d** $t = 4$, **e** $t = 5$, and **f** $t = 6$

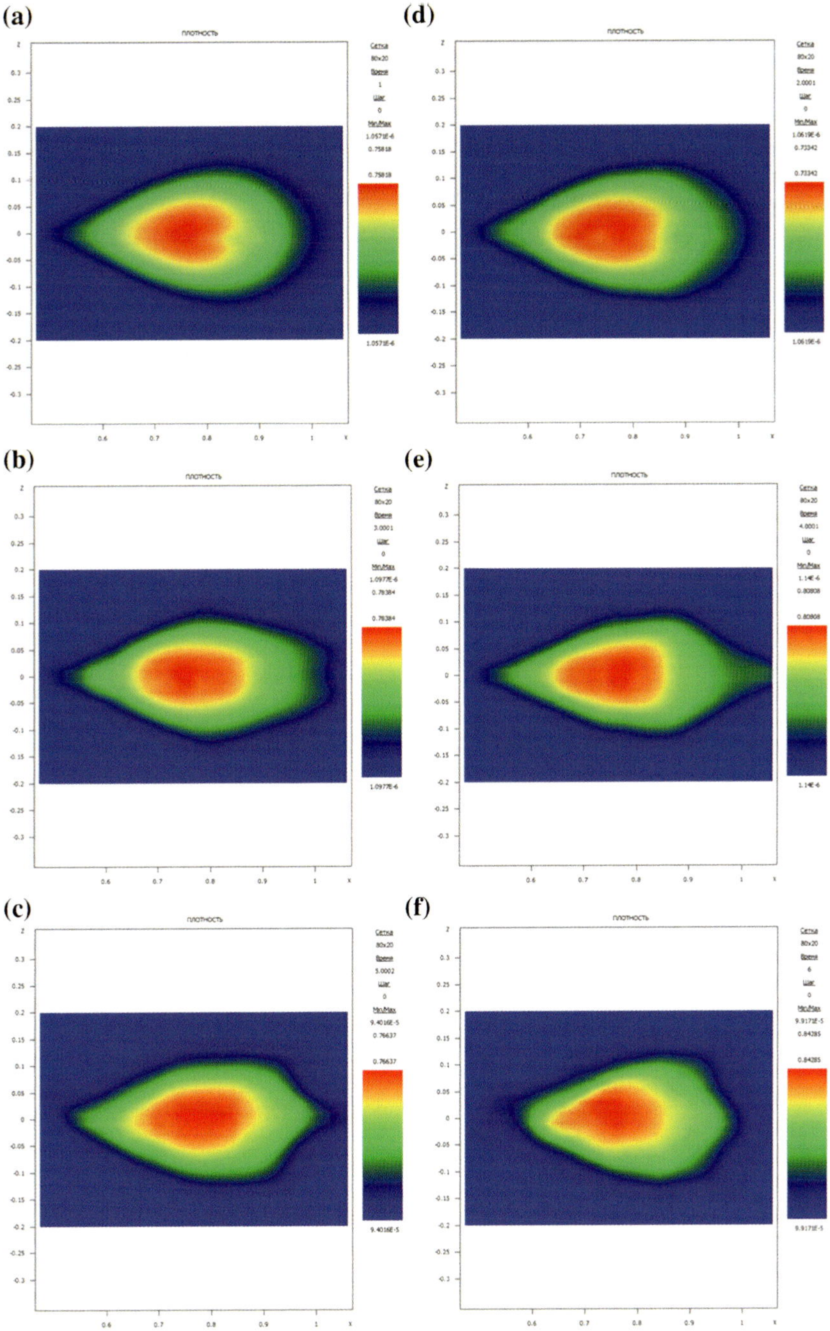

Fig. 3.6 Density fields in the plane $\varphi = 0$ of the accretion disk at different points in time: **a** $t = 1$, **b** $t = 2$, **c** $t = 3$, **d** $t = 4$, **e** $t = 5$, and **f** $t = 6$

case, the flows obtained in different models have a qualitative similarity in the sense that large spiral vortex structures leading to a loss of flow symmetry are formed in the flow, while the flow remains vortex. It is already known that the vortex flows have a significant influence on the evolution of accretion disks. Therefore, the authors consider their studying necessary not only to understand the properties and evolution of the vortex flows but also for research of the evolution of accretion disks, which are significantly influenced by the vortexes.

Another important result is a demonstration of the possibility to model the real astrophysical objects on the basis of supercomputers in real physical conditions including huge space-time areas. Calculations are carried out on the computational resources of the JSCC of the RAS as well as using the equipment of the Research Computing Center of the Lomonosov Moscow State University.

Acknowledgements The authors are grateful to A.G. Aksenov for helpful discussions of the setting of the initial field for the modeling of fast-rotating stellar accretion disks.

The work is performed in the framework of state assignments of the ICAD RAS and KIAM RAS.

References

1. Babakov, A.V., Popov, M.V., Chechetkin, V.M.: Mathematical simulation of a massive star evolution based on a gasdynamical model. Math. Model. Comput. Simul. **10**(3), 357–362 (2018)
2. Aksenov, A.G., Babakov, A.V., Chechetkin, V.M.: Mathematical simulation of the vortex structures in the fast rotation astrophysical objects. Comput. Math. Math. Phys. **58**(8), 1287–1293 (2018)
3. Babakov, A.V., Lugovsky, A.Y., Chechetkin, V.M.: Mathematical modeling of the evolution of compact astrophysical gas objects. In: Petrov I.B., Favorskaya, A.V., Favorskaya, M.N., Simakov, S.S., Jain, L.C. (eds.) Proceedings of the Conference on 50 Years of the Development of Grid-Characteristic Method, SIST, vol. 133, pp. 210–227. Springer (2019)
4. Lugovsky, A.Y., Popov, Y.P.: Roe–Einfeldt–Osher scheme as applied to the mathematical simulation of accretion disks on parallel computers. Comput. Math. Math. Phys. **55**(8), 1407–1418 (2015)
5. Boyarchuk, A.A., Bisikalo, D.V., Kuznetsov, O.A., Chechetkin, V.M.: Mass Transfer in Close Binary Stars. Taylor and Francis, London and New York (2002)
6. Shakura, N.I., Sunyaev, R.A.: Black holes in binary systems. Observational appearance. Astron. Astrophys. **24**, 337–355 (1973)
7. Belotserkovskii, O.M., Oparin, A.M., Chechetkin, V.M.: Turbulence: New approaches Nauka. Moscow (2003)
8. Velikhov, Y.P., Lugovsky, A.Y., Mukhin, S.I., Popov, Y.P., Chechetkin, V.M.: The impact of large-scale turbulence on the redistribution of angular momentum in stellar accretion disks. Astron. Rep. **51**(2), 154–160 (2007)
9. Lugovsky, A.Y., Mukhin, S.I., Popov, Y.P., Chechetkin, V.M.: The development of large-scale instability in stellar accretion disks and its influence on the redistribution of angular momentum. Astron. Rep. **52**(10), 811–814 (2008)
10. Velikhov, Y.P., Sychugov, K.R., Chechetkin, V.M., Lugovskii, A.Y., Koldoba, A.V.: Magneto-rotational instability in the accreting envelope of a protostar and the formation of the large-scale magnetic field. Astron. Rep. **56**(2), 84–95 (2012)

11. Lugovskii, A.Y., Filistov, E.A.: Numerical modeling of transient structures in the disks of spiral galaxies. Astron. Rep. **58**(2), 48–62 (2014)
12. Belotserkovskii, O.M., Severinov, L.I.: The conservative "flow" method and the calculation of the flow of a viscous heat-conducting gas past a body of finite size. Comput. Math. Math. Phys. **13**(2), 141–156 (1973)
13. Belotserkovskii, O.M., Babakov, A.V.: The simulation of the coherent vortex structures in the turbulent flows. Adv. Mech. **13**(3/4), 135–169 (1990)
14. Babakov, A.V.: Program package FLUX for the simulation of fundamental and applied problems of fluid dynamics. Comput. Math. Math. Phys. **56**(6), 1151–1161 (2016)
15. Osher, S., Solomon, F.: Upwind difference schemes for hyperbolic systems of conservations laws. Math. Comput. **38**, 339–374 (1982)
16. Chakravarthy, S., Osher, S.: A new class of high accuracy TVD schemes for hyperbolic conservation laws. AIAA Pap. **85**(0363), 1–11 (1985)
17. Einfeldt, B.: On Godunov type methods for gas dynamics. SIAM J. Numer. Anal. **25**, 294–318 (1988)
18. Kuznetsov, O.A.: Preprint No. 43. IPM RAN (Keldysh Institute of Applied Mathematics of Russian Academy of Sciences), Moscow (in Russian) (1998)
19. Lugovskii, A.Y., Chechetkin, V.M.: The development of large-scale instability in Keplerian stellar accretion disks. Astron. Rep. **56**(2), 96–103 (2012)
20. Babakov, A.V.: On the possibility of the numerical modeling of non-stationary vortex structures in a near wake. Comput. Math. Math. Phys. **28**(1), 173–180 (1988)
21. Abakumov, M.V., Mukhin, S.I., Popov, Y.P., Chechetkin, V.M.: Studies of equilibrium configurations for a gaseous cloud near a gravitating center. Astron. Rep. **40**(3), 366–377 (1996)

Chapter 4
Numerical Modeling of the Kolmogorov Flow in a Viscous Media

Svetlana V. Fortova and Alexey N. Doludenko

Abstract In this chapter, we consider the problem proposed by Kolmogorov to study the causes of turbulence. We consider the action of the external periodic field alone in one of the coordinates on a viscous conductive media in the two-dimensional case. We propose a numerical study of this problem based on the solution of the systems of Navier–Stokes equations. The calculations show that under certain conditions periodic vortex structures may appear in the liquid similar to the "parquet" mode in Kolmogorov task with the subsequent development of the self-similar regime.

4.1 Introduction

The study of the laminar flow stability with respect to small perturbations that always exist in nature is of particular interest both for theoretical studies and for practical applications [1]. This is explained by the fact that the explanation of diverse complex fluid motions and problem of the occurrence of large-scale eddy currents as well are associated with questions of stability [1, 2].

Despite the fact that three-dimensional turbulence is a diverse and essentially non-linear phenomenon, interest to two-dimensional turbulence attracts the attention of many researchers [3–15]. Kreichnan and Batchelor [16, 17] established the fact that two-dimensional turbulence is not a simplified model of three-dimensional turbulence, but has its own unique properties. According to [18], the essential difference between two-dimensional and three-dimensional turbulence is as follows: In three-dimensional case, motions are generated with scales smaller than the scale,

S. V. Fortova (✉)
Institute of Computer Aided Design of the RAS, 19/18 Vtoraya Brestskaya ul., Moscow 123056, Russian Federation
e-mail: sfortova@mail.ru

A. N. Doludenko
Joint Institute for High Temperatures of the RAS, 13 Str.2, Izhorskaya ul., Moscow 125412, Russian Federation
e-mail: adoludenko@gmail.com

L. C. Jain et al. (eds.), *Advances in Theory and Practice of Computational Mechanics*, Smart Innovation, Systems and Technologies 173,
https://doi.org/10.1007/978-981-15-2600-8_4

at which turbulence is excited (the "pumping" scale) [1, 2]. In this case, the energy is distributed in a direct cascade with Kolmogorov law –5/3 [18, 19] in the inertial range of the kinetic energy spectrum. In the two-dimensional case, nonlinearity leads to the appearance of motions with scales far exceeding the "pumping" scale with the appearance of large coherent structures [18]. Energy can be distributed in a reverse cascade (from small structures to large ones) with the –5/3 law of Kreichnan [16, 18] and is also transferred from the pump scale to small scales (direct cascade) with the –3 law due to dissipation of enstrophy. Such movements occur in the formation of cyclones and anticyclones, when the height of the atmosphere is about 10 km, the cyclone size is about 100 km, and consequently, the atmospheric flow can be referred to as quasi-two-dimensional [1].

The inverse cascade in two-dimensional turbulence was studied experimentally [5] and numerically [6, 7]. The peculiarities of these studies are the emergence of intensive large-scale movement including large eddies. In [8], a vortex dipole-stable coherent structure was obtained numerically in a square cell with periodic boundary conditions. In the field experiment [12], a stable coherent structure was also obtained.

In [2], the birth of a periodic self-oscillatory regime is considered as the first step in the transition from a laminar flow to a turbulent one. It is known [1] that in the flat case there is no strictly defined transition point: the laminar flow is transformed into completely chaotic over a certain transition region, where secondary, as a rule, oscillatory movements are superimposed on the main flow. It is of interest to obtain general flow patterns in transition regions, study the dynamics of their behavior, determine the critical values of flow parameters, as well as, find self-oscillating periodic flows and the possibilities of transition to chaos [20]. For geophysical applications, it is of interest to study the stability of such a class of flows, for which the scales of unstable perturbations are commensurate with the spatial scale of the main flow. Such a study, apparently, can provide a key to explaining the evolution of cyclones and anticyclones in the atmosphere and synoptic eddies in the ocean, etc. [1, 2, 18].

In this chapter, we consider the problem proposed by Kolmogorov to study the causes of turbulence in the two-dimensional case and that was experimentally investigated in [21]. It is a study of the plane flow of an incompressible fluid under the action of an external force, periodic in the transverse direction. In the linear formulation, this problem was studied in [22, 23], where the fact of stability loss of the main laminar flow with respect to spatially periodic perturbations with a long wavelength along the flow was proved. The problem of the non-linear development of perturbations and the occurrence of secondary stationary or periodic flows with loss of stability of the laminar flow was described in [20, 23]. However, the issues related to the existence of stable stationary or self-oscillatory regimes of the secondary flow, as well as, the possibility of transition to chaos are still largely open.

We propose a numerical study of the problem of the flat flow of a viscous slightly compressible fluid under the action of a periodic in the transverse direction force. The flow parameters that lead to the appearance of a "vortex parquet" and the subsequent loss of stability of the secondary flow have been determined.

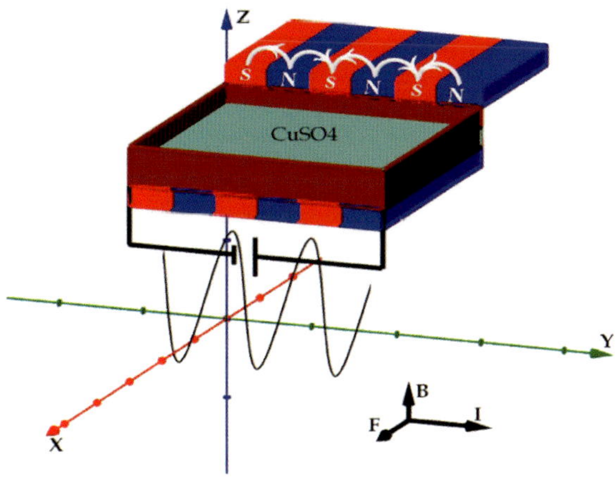

Fig. 4.1 Scheme of the experiment

The chapter is structured as follows. Description of the experiment is given in Sect. 4.2. Section 4.3 provides the problem statement and numerical method. Results of the conducted experiment are presented in Sect. 4.4. Section 4.5 concludes the chapter.

4.2 Description of the Experiment

Our numerical simulation is an interpretation of an experiment that can be found in [21]. Scheme of the experiment is shown in Fig. 4.1.

A flat horizontal rectangular cuvette was filled with an electrically conductive electrolyte aqueous solution (CuSO4). With the help of electrodes mounted on the longitudinal sidewalls of the cell, a constant electric current was passed through the electrolyte in the transverse direction. The cell with the electrolyte was mounted on a sheet of magnetoelastic rubber, which served as the source of an external magnetic field (see Fig. 4.1). By special magnetization, the magnetic field strength was created with a profile close to sinusoidal. Thus, Lorentz electromagnetic force acted on the moving fluid. As a result of this experiment, the authors managed to get a clear picture of the self-oscillating regime of vortex structures called "vortex parquet" or "Kolmogorov parquet." We propose a numerical study of the flat flow problem mainly of a viscous weakly compressible fluid under the action of a periodic force.

4.3 Problem Statement and Numerical Method

The problem statement and numerical method are considered in Sects. 4.3.1–4.3.2, respectively.

4.3.1 Problem Statement

Consider the problem of a flat flow of a viscous weakly compressible fluid under the action of an external periodic force directed along OX-axis, which equals to $\rho G \sin(ky)$. Here, $G = 0.01\ N/\text{kg}$ is Lorentz force equaled to the vector product of the strength of the current passed through the liquid, by the magnetic field strength, k is the wavenumber that specifies the period of the force (in our calculation $k = 1$). This G force came from the experiments with the conducting fluid, which were carried out by other researchers. The motion of the medium in this case is described by Navier–Stokes equations in the form of Eq. 4.1.

$$
\begin{aligned}
&\frac{\partial\rho}{\partial t} + \nabla\cdot(\rho\mathbf{V}) = 0\\
&\frac{\partial\rho u}{\partial t} + \nabla\cdot(\rho u\mathbf{V}) = -\frac{\partial p}{\partial x} + \rho G\sin ky + \mu\Delta u\\
&\frac{\partial\rho v}{\partial t} + \nabla\cdot(\rho v\mathbf{V}) = -\frac{\partial p}{\partial y} + \mu\Delta v\\
&p = \rho/\beta
\end{aligned}
\tag{4.1}
$$

Here, $\mathbf{V} = (u, v)^T$, where u, v are the components of the velocity vector, along OX- and OY- axes, respectively, P *is the* pressure, ρ is the density, β is the artificial compressibility factor.

The flow is investigated in a rectangular area with periodic boundary conditions. Size of this calculation domain equals to $8\pi \times 4\pi$ along OX and OY axes, respectively. Calculations were performed on a grid size of 200×100 and 800×400 cells along OX- and OY- axes, respectively.

In the calculation, the initial condition was immediately set as the sum of the main flow taken as $u = \sin(y)$ and superimposed small disturbances, which can be found in Eqs. 4.2–4.3.

$$u(t = 0) = 0.1\sin(y) + 0.001\sin(x/2) \tag{4.2}$$

$$v(t = 0) = 0.1\sin(y) + 0.001\sin(x/2) \tag{4.3}$$

It is known that for an incompressible fluid, the long-wave perturbations imposed on the main flow field are the most unstable. Therefore, such a superposition of the main flow velocity and small perturbations superimposed on the main flow are presented as the initial conditions for the velocity field. Such a statement of the initial conditions made it possible to obtain an analogue of the self-oscillatory regime.

Other used initial conditions can be represented as follows:

$$P(t = 0) = P_0 = 10^5\ \text{Pa},\ \rho = 1000\ \text{kg}/\text{m}^3,\ \ \mu = 2\ \text{Pas}.$$

Here μ is the viscosity of fluid.

4.3.2 Numerical Method

The calculation algorithm used for viscous medium modeling is based on MacCormack explicit method, which is of second order in accuracy, in time, and in space and well-proven in solving hyperbolic equations. Navier–Stokes equations are solved using the method of artificial compressibility [24]. In this case, the hyperbolic part of the equations is solved by MacCormack method, and the parabolic part is solved using standard finite difference method.

The spectral representation of the kinetic energy will be obtained by decomposition in a two-dimensional Fourier integral. Each of the velocity components can be expanded into a series of orthogonal harmonic functions:

$$v_i(x, y) = \sum_{k_x}\sum_{k_y}[v_i^{(1)}(k_x, k_y)\cos(k_x x)\cos(k_y y) + v_i^{(2)}(k_x, k_y)\cos(k_x x)\sin(k_y y) \qquad i = 1, 2$$
$$+ v_i^{(3)}(k_x, k_y)\sin(k_x x)\cos(k_y y) + v_i^{(4)}(k_x, k_y)\sin(k_x x)\sin(k_y y)], \tag{4.4}$$

where εv_i is one of the velocity components , k_x and k_y are the wave vector components along OX and OY, respectively, $v_i^{(j)^{(i)}}(k_x, k_y)$ $j = 1 \div 4$ are the Fourier coefficients obtained as

$$v_i^{(1)}(k_x, k_y) = \frac{1}{\pi}\int_0^{2\pi}\int_0^{2\pi} v_i(x, y)\cos(k_x x)\cos(k_y y)dxdy, i = 1, 2,$$

$$v_i^{(2)}(k_x, k_y) = \frac{1}{\pi}\int_0^{2\pi}\int_0^{2\pi} v_i(x, y)\cos(k_x x)\sin(k_y y)dxdy, i = 1, 2,$$

$$v_i^{(3)}(k_x, k_y) = \frac{1}{\pi}\int_0^{2\pi}\int_0^{2\pi} v_i(x, y)\sin(k_x x)\cos(k_y y)dxdy, i = 1, 2,$$

$$v_i^{(4)}(k_x, k_y) = \frac{1}{\pi}\int_0^{2\pi}\int_0^{2\pi} v_i(x, y)\sin(k_x x)\sin(k_y y)dxdy, i = 1, 2. \tag{4.5}$$

Value

$$\varepsilon(k_x, k_y) = \frac{[v_1(k_x, k_y)]^2 + [v_2(k_x, k_y)]^2}{2}, \textit{where}$$

$$v_i(k_x, k_y) = \sqrt{\left[v_i^{(1)}(k_x, k_y)\right]^2 + \left[v_i^{(2)}(k_x, k_y)\right]^2 + \left[v_i^{(3)}(k_x, k_y)\right]^2 + \left[v_i^{(4)}(k_x, k_y)\right]^2}, i = 1, 2 \quad (4.6)$$

will be the desired image of the kinetic energy in the space of wave numbers.

So, calculating Fourier coefficients (Eq. 4.5), and, further, calculating Eq. 4.6, we obtain a quantity $\varepsilon(k_x, k_y)$ depending on k_x and k_y. Going through all k_x and k_y in both directions, we obtain the energy spectrum.

4.4 Results

We will give the views of arising characteristic flows of viscous fluid. From Fig. 4.2, it can be seen that a "parquet" of not so big eddies first arises from the initial state with the subsequent development of two relatively large vortices.

In the same Fig. 4.2, one can see the series of the vorticity magnitude taken in different moments of time. It can be seen that from almost initial state (time = 60, let name it phase 1) vortices form a parquet-like pattern (time = 200, let name it phase 2), which over time combine at first into four vortices (time = 400, let name it phase 3), and then into two vortices (time = 1000, 3140, let name it phase 4), which rotate in different directions. Each of the pictures corresponds to more or less stable pattern. These quasi-stable regimes exist on the time moments between peaks of kinetic energy and enstrophy, which one can see in Fig. 4.3a. Thus, each peak separates arising quasi-stationary flows. At the moments of local peaks, transient regimes occur transforming one quasi-stationary state into another.

Turning to the consideration of the viscous case in detail, we can pay attention to Fig. 4.3. One can see the graphs of kinetic energy and enstrophy on time normalized on the maximum value of these parameters. The enstrophy can be described as the integral of the square of the vorticity. In Fig. 4.3a, there are graphs for the case with the region size equals to 200 × 100 computational cells. In Fig. 4.3b, there are graphs for the case with the region size equals to 800 × 400 computational cells. Before moment of time = 400 the behavior of fluid in two of these cases remains similar. Even the time of transition between phase 1 and phase 2 is the same and equals to approximately 70. Time of transition to the phase 3 is almost the same ($t \approx 300$) also but after the phase 3 in more rough mesh it goes into stable phase 4 during the period of time from ≈520 to 1000. Not such a picture is observed in the case of a finer grid. In this case, phase 3 seems to be stable, but nonetheless goes into a truly stable phase 4 in the time zone from 1000 to 1500 (Fig. 4.2).

In the zone of time which equals to 1000 for rough mesh and 1500 for the finer mesh after a long transient regime, a system consisting of two vortices rotating in opposite directions begins to form. The conditional left vortex rotates counterclockwise, and the conditional right vortex rotates clockwise. After that, the system of two vortices goes into the self-similar state. Each of the vortices from time to time forms a "ring." Further, this "ring" is divided into two vortices rotating in one direction and

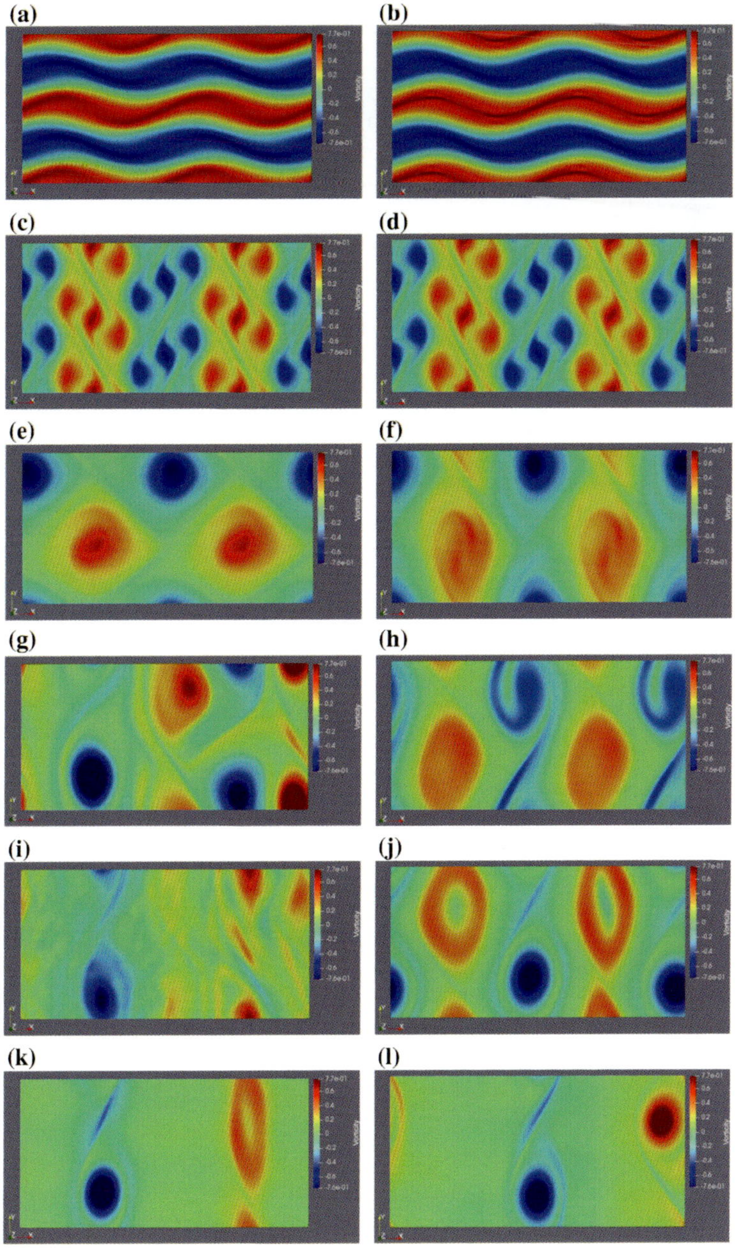

Fig. 4.2 Vorticity magnitude taken in different moments of time for different regions: **a** time = 60, size = 200 × 100 cells, **b** time = 60, size = 800 × 400 cells, **c** time = 200, size = 200 × 100 cells, **d** time = 200, size = 800 × 400 cells, **e** time = 400, size = 200 × 100 cells, **f** time = 400, size = 800 × 400 cells, **g** time = 600, size = 200 × 100 cells, **h** time = 600, size = 800 × 400 cells, **i** time = 1000, size = 200 × 100 cells, **j** time = 1000, size = 800 × 400 cells, **k** time = 3140, size = 200 × 100 cells, **l** time = 2640, size = 800 × 400 cells

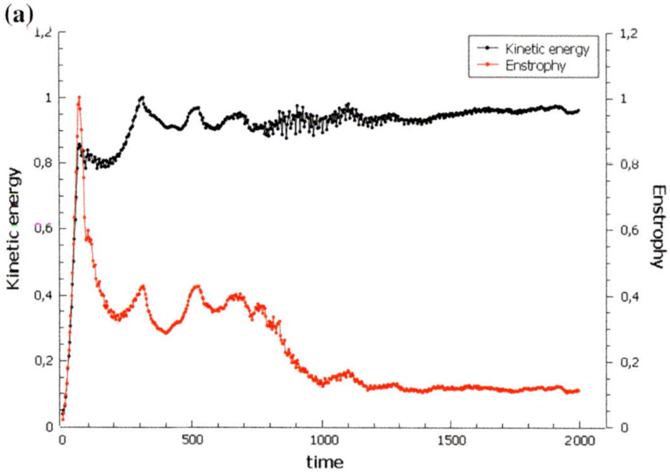

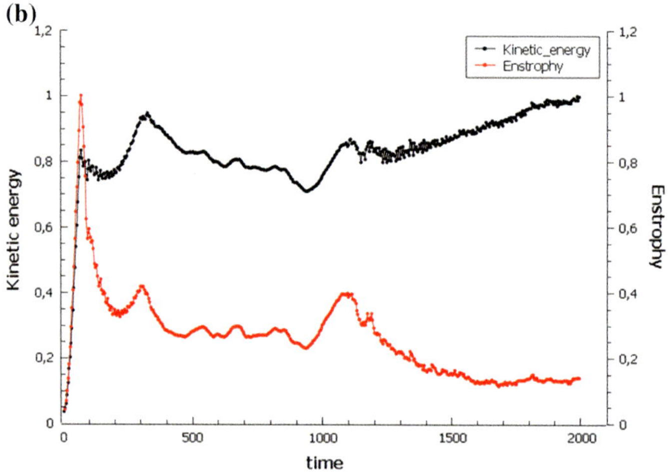

Fig. 4.3 Graphs of kinetic energy and enstrophy on time normalized on the maximum value of these parameters: **a** size of calculated region equals to 200 × 100 cells, **b** size of calculated region equals to 800 × 400 cells

located close to each other. As a result of their interaction, a larger vortex is formed, which later, in its turn, forms a "ring." The process is repeated cyclically throughout the observed time up to 12,800 time steps (for rough mesh). This self-similar mode is observed with the minimum value of enstrophy. It can be also seen as a more pronounced increase in kinetic energy in phase 4 for a finer grid.

At first glance, the behavior of a turbulent system, when energy enters, should tend from order to chaos. This is how most systems behave. Over time, large vortices should disintegrate into small ones. And this process is indeed observed when studying 3D turbulence. However, in 2D case, everything happens the other way around. The entry of energy into a chaotic mixture of small vortices (phase 2) in a

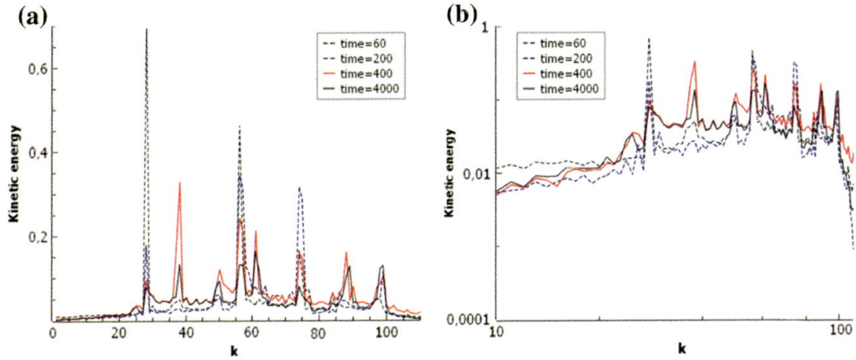

Fig. 4.4 Kinetic energy spectrum at different moments of time, 200 × 100 cells: **a** linear scale, **b** logarithmic scale

two-dimensional system leads to the fact that vortices rotating in the same direction will form more and more large vortices (phase 3) with time. Those, in turn, interact with each other until a pair of stable vortices remains in the system (phase 4). In this case, the system becomes more orderly rather than chaotic. That means energy has to flow from small structures to large ones with the –5/3 law of Kreichnan [16, 18].

If we consider the kinetic energy spectrum in our case (Fig. 4.4), then this kind of spectrum is not clearly observed. In Fig. 4.4, one can find four graphs corresponding to the above phases of fluid flow. Thus, we can observe different peaks on the kinetic energy graphs (these peaks are better seen in Fig. 4.4 with linear scale).

Over time, some peaks disappear, for example, in the region of $k = 28$. Others decrease in amplitude, for example, in the region of $k = 56, 74$. Still others are formed in new locations, for example, peaks in the region of $k = 38, 90$. All this says that there are different modes that carry most of the kinetic energy. And the kinetic energy flows from one mode to another over time. All this happens because we observe not the uniform turbulence, but phases passing into each other with more or less regular structures (vortices). Moreover, during the transition from one phase to another, the structures are combined or, in other words, they are enlarged.

4.5 Conclusions

The result of this research is a direct numerical simulation of the vortex flow formation regime in a layer of weakly compressible medium based on Navier–Stokes equations. Namely, a small perturbation of the velocity components leads to the appearance of a "vortex parquet" and, further, the two rotating vortexes. In addition, when modeling viscous fluid, several transient modes were found separated from each other by maxima of kinetic energy or enstrophy. In the case of a finer computational grid, phase 3 turned out to be more extended in time, and the transition to the self-similar regime took place somewhat later than in the case of a coarser grid. Besides that, the self-similar regime is accompanied by the minimum value of enstrophy.

References

1. Belocerkovskij, O.M., Oparin, A.M.: Numerical Experiment: From Order to Chaos. Nauka, Moscow (in Russian) (2000)
2. Landau, L.D., Lifshicz, E.M.: Continuum Mechanics. Gostex-teorizdat (In Russian) (1953)
3. Boffetta, G., Ecke, R.E.: Two-dimensional turbulence. Rev. Fluid Mech. **44**, 427–451 (2012)
4. Kraichnan, R.H., Montgomery, D.: Two-dimensional turbulence. Rep. Prog. Phys. **43**(5), 547–619 (1980)
5. Sommeria, J.: Experimental study of the two-dimensional inverse energy cascade in a square box. Fluid Mech. **170**, 139–168 (1986)
6. Smith, L.M., Yakhot, V.: Finite-size effects in forced two-dimensional turbulence. J. Fluid Mech. **274**, 115–138 (1994)
7. Boffetta, G., Celani, A., Vergassola, M.: Inverse energy cascade in two-dimensional turbulence: deviations from Gaussian behavior. Phys. Rev. E Stat. Phys. Plasmas Fluids Relat. Interdiscip. Topics. E **61**(1), R29–R32 (2000)
8. Chertkov, M., Connaughton, C., Kolokolov, I., Lebedev, V.: Dynamics of energy condensation in two-dimensional turbulence. Phys. Rev. Lett. **99**(8), 084501.1–084501.4 (2007)
9. Falkovich, G.: Symmetries of the turbulent state. J. Phys. A: Math. Theor. 42(12), 123001.1–123001.18 (2009)
10. Kolokolov, I.V., Lebedev, V.V.: Profile of coherent vortices in two-dimensional turbulence. JETP Lett. **101**(3), 164–167 (2015)
11. Kolokolov, I.V., Lebedev, V.V.: Velocity statistics inside coherent vortices generated by the inverse cascade of 2-D turbulence. J. Fluid Mech. **809**, R2.1–R2.11 (2016)
12. Infeld, E., Rowlands, G.: Nonlinear Waves, Solitons and Chaos, Cambridge UP (2000)
13. Xia, H., Shats, M., Falkovich, G.: Spectrally condensed turbulence in thin layers. Phys. Fluids **21**(12), 125101.1–125101.11 (2009)
14. Laurie, J., Boffetta, G., Falkovich, G., Kolokolov, I., Lebedev, V.: Universal profile of the vortex condensate in two-dimensional turbulence. Phys. Rev. Lett. **113**, 254503.1–254503.5 (2014)
15. Fortova, S.V., Oparina, E.I., Belotserkovskaya, M.S.: Numerical simulation of the Kolmogorov flow under the influence of the periodic field of the external force. J. Phys.: Conf. Ser. **1128**, 012089.1–012089.5 (2018)
16. Kraichnan, R.H.: Inertial ranges in two-dimensional turbulence. Phys. Fluids **10**(7), 1417–1423 (1967)
17. Batchelor, G.K.: Computation of the energy spectrum in homogeneous two-dimensional turbulence. Phys. Fluids **12**(12), II-233–II-239 (1969)
18. Frik, P.G.: Turbulence: Models and Approaches. Lecture Course. Part 2. Perm (in Russian) (1999)
19. Kolmogorov, A.N.: Local structure of turbulence in incompressible fluid at very high Reynolds numbers. Dokl. Akad. Nauk SSSR **30**(4), 299–303 (in Russian) (1941)
20. Belocerkovskij, S.O., Mirabel, A.P., Chusov, M.A.: On the construction of a supercritical regime for a plane periodic flow. Izv. Acad. Sci., USSR, Atmos. Oceanic Phys. **14**(1), 11–20 (in Russian) (1978)
21. Gledzer, E.B., Dolzhanskij, F.V., Obuxov, A.M.: Hydrodynamic Type Systems and Their Application. Nauka, Moscow (in Russian) (1981)
22. Meshalkin, L.D., Sinaj, Ya.G.: Investigation of the stationary solution stability of one system of equations of an incompressible viscous fluid plane motion. J. Appl. Math. Mech. **25**(6), 1140–1143 (in Russian) (1961)

23. Yudovich, V.I.: On the instability of parallel flows of a viscous incompressible fluid with respect to spatially periodic perturbations. Zh. Vychisl. Mat. Mat. Fiz. 242–249 (in Russian) (1966)
24. Anderson, D., Tannehill, J, Pletcher, R.: Computational Fluid Mechanics and Heat Transfer. Mir, Moscow (in Russian) (1990)

Chapter 5
On Structures of Supersonic Flow Around Plane System of Cylindrical Rods

Sergey V. Guvernyuk and Fedor A. Maksimov

Abstract The chapter presents the results of numerical simulation of two-dimensional laminar flows near a regular system of cylinders, forming a plane lattice perpendicular to the velocity vector of the oncoming supersonic flow. A multiblock computing technology is applied using local curvilinear grids adapted to the surface of bodies and having finite areas of overlap with a global rectangular grid. The viscous boundary layers are resolved on the local grids using Navier–Stokes equations. The interaction of shock-wave structures and aerodynamic wakes behind the elements of the lattice is described within Euler equations. With a sequential increase and decrease in Mach number of the oncoming flow, several rearrangements of the flow structure near the grid are found. A multiple hysteresis was revealed, which is expressed in the fact that the flow structure and aerodynamic loads on the lattice elements depend not only on Mach number but also on the history of its change.

5.1 Introduction

The study of a supersonic flow around the bodies containing permeable structures is important for a number of technical applications [1–3]. A supersonic flow around a lattice of cylinders is of interest as a model problem, by the example of which it is convenient to research the interrelation of the external large-scale and local near-wall flows near the permeable bodies, such as mesh or perforated screens, distributed systems of solid particles, etc. Very complex interaction of local compression shocks

---

S. V. Guvernyuk · F. A. Maksimov (✉)
Institute of Mechanics, Lomonosov Moscow State University, 1, Michurinskii prt., Moscow 119192, Russian Federation
e-mail: f_a_maximov@mail.ru

S. V. Guvernyuk
e-mail: guv@mail.ru

F. A. Maksimov
Institute of Computer Aided Design of the RAS, 19/18, Vtoraya Brestskaya ul., Moscow 123056, Russian Federation

L. C. Jain et al. (eds.), *Advances in Theory and Practice of Computational Mechanics*, Smart Innovation, Systems and Technologies 173,
https://doi.org/10.1007/978-981-15-2600-8_5

occurs among themselves and with the aerodynamic wakes behind the lattice elements [4–6] in the case of sufficiently rarefied grids. Moreover, different schemes of the resulting flows of such interaction are equally possible for some combinations of the problem parameters. This indicates a possibility of the parametric hysteresis.

Hysteresis of the flow structure behind an infinite lattice of cylinders is revealed in [5] under the conditions, when both regular and Mach schemes of intersection of the oblique shocks propagating from the neighboring lattice elements, are equally possible. Two types of hysteresis were identified by the lattice permeability in the study [6], at fixed Mach number $M = 6$. One of them is associated with the destruction of the collective flow around the lattice elements. The second one is connected with the rearrangement of the near wake behind the lattice elements under the influence of the local shock waves from the neighboring lattice elements.

The chapter considers the supersonic flow around a geometrically unchanged lattice of cylinders (permeability 80%) with a variation in Mach number in the direction of increasing and decreasing in the range from 2.0 to 4.5.

The chapter is structured in the following manner. Section 5.2 provides a formulation of problem. Multiblock computing technology is discussed in Sect. 5.3. The calculation results for lattice of 10 elements and infinite lattice with periodical conditions are represented in Sects. 5.4 and 5.5, respectively. Conclusions are given in Sect. 5.6.

5.2 Formulation of Problem

The flow around the grid of circular cylindrical rods with axes parallel to each other and lying in a plane, perpendicular to the direction of a uniform supersonic stream of a viscous perfect gas, is considered. It is assumed that the length of the rods is larger than the transverse size of the lattice. Therefore, a flow field up to sufficiently large distances in front and behind the plane lattice does not depend on the length of the rods and is described by the system of two-dimensional Navier–Stokes equations with boundary conditions of adhesion on the surface of the lattice elements.

The determining dimensionless parameters of the problem are: M is Mach number of the unperturbed oncoming flow, Re is Reynolds number based on cylinders diameter d, γ is the ratio of gas heat capacities (adiabatic exponent), n is the number of rods in the lattice, h/d is the relative period of the lattice. The geometric permeability of the lattice is defined as $\sigma = (h - d)/d$, $0 < \sigma < 1$. The characteristic mode of a supersonic flow around a permeable screen is a flow with a smooth detached shock wave in front of the screen at small and moderate values of σ [1, 6, 7]. We will refer to such a flow as mode ***A***. The head shock wave ceases to be common for the entire lattice and transforms into a system of local shock waves in the vicinity of cylinders [6] for sufficiently large σ and M. This flow will be referred to as mode ***B***.

The task of flow around a single lattice element cannot be set without taking into account the location of this element relative to the edge of the lattice in the case of mode ***A***. On the other hand, a flow along the most elements of the lattice is local in mode ***B***, independent of the location of the element relative to the edge of the lattice. This allows us to use a simplified formulation of the problem considering a flow around any of the rods as a fragment of a periodic flow near the infinite system of cylinders [5, 6]. We use both of the described formulations of the problem in this chapter.

A uniform supersonic flow is specified before the grid at the input boundary of the rectangular computational domain. The lateral boundaries of the computational domain are taken at a sufficiently large distance from the edges of the lattice and non-reflective boundary conditions are set on them when calculating the joint flow around all the elements of the lattice (mode ***A***). The lateral boundaries pass along the lines of symmetry between the rods at a distance of lattice period h and conditions for the periodicity of the flow are set on them when calculating in mode ***B***.

5.3 Multiblock Computing Technology

System of overlapping grids is used for numerical simulation of the flow: one global and many local ones (by the number of elements in the lattice).

Description of the far field of the flow is carried out on a uniform global grid with rectangular cells. This grid, strictly speaking, does not allow describing physical dissipative processes in the wake. The application of Navier–Stokes equations with a limited number of nodes used in real calculations on the global grid cannot be justified. Therefore, Euler equations are applied (the dissipative term is assumed to be zero) on this grid in numerical simulation of the flow. The local grids are adapted to the surface of the lattice elements and have an exponential refinement, which makes it possible to adequately simulate a viscous near-wall flow near these elements. In this chapter, Navier–Stokes equations are used in the thin-layer approximation, i.e., only the second derivatives along the normal to the surface of the body are taken into account when calculating the dissipative term, as in the theory of the boundary layer.

The nonstationary Navier–Stokes equations in a thin-layer approximation for a two-dimensional plane flow of compressible gas in a dimensionless vector form in a curvilinear coordinate system $\xi = \xi(x, y)$ and $\eta = \eta(x, y)$ are as follows:

$$\frac{\partial}{\partial \tau}\frac{\mathbf{U}}{J} + \frac{\partial}{\partial \xi}\frac{\xi_x\mathbf{E} + \xi_y\mathbf{F}}{J} + \frac{\partial}{\partial \eta}\frac{\eta_x\mathbf{E} + \eta_y\mathbf{F}}{J} = \frac{\partial}{\partial \eta}\frac{\mathbf{S}}{J},$$

$$\mathbf{U} = \begin{Bmatrix} \rho \\ \rho u \\ \rho v \\ e \end{Bmatrix},\ \mathbf{E} = \begin{Bmatrix} \rho u \\ \rho u^2 + p \\ \rho uv \\ (e + p)u \end{Bmatrix},\ \mathbf{F} = \begin{Bmatrix} \rho v \\ \rho uv \\ \rho v^2 + p \\ (e + p)v \end{Bmatrix},\ \mathbf{S} = \frac{\mu}{\mathrm{Re}}\begin{Bmatrix} 0 \\ m_1\frac{\partial u}{\partial \eta} + m_2\eta_x \\ m_1\frac{\partial v}{\partial \eta} + m_2\eta_y \\ m_3 \end{Bmatrix},$$

$$m_1 = \eta_x^2 + \eta_y^2,\ m_2 = \frac{1}{3}\left[\eta_x\frac{\partial u}{\partial \eta} + \eta_y\frac{\partial v}{\partial \eta}\right],$$

$$m_3 = m_1\left[\frac{\gamma}{\gamma-1}\frac{1}{\Pr}\frac{\partial T}{\partial \eta} + \frac{\partial}{\partial \eta}\frac{u^2+v^2}{2}\right] + m_2\left[\eta_x u + \eta_y v\right].$$

Here, t is the time, ρ designates the density, (u, v) are the components of the velocity vector V in the respective directions (x, y) of Cartesian coordinate system, p is the pressure, and e denotes the total energy of the unit gas volume, which for a perfect gas can be represented as $e = \rho\left(\varepsilon + \frac{1}{2}\left(u^2+v^2\right)\right)$, where $\varepsilon = \frac{1}{\gamma-1}\frac{p}{\rho}$ is the internal gas energy, and γ is the adiabatic exponent.

The dimensionless variables are defined through dimensional quantities, which are indicated by Eq. 5.1.

$$t = \sqrt{\frac{p'_o}{\rho'_o}}\frac{t'}{L'}\quad \mathbf{X} = \frac{\mathbf{X}'}{L'}\quad \mathbf{V} = \sqrt{\frac{\rho'_o}{p'_o}}\mathbf{V}'\quad \rho = \frac{\rho'}{\rho'_o}\quad p = \frac{p'}{p'_o}\quad T = \frac{T'}{T'_o}\quad \mu = \frac{\mu'}{\mu'_o} \tag{5.1}$$

The subscript $_o$ in Eq. 5.1 means the value of the parameter in an undisturbed flow. Here, L' is the characteristic dimension, $X = (x, y)$, and $V = (u, v)$.

It is assumed that the Prandtl number $\Pr = \frac{\mu c_p}{\lambda}$ is the constant, c_p is the heat capacity ratio, λ stands for the coefficient of heat conductivity, μ is the viscosity ratio. $\mathrm{Re} = \frac{\sqrt{p'_o\rho'_o}L'}{\mu'_o}$ is Reynolds number. The system of differential equations is supplemented by the equation of state $P = \rho RT$, where T is the temperature and R denotes the gas constant ($p = \rho T$ in dimensionless form).

The coefficients of the transformation matrix can be calculated by the following formulas: $\xi_x = J\frac{\partial y}{\partial \eta}$, $\xi_y = -J\frac{\partial x}{\partial \eta}$, $\eta_x = -J\frac{\partial y}{\partial \xi}$, and $\eta_y = J\frac{\partial x}{\partial \xi}$. Here, J is the Jacobian of the transformation, which is determined by the formula $J^{-1} = \frac{\partial x}{\partial \xi}\frac{\partial y}{\partial \eta} - \frac{\partial x}{\partial \eta}\frac{\partial y}{\partial \xi}$. The use of a generalized transformation makes it possible to construct a uniform grid in the form of a unit square. The coefficients of the transformation matrix for a given distribution of nodes in the physical domain are calculated using the central differences.

In deriving the reduced system of equations, it is assumed that the coordinate lines, $\xi = const$, are oriented along the normal to the surface of the body, and the derivatives of the direction η actually correspond to the derivatives along the local normal to the body surface. Due to this, the second derivatives along the normal to the body surface are taken into account when calculating the dissipative term.

Figure 5.1a presents an example of a fragment of an external computational grid with uniform rectangular cells superimposed by a body domain and the contour of the external boundary of the grid near the body. Figure 5.1b presents a magnified fragment of the calculation area near the edge of the body with a grid. First, coordinate lines $\xi = const$ are constructed (the conformal mapping ensures the orthogonality of these lines to the body contour), and second, the grid nodes are arranged along these lines using exponential refinement taking into account the minimum distance between the body node and the nearest node to the body and the distance from the body to the outer boundary of the computational domain. On considering the grid, the

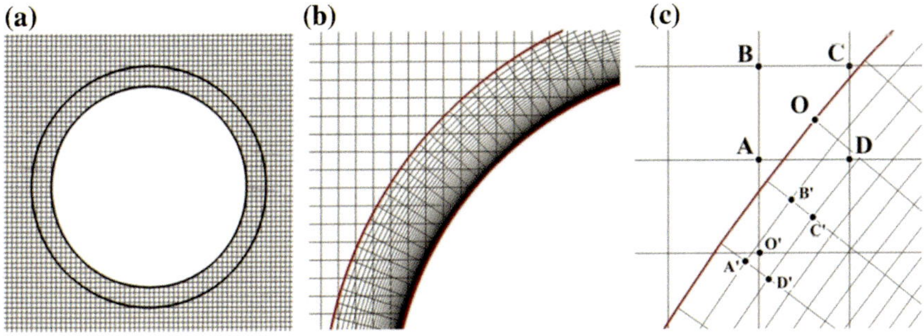

Fig. 5.1 Construction of computational grid: **a** fragment of an external computational grid, **b** magnified fragment of calculation area, **c** interpolation of values

outer grid is superimposed by a set of n bodies, near each of which the corresponding curvilinear grid is constructed.

The solution is obtained by the relaxation method. An explicit second-order approximation scheme [8] is used. The specific feature of calculations using multi-block technology is integration with a common time step on the outer computational grid, i.e., the minimum integration time step is chosen from the stability condition over the entire computational domain. This is not a significant limitation because the grid is uniform. At the same time, integration on the grids near the bodies involves the use of the local time step, i.e., the choice of the time step at each grid nose is determined by the local conditions. This results in a faster disturbance propagation and, therefore, relaxation.

In order to tie together the solutions on the outer grid and the grids around the bodies into a single whole, after completing the integration step, the values of the gas-dynamic functions on the outer boundary L of the grid near the body are determined by the interpolation from the solution obtained on the outer grid. Since a two-step difference scheme [8] is used, a similar procedure is also performed for the node layer in the vicinity of the boundary L. At the same time, the solution at all nodes at the outer grid that happens to be inside the domain of the solution determination near the body is replaced by the solution obtained on this grid [9].

When recalculating the values of gas-dynamic functions from one grid to another, interpolation is used. Interpolation is implemented in the following form (in the considered two-dimensional case). At first, we define the cell $ABCD$, in which point O is located (Fig. 5.1c). Then values of the functions at point O are determined by the values of the functions at nodes A, B, C, and D. The value of function f at node O can be obtained from its values at all three nodes, for definiteness, let them be node B, A, and D, by the interpolation formula:

$$f_O = f_A + \alpha \cdot (f_B - f_A) + \beta \cdot (f_D - f_A),$$

where $\alpha = \frac{|AO \times AB|}{|AD \times AB|}$ and $\beta = \frac{|AO \times AD|}{|AB \times AD|}$.

In order to take into account the value of the function at node C, we can similarly express f_O in terms of its values at a point of another triple of nodes, for instance, D, C, and B. The final expression for f_O is taken as an arithmetic average of the values for the four chosen variants of the corner point. The interpolation coefficients are determined for all nodes on the contour L and adjacent to it.

For nodes of the uniform grid that are inside the grids around the bodies, for example, the point O' that happens to be in the cell $A'B'C'D'$ (Fig. 5.1c), the same procedure is performed, and the corresponding interpolation coefficients are determined.

In addition to the area, in which solutions are conjugated on the outer uniform grid and grids near bodies, the necessary boundary conditions ought to be set at other boundaries. For grids near bodies, this is the boundary corresponding to the surface of the body, on which the no-slip condition and the given surface temperature are set.

In calculation, the grids with the following numbers of cells are used: the local grid near the bodies includes 180×40 cells, the outer grid around body system involves 2000×2400 cells, and the outer grid in the periodic conditions about one lattice element consists of 8000×800 cells.

5.4 Calculation Results for Lattice of 10 Elements

The calculations were performed for fixed $\gamma = 1.4$, $Re = 10^5$, $n = 10$, and $h/d = 5$ ($\sigma = 80\%$) with variation of M in the range $2.0 \leq M \leq 4.5$. The symmetry of the flow around the lattice was assumed, and in fact only the upper half of the flow region with five cylinders located in it was calculated.

Figures 5.2, 5.3, 5.4, and 5.5 show the flow patterns obtained according to the scenario with a sequential increase in Mach number and the use of the solutions from the previous step in the parameter M as the initial field. The visualization of the flow fields is represented by the levels of the density gradient modulus. The coordinate axes are normalized to the diameter of the elements of lattice d. When constructing all the drawings, the same variant of the palette settings is used, therefore, at higher Mach numbers, the drawings are darker. First, up to $M = 2.4$, there is a smooth detached shock wave (Fig. 5.2) before the lattice as before a solid screen. The flow between this shock and the lattice is subsonic, and the sound speed is reached (mode A) in the local gap between the adjacent rods.

Then, when $M = 2.5$ is reached, a sudden qualitative change in the flow pattern occurs: the smoothness of the bow shock wave is disturbed, numerous branch points appear, and a characteristic configuration of Mach intersection of shock waves is formed between the neighboring lattice elements (Fig. 5.3). The reflected shock waves interact with the subsonic region of the near wake, as a result of which pressure rises there and the local areas of the separated flow behind the cylinders increase substantially. A similar phenomenon was observed with increasing lattice permeability [6]. This mode is characterized by the fact that all elements of the finite lattice, except for the extreme ones, are flown around in almost the same way. As Mach number

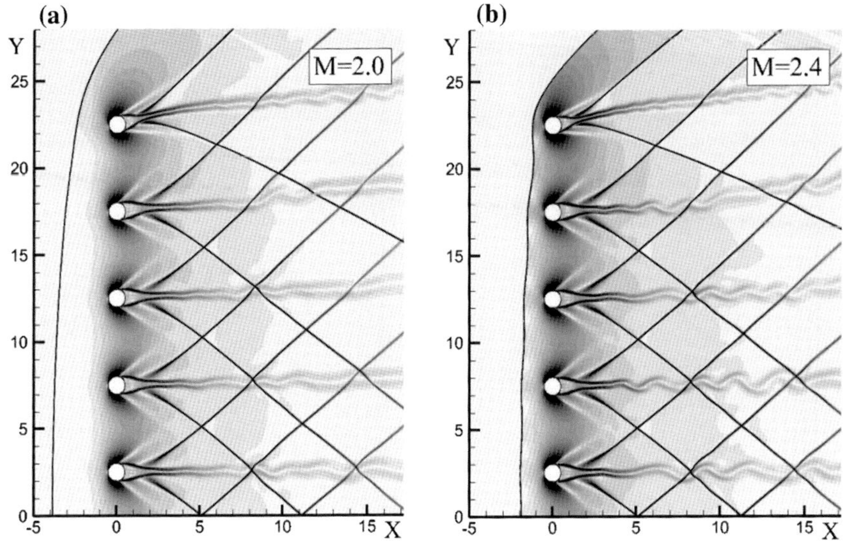

Fig. 5.2 Flow around lattice as M increases, mode A: **a** $M = 2.0$, **b** $M = 2.4$

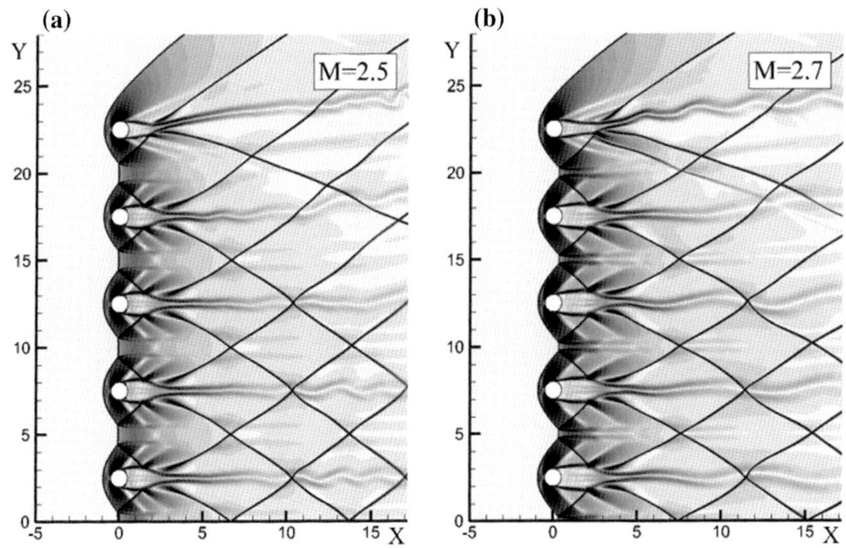

Fig. 5.3 Flow around lattice as M increases, mode B: **a** $M = 2.5$, **b** $M = 2.7$

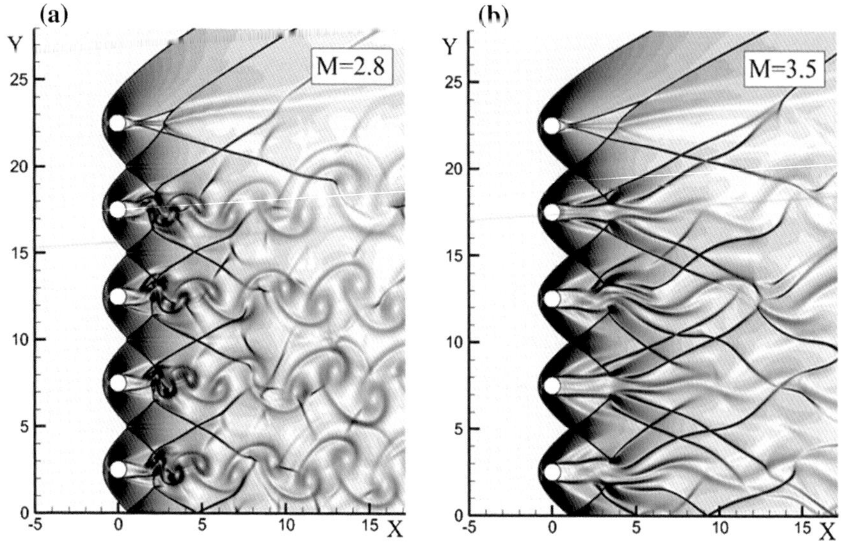

Fig. 5.4 Flow around lattice as M increases, mode B: **a** $M = 2.8$, **b** $M = 3.5$

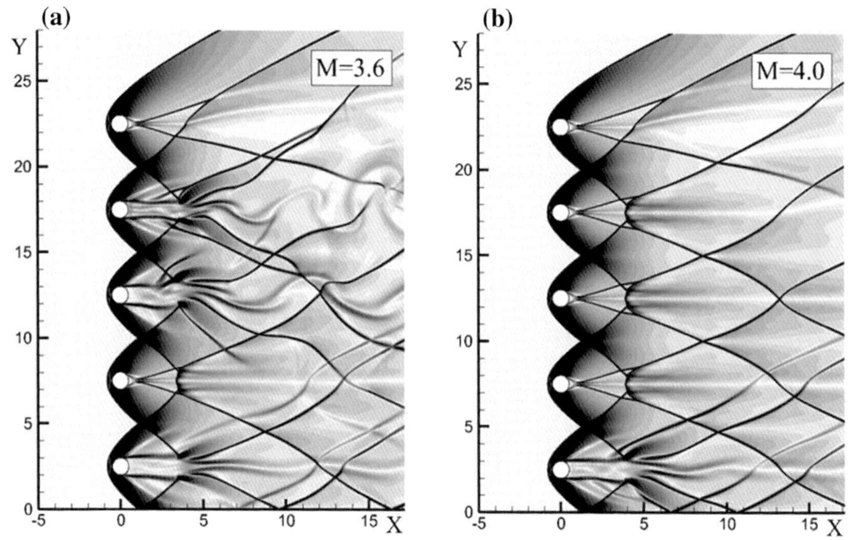

Fig. 5.5 Flow around lattice as M increases, mode B: **a** $M = 3.6$, **b** $M = 4.0$

increases, the reflected shock waves shift downstream, and the length of the separation areas behind the cylinders increases. In this case, the wake behind the lattice as a whole has a fairly regular appearance, which persists up to $M = 2.7$ (Fig. 5.3).

However, when reaching the value $M = 2.8$ (Fig. 5.4), the flow with a stationary separation area behind each cylinder and a regular wake, as in Fig. 5.3, collapses.

Apparently, with such an extended separation region behind the cylinders, the stationary balance of mass and momentum fluxes at the boundaries of the separation regions ceases to be ensured, which leads to periodic self-oscillations in the position and size of these regions and to a significant change in the structure of the entire flow (Fig. 5.4). The visualized traces behind the elements of the lattice look like Karman vortex street, however, it should be borne in mind that this happens against the background of the generally supersonic flow behind the lattice.

The unsteady flow pattern is qualitatively preserved for all internal elements of the lattice, when $2.8 \leq M \leq 3.5$. At the same time, a transition from Mach to a regular branching diagram of the bow shock occurs (Fig. 5.4). Then, starting with $M = 3.6$, abrupt transitions to patterns with independent flow around lattice elements begin to occur. The separation region sharply decreased near one of the internal elements of the lattice (the second element from the plane of symmetry in Fig. 5.5) and became the same as near the extreme element when $M = 3.6$.

The same thing happened near all elements, except for the one closest to the plane of symmetry (Fig. 5.5), when $M = 4.0$. Finally, when $M = 4.5$, all lattice elements are being flowed around without mutual influence on each other. The local flow is the same as near a single cylinder in a supersonic flow. Shock waves from neighboring lattice elements fall only on the supersonic part of the wakes, which affects the flow characteristics downstream, but does not change the local flow around the elements themselves. A similar flow around the lattice took place, when $M = 4.0$, while a uniform field of the incoming flow was taken as the initial field (Fig. 5.6).

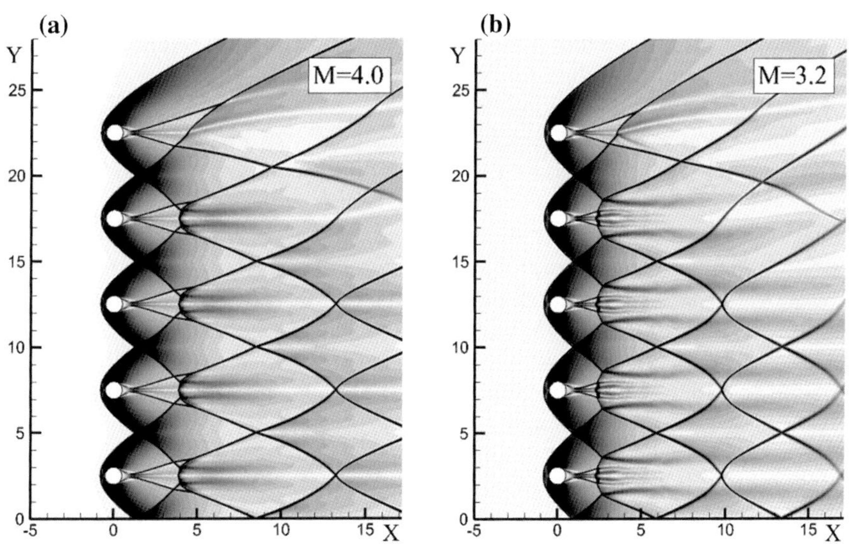

Fig. 5.6 Flow around lattice as M increases, mode B: **a** $M = 4.0$, **b** $M = 3.2$

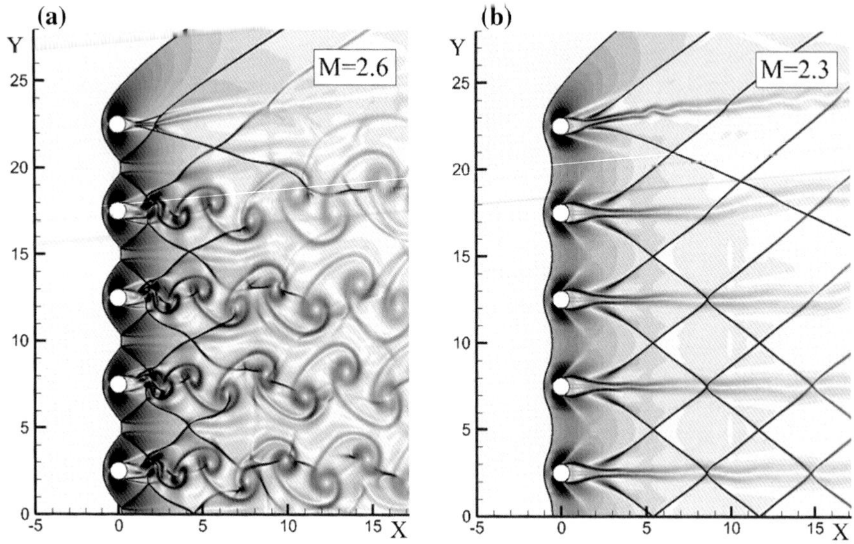

Fig. 5.7 Flow around lattice as M decreases: **a** mode B, $M = 2.6$, **b** mode A, $M = 2.3$

The following are examples of the results of a series of calculations performed in a different scenario: with a sequential decrease in Mach number from 4.5 to 2.0 (Figs. 5.6 and 5.7). All the flow patterns described above were also observed in this case, but with a certain shift of the boundaries of the transition from one flow pattern to another. The flow pattern is implemented without the influence of neighboring elements on the flow around each other (Fig. 5.6) in the range of $4.0 \geq M \geq 3.2$.

When $3.1 \geq M \geq 2.6$, a flow pattern with a nonstationary wake behind the cylinders and a significant effect of their interference on the parameters of the local separated flows in the bottom area behind the lattice elements (Fig. 5.7, $M = 2.6$) is observed.

When $2.5 \geq M \geq 2.3$, the flow is stabilized in the bottom area behind the cylinders, while the length of the region of the separated flow is abnormally large (Fig. 5.7, $M = 2.3$) as compared to other modes. The bottom pressure is increased on the lattice elements. The latter means a decrease in the aerodynamic drag. When $2.2 \geq M \geq 2.0$, the flow around the grid returns to mode A (as in Fig. 5.2).

It should be stressed that a transition from one flow pattern around the lattice to another occurs in a threshold manner in both considered scenarios of Mach number changes. The transition takes place so that there are ranges of values of M, for which different flow patterns can exist. The selection of an actual flow pattern is determined by the pre-history of the change in Mach number.

The implemented flow patterns affect the aerodynamic drag F_x of the grid. According to the calculation results, an estimate was obtained for the dependence of the aerodynamic drag coefficient $c_x = 2F_x/\gamma\, p_\infty M^2 S_x$ (p_∞ is the static pressure in the undisturbed flow and S_x designates the drag area) on Mach number, Fig. 5.8. The

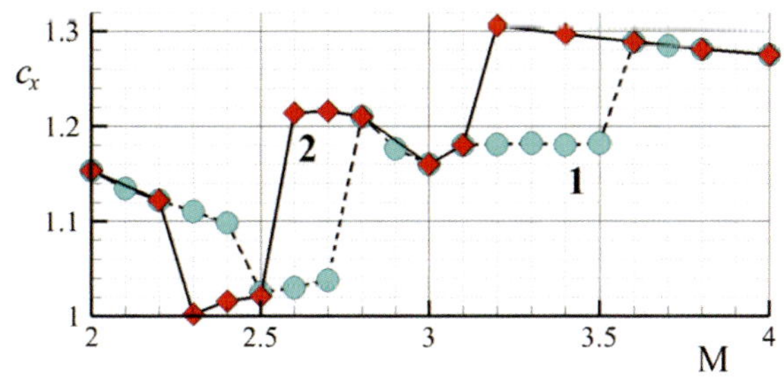

Fig. 5.8 Dependence of aerodynamic drag coefficient of lattice element on Mach number, curve 1—$c_x = f(M)$ with consistent increase in M, curve 2—with decrease in M

unsteadiness of the wake is weakly reflected in the oscillations of c_x (maximum deviation of the average value is not more 6%).

In accordance with the four qualitatively different flow patterns listed above, four levels of drag coefficient values are realized, the transitions between which are characterized by three double-valued solution intervals.

5.5 Calculation Results for Infinite Lattice with Periodical Conditions

As it was noted above, flows on mode ***B*** can be investigated under a simplified formulation of the problem, considering one lattice element with periodicity conditions on the side boundaries. This allows one to significantly reduce the computational domain and increase the level of the solution refinement. On the other hand, the correctness of the periodicity conditions in the case of nonstationary flow patterns is not obvious, since the periodicity condition may impose restrictions on the nature of the disturbance propagation in the flow region. Figure 5.9 shows the calculation results for a single lattice element with periodicity conditions on the side boundaries.

The oncoming supersonic flow is directed from left to right in each of the presented flow patterns. The distribution of the density gradient modulus is presented. The upper and lower boundaries of the computational domain correspond to the period of the infinite lattice. Transitions between the same flow patterns that were identified above in the case of a lattice with a finite number of elements for mode *B* are observed.

Figure 5.10 shows a summary graph of the dependence of the drag coefficient c_x of a single lattice element. Line 1′ corresponds to the scenario with a sequential increase in M and line 2′ indicates the scenario with a decrease in M for an infinite lattice. The data of Fig. 5.8 obtained for a lattice of 10 elements (lines 1 and 2) are shown in the same place in Fig. 5.10.

The calculation results using a finite-size lattice model and a simplified one for a single lattice element with periodicity conditions do not qualitatively differ, but there are some differences in estimating the hysteresis boundaries and the value of aerodynamic drag in the mode with the formation of Karman vortex street. This may be due to a virtually different resolution of the computational grid in different models.

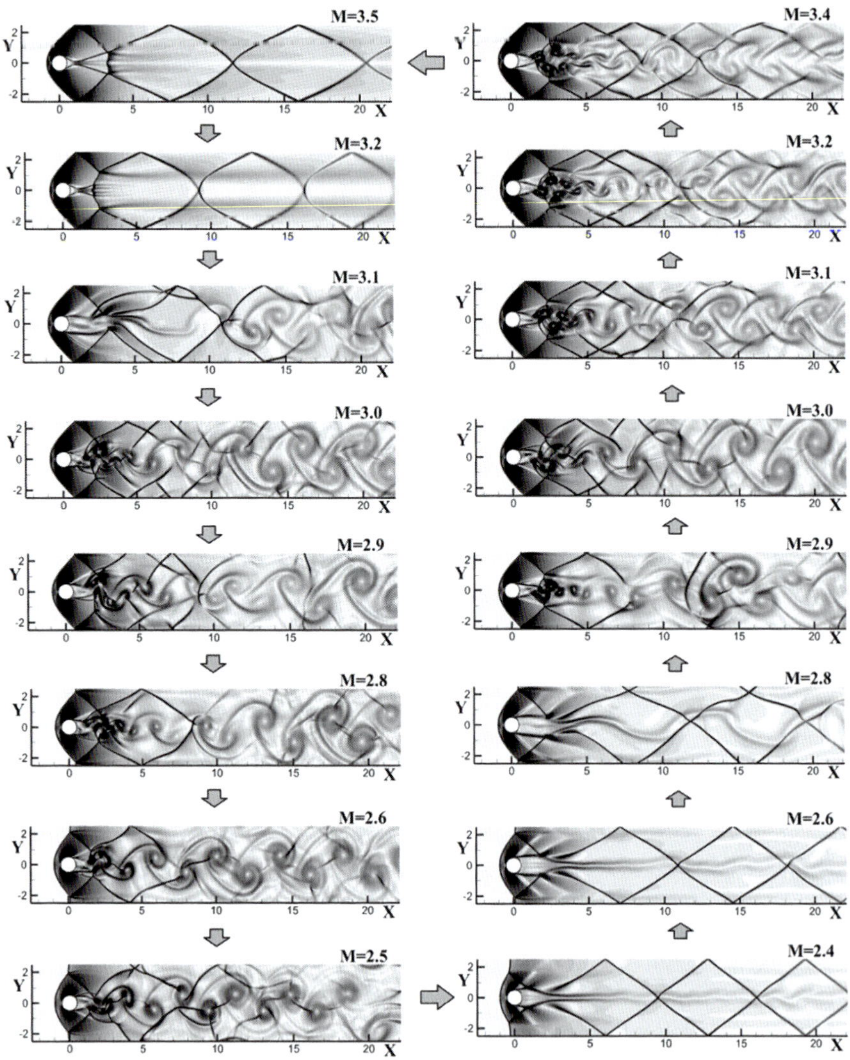

Fig. 5.9 Evolution of flow structures around lattice element with continuous decrease and increase in *M*

Fig. 5.10 Aerodynamics drag coefficient of lattice element with successive increase (line 1, 1′) and decrease (line 2, 2′) in Mach number *M*

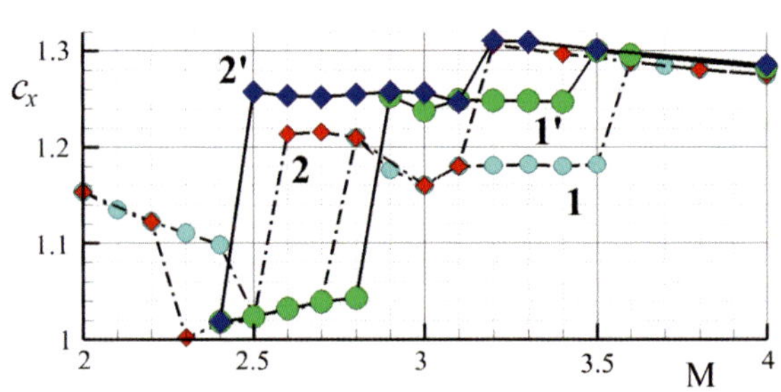

However, it can also be associated with the presence of an extreme element in a finite-size lattice, which affects indirectly all the lattice elements, which is not taken into account in the simplified formulation of the problem with periodic boundary conditions on the side boundaries.

5.6 Conclusions

The multiblock computing technology is used to calculate supersonic flow around a system of cylindrical rods that form a plane lattice of finite width at various Mach numbers and various scenarios of its change. Four flow patterns were revealed, the new of which was a pattern with nonstationary periodic self-oscillations of the flow in the near wake behind the lattice elements. The implemented flow patterns and the boundaries of transition from one flow pattern to another depend on the pre-history of the change in Mach number.

Using an example of the flow around the finite lattice of cylindrical rods with a permeability of 80%, three ranges of flow ambiguity and the corresponding hysteresis of Mach number characteristics were revealed. The first type is associated with restructuring between the collective and local flow regimes of the lattice elements. The second type is conditioned by the restructuring of the near wake behind the cylindrical elements as a result of interaction with local shock-wave systems from neighboring lattice elements. The third type of hysteresis is associated with the occurrence of a nonstationary periodic flow in the wake of the lattice elements.

Acknowledgements The work was carried out with a partial financial support of the Russian Foundation for Basic Research (Project No. 19-01-00242) and partially under the state task of the ICAD RAS.

The calculations were carried out on MVS-100 K at Interdepartmental supercomputer center of the Russian Academy of Sciences.

References

1. Guvernyuk, S.V.: On hypersonic flow around bodies with wire screens. Gas Wave Dyn. **4**, 236–242 (in Russian) (2005)
2. Fomin, V.M., Mironov, S.G., Serdyuk, K.M.: Reducing the wave drag of bodies in supersonic flows using porous materials. Tech. Phys. Lett. **35**(2), 117–119 (2009)
3. Kirilovskiy, S.V., Maslov, A.A., Mironov, S.G., Poplavskaya, T.V.: Application of the skeleton model of a highly porous cellular material in modeling supersonic flow past a cylinder with a forward gas-permeable insert. Fluid Dyn. **53**(3), 409–416 (2018)
4. Maksimov, F.A.: Supersonic body wrapping. Comput. Res. Model. **5**(6), 969–980 (in Russian) (2013)
5. Kudryavsev, A.N., Epshtein, D.B.: Hysteresis phenomenon in supersonic flow past a system of cylinders. Fluid Dyn. **47**(3), 395–402 (2012)

6. Guvernyuk, S.V., Maksimov, F.A.: Supersonic flow past a flat lattice of cylindrical rods. Comput. Math. Math. Phys. **56**(6), 1012–1019 (2016)
7. Guvernyuk, S.V., Savinov, K.G., Ul'yanov, G.S.: Supersonic flow round blunt perforated screens. Fluid Dyn. **20**(1), 124–129 (1985)
8. Maksimov, F.A., Churakov, D.A., Shevelev, Y.D.: Development of mathematical models and numerical methods for aerodynamic design on multiprocessor computers. Comput. Math. Math. Phys. **51**(2), 284–307 (2011)
9. Isaev, S.A., Baranov, P.A., Usachov, A.E.: Multiblock Computing Technologies in VP2/3 Aerothermodynamics Package. LAP LAMBERT Academic Publishing, Saarbryuken (in Russian) (2013)

Chapter 6
Limiting Functions Affecting the Accuracy of Numerical Solution Obtained by Discontinuous Galerkin Method

Marina E. Ladonkina, Olga A. Nekliudova and Vladimir F. Tishkin

Abstract In the numerical solution of hyperbolic systems of equations, Galerkin method with discontinuous basic functions is proved to be very reliable. However, to ensure the monotony of the solution obtained by this method, it is necessary to use a smoothing operator, especially if the solution contains strong discontinuities. In this chapter, we consider the classic Cockburn limiter, a moment limiter that preserves the high order of the scheme, well-proven smoothing operator based on Weighted Essentially Non-Oscillatory (WENO) reconstruction, the smoothing operator of a new type based on averaging solutions, taking into account the rate of change of the solution and the rate of change of its derivatives and slope limiter, preserving the positivity of pressure. A comparison was made of the actions of these limiters on a series of test problems. Numerical results show that using discontinuous Galerkin method and applying moment limiter, slope limiter, WENO limiter, or limiter based on averaging allows to obtain a high order of accuracy on smooth solutions, as well as the clear, non-oscillating profiles on shock waves provided with appropriate constants for the correct determinations of defective cells. In addition, slope limiter, WENO limiter, and averaging limiter are simple enough to implement and be generalized on the multidimensional unstructured grids.

M. E. Ladonkina (✉) · O. A. Nekliudova · V. F. Tishkin
Keldysh Institute of Applied Mathematics of RAS, 4, Miusskaya Square, Moscow 125047, Russian Federation
e-mail: ladonkina@imamod.ru

O. A. Nekliudova
e-mail: nek_olga@mail.ru

V. F. Tishkin
e-mail: v.f.tishkin@mail.ru

L. C. Jain et al. (eds.), *Advances in Theory and Practice of Computational Mechanics*, Smart Innovation, Systems and Technologies 173,
https://doi.org/10.1007/978-981-15-2600-8_6

6.1 Introduction

In the last two decades, Galerkin method with discontinuous basic functions has been actively developed. This method was proposed more than fifty years ago to solve the neutron transport equation [1]. Nowadays, an enormous amount of publications appears annually on the study and application of Discontinuous Galerkin (DG) method for solving a wide class of applied problems.

Galerkin method with discontinuous basic functions is one of the numerical methods of high accuracy. This method has a number of advantages inherent in both finite-element and finite-difference approximations. Notably, it provides a given order of accuracy even on unstructured meshes [2] and can be used for meshes with an arbitrary cell shapes. This is especially relevant for solving complex multi-scale tasks. It is well known that there are two approaches to improve the accuracy of the resulting solution. One of them is to grind up the grid in the areas of the existing features of the solution, while the second approach is to increase the order of accuracy of the scheme. The discontinuous Galerkin method allows one to implement both approaches at once. Thus, the order of accuracy using the high-order polynomials is increased making a local grid refinement (the so-called hp-adaptation) [3, 4] at the same time. This is especially relevant for solving complex multi-scale problems.

The choice of a grid is one of the important issues of method implementation. The undoubted advantage of discontinuous Galerkin method is a possibility of its application on the grids of arbitrary structure. There are successful software implementations of Rectangular Mesh Generator (RMG) for solving three-dimensional problems on unstructured grids containing as elements of only one type tetrahedral [5, 6] or hexahedral [4, 7], as well as, for grids with different types of cells [8].

However, DG method has some implementation difficulties. First of all, to ensure the monotony of solution obtained by this method, it is necessary to introduce so-called slope limiters, especially if the solution contains strong discontinuities. Cockburn limiter [2] is the most widely used one on the tetrahedral grids. The concept of this limiter can be implemented in the multidimensional problems with arbitrary structured grids. However, this limiter, as all Total Variation Diminishing (TVD) limiters, reduces the accuracy of the resulting solution [9].

Recently, various approaches have been developed to solve this problem. One of them is to create a limiter of a higher order of accuracy. It was proposed in [10]. However, this limiter works well only on the structured grids. Another approach of creating a limiter with a higher order of accuracy consists of using WENO limiter [11]. In [12], the limiters were proposed that do not resort to the use of min mod procedures and, accordingly, do not reduce the accuracy of the solution, which is a great advantage of these limiters. In [13], a smoothing operator of a new type was developed based on averaging of solutions, taking into account the rate of change of the solution and the rate of change of its derivatives. It was shown that the application of the proposed smoothing operator is not inferior to WENO limiter, and in some cases exceeds the accuracy of the obtained solution, which is confirmed by numerical studies.

Another fundamentally different way of monotonizing a solution is based on introducing the artificial viscosity into a numerical scheme [14]. This approach involves a use of empirical constants, which complicates its application to solve real problems. In [12], a modification of DG method for two-dimensional case was proposed providing the possibility of a smooth transition from a high-order accuracy scheme to a first-order monotonic scheme in flow singularity regions. The most simple and effective implementation of the limiter was proposed in [8], but this limiter has its flaws, since its implementation does not guarantee the suppression of nonphysical oscillations.

Nowadays, a number of papers have shown that the use of limiters can adversely affect the accuracy of the numerical solution [9, 15]. Therefore, investigations of preserving the order of accuracy of the solution and ensuring its monotony remain relevant.

In this chapter, we consider a model approach for DG method implementation. Approximate solution is defined as the projection of a vector of unknown variables onto a space of polynomials of degree p with time-dependent coefficients and coefficients of polynomials as unknowns. Also, we explore the limiting functions that are most often used in this approach for real calculations.

In Sect. 6.2, we provide a description of DG method for Euler equations, as well a description of tested limiters. Section 6.3 presents the test results and research on solution of Einfeldt problem. Section 6.4 contains the conclusions for this chapter.

6.2 Discontinuous Galerkin Method for Euler Equations

We study Euler equation written in a conservative form:

$$\partial_t \mathbf{U} + \nabla \cdot \mathbf{F}(\mathbf{U}) = 0. \tag{6.1}$$

It is supplemented with suitable initial-boundary conditions, the form of which depends on the specific problem and will be specified later. The conservative variables $\mathbf{U}$ and the components of the stream function $\mathbf{F(U)}$ are given as

$$\mathbf{U} = \left\{ \begin{matrix} \rho \\ \rho\, u \\ E \end{matrix} \right\}, \quad \mathbf{F}_x(\mathbf{U}) = \left\{ \begin{matrix} \rho\, u \\ \rho\, u^2 + p \\ (E + p)u \end{matrix} \right\}, \tag{6.2}$$

where ρ is the density of the fluid, u is the velocity, p is the pressure, ε is the specific internal energy, and $E = \rho\big(\varepsilon + u^2/2\big)$ is the total energy per volume unit. To determine the pressure p, we use the ideal gas state equation:

$$p = (\gamma - 1)\rho\varepsilon,$$

where γ is the adiabatic index.

To apply the discontinuous Galerkin method, we cut the area, where the solution is sought with a grid $0 = x_{1/2} \le x_{3/2} \le \dots \le x_{N+1/2} = L$ with grid spacing $\Delta x_i = (x_{i+1/2} - x_{i-1/2})$. At each interval $x_{i-1/2} \le x \le x_{i+1/2}$, we search an approximate solution of the system of equations (Eq. 6.1) in the form of a projection of a vector of conservative variables $\mathbf{U} = (\rho, \rho u, E)$ onto the space of polynomials $P(x)$ of degree p in the basis $\{\phi_k(x)\}$ with coefficients depending on time. Then, the solutions will be

$$\mathbf{U}_h(x,t) = \sum_{k=0}^{p} \mathbf{U}_k(t)\phi_k(x),$$

where p is the degree of polynomials, and $\phi_k(x)$ is the corresponding basic function.

In this chapter, we will use Taylor basis.

An approximate solution of the system (Eq. 6.1) in the discontinuous Galerkin method is sought as a solution of the following system [2]:

$$\begin{gathered}\int\limits_{I_i} \partial_t \mathbf{U}_h(x,t)\cdot\phi_k(x)\mathrm{d}x - \int\limits_{I_i} \mathbf{F}(\mathbf{U}_h(x,t))\partial_x\phi_k(x)\mathrm{d}x + \\ +\mathbf{F}_{i+1/2}\phi_k(x^l_{i+1/2}) - \mathbf{F}_{i-1/2}\phi_k(x^r_{i-1/2}) = 0,\end{gathered} \tag{6.3}$$

where $i = 0, \dots, N$, $k = 0, 1, 2$.

In Eq. 6.3, $\mathbf{U}_h(x,t) = (\rho_h(x,t)\rho u_h(x,t)E_h(x,t))^T$ is the solution vector; $\phi_k(x^l_{i+1/2})$, $\phi_k(x^r_{i-1/2})$ are the basis functions with the number k on the interval I_i calculated in points $x_{i+1/2}$,$x_{i-1/2}$; and $\mathbf{F}_{i+1/2}$, $\mathbf{F}_{i-1/2}$ are the discrete flows, which are monotonic functions of two variables:

$$\mathbf{F}_{i+1/2} = \Phi(\mathbf{U}_h(x^l_{i+1/2}, t), \mathbf{U}_h(x^r_{i+1/2}, t)),$$
$$\mathbf{F}_{i-1/2} = \Phi(\mathbf{U}_h(x^l_{i-1/2}, t), \mathbf{U}_h(x^r_{i-1/2}, t)),$$

for which the condition of approval is satisfied:

$$\Phi(\mathbf{U}_h(x_i,t), \mathbf{U}_h(x_i,t)) = \mathbf{F}(\mathbf{U}_h(x_i,t)).$$

In this chapter, Rusanov–Lax–Friedrichs flows [16, 17] and Godunov flow [18] are used as a numerical flux.

Hereinafter, Cockburn limiter, moment limiter, limiter based on WENO reconstruction, limiter based on averaging the solution, and slope limiter are considered in Sects. 6.2.1–6.2.5, respectively.

6.2.1 Cockburn Limiter

A limiter is a certain operator acting on an approximate solution function on each interval $x_{i-1/2}, x_{i+1/2}$. According to [2], we denote the action of this operator on the function u by $\Lambda\,\Pi_h u$.

Cockburn limiter is described in detail in [2]. For a linear function $u = u_0^i + u_1^i \phi_1$, the action of the limiting operator can be written as

$$\Lambda\,\Pi_h u = u_0^i + \tilde{u}_1^i \phi_1, \tag{6.4}$$

$$\tilde{u}_1^i = \min\ \mathrm{mod}\ \left(u_1^i, \alpha\left(u_0^{i+1} - u_0^i\right), \alpha\left(u_0^i - u_0^{i-1}\right)\right), \tag{6.5}$$

$$\min\ \mathrm{mod}\ (a_1, \ldots, a_N) = \begin{cases} \mathrm{sign}(a_1) \min\limits_{1\le j\le N} \left|a_j\right| \text{ if } \mathrm{sign}(a_1) = \mathrm{sign}(a_2) = \cdots = \mathrm{sign}(a_N) \\ 0 \quad \text{otherwise} \end{cases}. \tag{6.6}$$

To use Cockburn limiter, in the case when the order of the polynomial is $p > 1$, we project the function u onto the space of linear polynomials u^l. Further, after applying the limiter (Eqs. 6.4–6.6) to the linear function u^l, the coefficients at the linear terms $\tilde{u}_1^l$ and u_1^l are compared, and, if they are equal, the original function is not being changed after the action of the limiter. Otherwise, the result of the limiter will be a linear function. When applying this limiter, the choice of the parameter $1 \le \alpha \le 2$ plays an important role. When $\alpha = 1$, we get the most "hard" limiter ensuring the monotony of the solution, when $\alpha = 2$, we get the "less strict" limiter $\Lambda\,\Pi_h$.

6.2.2 Moment Limiter

The next type of limiter investigated in this chapter is the "momentum" limiter described in [10]. This limiter is characterized by the fact that it preserves the highest possible order of the scheme.

The main idea of the method is that the solution is limited by limiting its coefficients. The coefficient $\tilde{u}_k^i$ corresponds to the kth derivative of the solution, and it is compared with the alternative approximation of the kth derivative in terms of the right and left differences of the $(k-1)$th derivative.

Starting with the highest coefficients $k = p$, we replace $\tilde{u}_k^i$ in Eq. 6.7.

$$\tilde{\tilde{u}}_k^i = \min\ \mathrm{mod}\ \left(\tilde{u}_k^i,\ \alpha_k\left(\tilde{u}_{k-1}^{i+1} - \tilde{u}_{k-1}^i\right),\ \alpha_k\left(\tilde{u}_{k-1}^i - \tilde{u}_{k-1}^{i-1}\right)\right). \tag{6.7}$$

The limiter is triggered, when $\tilde{\tilde{u}}_k^i \ne \tilde{u}_k^i$. In the case of $\tilde{\tilde{u}}_k^i = \tilde{u}_k^i$, limiting is terminated, otherwise the coefficient $\tilde{u}_{k-1}^i$ is limited continuing as long as either $k =$

1 or the condition $\tilde{\tilde{u}}_k^i - \tilde{u}_k^i$ is fulfilled. In the limiter (Eq. 6.7), there is a parameter α_k, which value depends on the order of the coefficient k. In [17], it was shown that the range of parameter α_k variation is bounded below by a number $\frac{1}{2(2k-1)}$. It should be noted that in the case of $p = 1$, the limiter completely coincides with the min mod limiter.

6.2.3 Limiter Based on WENO Reconstruction

In [11], a limiter was proposed for the discontinuous Galerkin method based on WENO reconstruction, which allows to preserve the high accuracy of the method and does not distort the solution profile. At the first stage, the problem cells should be identified, i.e., those cells in which limiting may be required. At the next stage, the numerical solution in the problem cells is replaced with the reconstructed one, with the polynomials obtained during the reconstruction retaining the original integral average value in the cell and a high order of accuracy, but less prone to oscillations. To identify the problem cells, we will use Total Variation Bounded (TVB) min mod [11] limiter, where the min mod function is defined by Eq. 6.6 or through the function converted by TVB min mod (Eq. 6.8):

$$\tilde{m}(a_1, \ldots, a_N) = \begin{cases} a_1 & \text{if } |a_1| \le Mh^2 \\ \min \operatorname{mod}(a_1, \ldots, a_N) & \text{otherwise} \end{cases}, \tag{6.8}$$

where parameter M is chosen according to the solution of the problem.

The main idea of WENO limiter is that a new polynomial is built on the problem cell, which is a convex combination of the original polynomial and polynomials on neighboring cells with the necessary corrections to preserve the integral average in the cell. In the calculations presented below, we used the coefficients indicated in [11].

6.2.4 Limiter Based on Averaging the Solution

In [13], a different approach for limiter building was proposed. Let us consider a new polynomial:

$$P_j^{sm} = \bar{P}_j + P_j^0\left(1 - \max\left(\alpha_{j+1/2}, \alpha_{j-1/2}\right)\right), \quad P_j^0 = P_j - \bar{P}_j.$$

The criterion for selecting coefficients $\alpha_{j+1/2}, \alpha_{j-1/2}$ is thoroughly described in [13]. First, consider the order of deviation of neighboring polynomials from the arithmetic mean of their integral means. As one of the coefficients, we will choose:

$$\mu_{j+1/2} = \int\limits_{x_{j-1/2}}^{x_{j+3/2}} (P_j - P_{j+1})^2 \mathrm{d}x \Bigg/ \left(\int\limits_{x_{j-1/2}}^{x_{j+3/2}} \left(P_j - P_{j+1/2}^{av}\right)^2 \mathrm{d}x + \int\limits_{x_{j-1/2}}^{x_{j+3/2}} \left(P_{j+1} - P_{j+1/2}^{av}\right)^2 \mathrm{d}x \right),$$

$$P_{j+1/2}^{av} = \tfrac{1}{2}(\bar{P}_j + \bar{P}_{j+1}).$$

This coefficient does not reduce the order of the polynomial obtained.

Note that in the discontinuous Galerkin method, we can determine the coefficient $\beta_{j+1/2}$ in terms of the rate of change of the solution and its derivatives. We make the coefficient:

$$\alpha_{j+1/2} = \mu_{j+1/2}\beta_{j+1/2}\frac{c_{j+1/2} \cdot \Delta t}{\Delta x},$$

proportional to the time step, but not violating the dimension. There is a multiplier $\frac{1}{\Delta x}$ in the coefficient, which theoretically can reduce the accuracy of the solution, so replacing it with a multiplier $\frac{\rho_1}{\bar{\rho}}$, where $\rho_h(x)$ is the density, $\rho_h(x) = \sum_{k=0}^{p} \rho_k(x)\phi_k(x)$, $\bar{\rho}$ is the integral average, and ρ_1 the first derivative of $\rho_h(x)$ theoretically allows us to maintain the order of accuracy of the scheme:

$$\alpha_{j+1/2} = \mu_{j+1/2}\beta_{j+1/2}\frac{c_{j+1/2}\Delta t}{\bar{\rho}/\rho_1}.$$

6.2.5 Slope Limiter

We describe a design of slope limiter according to [8]. To limit the polynomials in the cell with the number *K*, it is necessary to go through all the neighboring cells and calculate the integral average pressure value in each of them $\bar{p}_i$, $i = 1, N$, where N is the number of neighboring cells and choose the maximum and minimum pressure values $p_{\max} = \max(\bar{p}_i)$, $p_{\min} = \min(\bar{p}_i)$, $i = 1, N$. Let us denote $p^+ = (1+\varepsilon)p_{\max}$, $p^- = (1-\varepsilon)p_{\min}$, where ε is some small positive constant. We calculate the pressure in all quadrature points of the limited cell. Find the maximum and minimum values $p_{\max}^K$, $p_{\min}^K$. If at any quadrature point the pressure value exceeds the value p^+, we multiply the original polynomials by some positive value α, so that the pressure value at this quadrature point does not exceed p^+. Similarly, we proceed with the condition on p^-. To get limited pressure values $\hat{p} = (1-\alpha)\bar{p} + \alpha p$, we recalculate the coefficients of all the original polynomials:

$$\hat{\mathbf{U}} = (1-\alpha)\bar{\mathbf{U}} + \alpha\mathbf{U} = (1-\alpha)\bar{\mathbf{U}} + \alpha \sum_{j=0,p} \mathbf{U}_j\varphi_j,$$

where $\mathbf{U} = (\rho \;\; \rho u \;\; E)^T$ is the vector of the original polynomials, and $\bar{\mathbf{U}}$ is the vector of integral means of conservative variables. In addition, the same procedure is done separately for density.

6.3 Numerical Experiments

Let us consider a series of test problems of one-dimensional unsteady gas dynamics. Despite the simplicity of the statement, these problems reflect all the features of gas-dynamic flows. The initial distribution of density, velocity, and pressure values are presented in Table 6.1.

Test 1 (Sod problem). As a result of this test, a shock wave arises moving into the low-pressure region and a rarefaction wave appears expanding into the high-pressure region and a contact discontinuity (Fig. 6.1a).

Test 2 (Lax problem). In this problem, the same configuration of the solution arises as in Sod problem, consisting of a shock wave and contact discontinuity with large differences in gas-dynamic parameters than in the previous problem but, unlike Sod, a less intense rarefaction (Fig. 6.1b).

Test 3 (supersonic shock tube). The configuration of the solution of this problem is similar to the two previous ones. The shock wave, rarefaction wave, and contact discontinuity arise. However, a solution of this problem allows us to estimate the operation of computational schemes in the emerging regions of supersonic flows (Fig. 6.1c).

Test 4 (Mach 3 problem). The solution is a contact discontinuity and two rarefaction waves (Fig. 6.1d).

Test 5 (Einfeldt problem). As a result of this solution, two symmetric rarefaction waves arise propagating in the opposite directions and fixed contact discontinuity (Fig. 6.1).

Test 6 (shock entropy wave interaction) is the interaction of a shock wave with an entropic perturbation. Mach number of the shock wave moving along the X-axis is $M = 3.5$. After the passage of the shock wave behind the front, a complex flow is formed, in which a series of smaller amplitude shock waves forms over time (Fig. 6.1e).

Test 7 (Woodward–Colella blast waves). This problem is a model of the interaction of two shock waves and is one of the generally accepted tests for testing the operability of numerical methods for solving gas dynamics problems. At the initial moment of time, the density $\rho = 1$, velocity $u = 0$, the pressure is distributed as follows: $p =$

Table 6.1 Initial distribution of density, velocity, and pressure

No.	Parameters to the left			Parameters to the right			Time
	Density	Velocity	Pressure	Density	Velocity	Pressure	
1	1	0	1	0.125	0	0.1	2.0
2	0.445	0.698	3.528	0.5	0	0.571	1.3
3	1	0	1	0.02	0	0.02	0.15
4	3857	0.920	10.333	1	3.55	1	0.09
5	1	−2	0.4	1	2	0.4	0.15
6	3.857143	2.629369	10.3333	$1 + 0.2\sin(5x)$	0	1	1.8

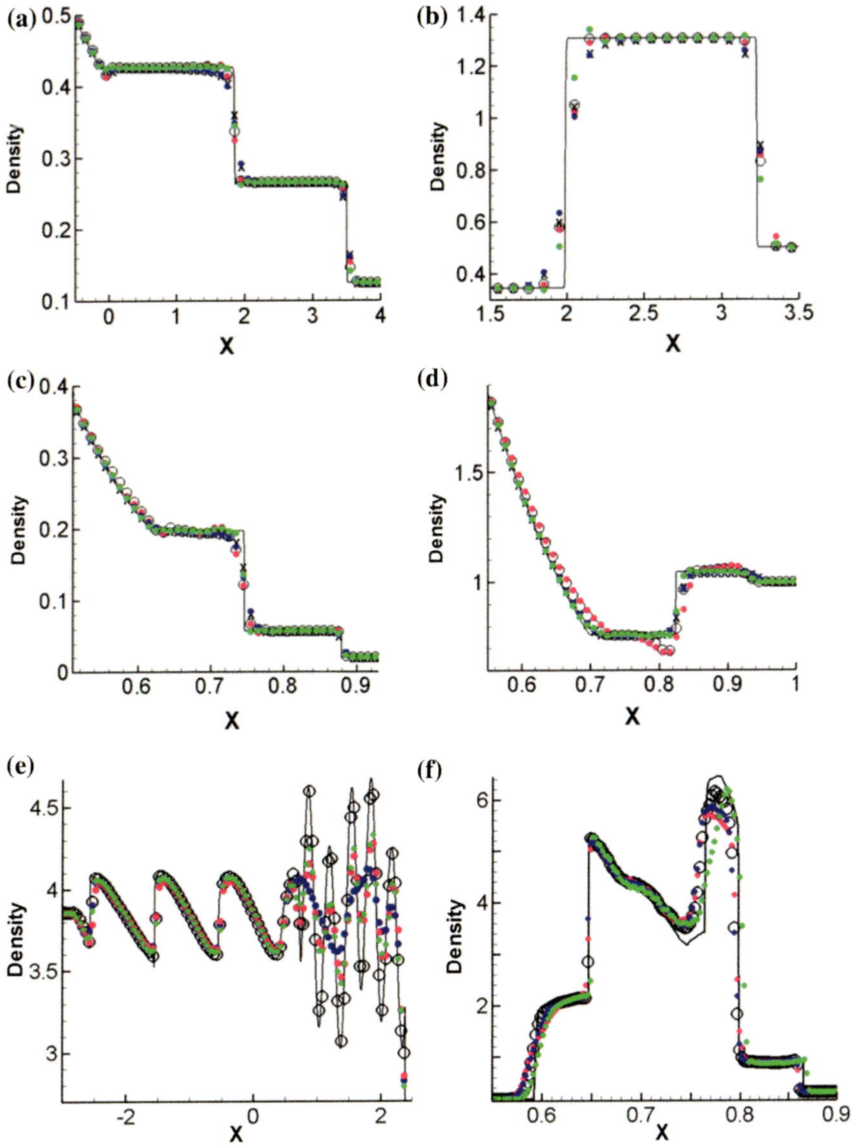

Fig. 6.1 Density: **a** Test 1, **b** Test 2, **c** Test 3, **d** Test 4, **e** Test 6, and **f** Test 7

103, at $0 \le x \le 0.1$, $p = 102$, at $0.1 \le x \le 0.9$, and $p = 102$, at $0.9 \le x \le 1$. Estimated time $t = 0.038$ (Fig. 6.1f).

In Tests 1, 2, and 6, the computational domain is $-5 \le x \le 5$, while in Tests 3, 4, and 5, the computational domain is $0 \le x \le 1$. The positions of the point of discontinuity are $-x_0 = 0$, $-x_0 = 0.5$, and $x_0 = -4$ in Tests 12, Tests 3–5, and Test

6, respectively. Calculations of Tests 1–5, Test 6, and Test 7 were carried out on the grids of 100 cells, 200 cells, and 400 cells, respectively. In all calculations, gas is assumed to be ideal, with an adiabatic exponent $\gamma = 1, 4$.

Note that for each limiter there are configurable parameters. We carry out a series of calculations with the Cockburn limiter, moment limiter, and slope limiter with parameters corresponding to weak limiting. Let us determine the optimal parameters for WENO and Average Solution Meaning (AvSM) limiters. To identify defective cells, you ought to define the parameter M in TVB min mod limiter. The first series of calculations was carried out with the parameter $M = 0.01$. This parameter with an excess determines the cells, in which it is necessary to carry out limiting. The next step was to determine the critical value of the parameter M so that its further increase would not affect the quality of the limitation that is, the calculation would be performed without a limiter. The next series of calculations consisted in determining the critical value of the parameter M. Thus, with an increase of the parameter M in all tests, more distinct profiles are observed in the areas of solution discontinuities, but at the same time, the appearance of oscillations is observed.

We present the results of calculations obtained with optimal parameters.

When solving problems 1–5, it is impossible to unambiguously determine the advantages of using one or another limiter. It should be noted that the slope limiter (green dots, Fig. 6.1) shows the most stable behavior. Although oscillations are present in Test 3, it is easy to suppress them by decreasing the epsilon parameter. Tests 4 and 5 turned out to be the most difficult for a numerical solution using WENO (red dots, Fig. 6.1) and AvSM (black rounds, Fig. 6.1) limiters. Both represent the decay of a gas-dynamic discontinuity in the form of a contact discontinuity and two rarefaction waves. When solving Test 4, the best results were obtained using Cockburn limiter (black crosses, Fig. 6.1), moment limiter (blue dots, Fig. 6.1), and slope limiter (green dots Fig. 6.1). However, slope limiter did not cope well with the solution of Test 5. For problem 7, errors were calculated in L2 norm with respect to the "reference" solution from [19]. Errors in L2 norm have the following values: 2.09*d*-5, 1.04*d*-5, 1.25*d*-5, 3.94*d*-5, and 4.09*d*-5 for Cockburn limiter, moment limiter, slope limiter, WENO limiter, and AvSM limiter, respectively.

When solving more complex problems 6 and 7, the use of limiters based on averaging gives the best result with a coefficient $k \sim \Delta x$. In [19], it was shown that in some cases, the coefficient $k \sim \rho_1/\bar{\rho}$ when using quadratic functions makes the limiter more dissipative.

It is worthwhile to dwell separately on solving problem 5. As is known, this problem is often used when testing numerical methods, and one of the indicators of a well-functioning scheme is the accuracy of the transfer of the contact discontinuity region. In almost all calculations, a nonphysical surge of internal energy is observed. We considered several numerical schemes for solving this problem. We present a series of calculations performed according to the first-order Godunov scheme (Fig. 6.2b) and the third-order discontinuous Galerkin method with quadratic basis functions using moment limiter [10] with coefficients corresponding to "hard" limiting (Fig. 6.2a).

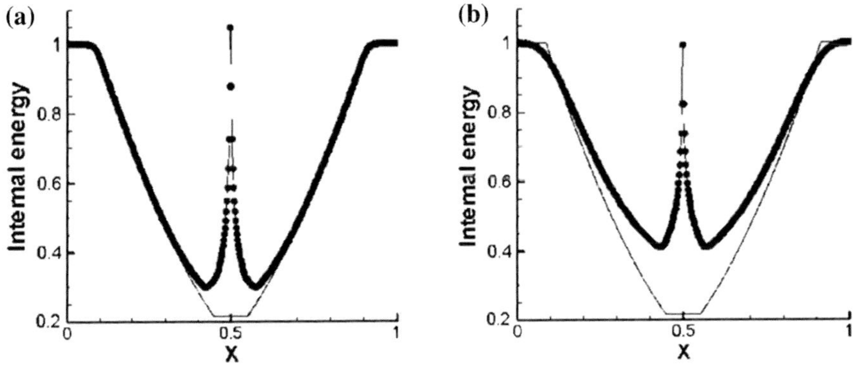

Fig. 6.2 Internal energy. Test 5. 500 points: **a** DG, $P = 2$, Godunov flux, moment limiter; **b** Godunov scheme, $P = 0$

Adding the pseudo-heat conductivity to the numerical approximation of the flow (Eq. 6.5), we obtain a clear decrease in the entropy release (Fig. 6.3).

Now let us pay attention to the definition of the initial data. In this problem, the break point $x_0 = 0.5$ falls on the boundary between the cells. In the region to the left of the discontinuity point $x \leq x_0$, the velocity is $u = -0.2$, and in the region on the right, $u = 0.2$. We choose a grid so that the point $x_0 = 0.5$ falls inside the cell approximating the speed in this cell. Then we get $u = 0$. In Fig. 6.4, it is clearly seen that there is practically no entropy surge.

Doing the same thing and by approximating the velocity gap in two cells, it is possible to obtain a solution corresponding to the physical one. Thus, it was established that the defect arising in the numerical solution of problem 5 is connected not only with the choice of the numerical scheme but also with the method of specifying the initial data.

Figure 6.5 shows the results of calculations of problem 5 with a corrected initial

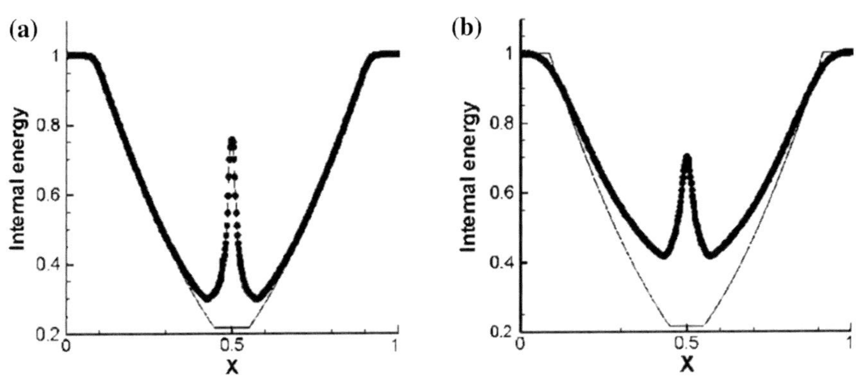

Fig. 6.3 Internal energy including pseudo-heat conductivity. Test 5: **a** moment limiter, $P = 2$; **b** Godunov scheme, $P = 0$

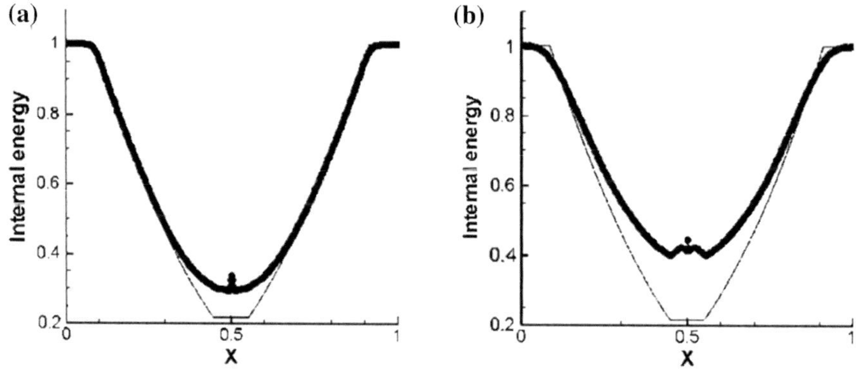

Fig. 6.4 Internal energy. Test 5. 501 points: **a** moment limiter, $P = 2$; **b** Godunov scheme, $P = 0$

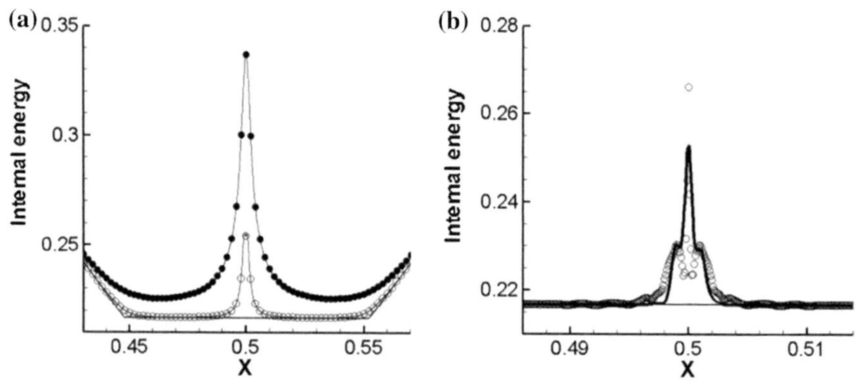

Fig. 6.5 Internal energy. Test 5. 50,000 points: **a** Godunov scheme $P = 0$ (black dots), moment limiter, $P = 2$ (white circles); **b** WENO limiter (white circles), AvSM limiter (solid line)

velocity profile. As can be seen from the graphs, we managed to improve the quality of the result.

Note that in this case, the addition to the numerical scheme of the coefficient corresponding to the pseudo-thermal conductivity leads to an additional smoothing of the solution. The results of Test 5 without pseudo-heat conductivity and with pseudo-heat conductivity are depicted in Fig. 6.6.

According to the results of the study, it is worth noting that all the limiters under study showed themselves quite well in solving complex problems of one-dimensional unsteady gas dynamics. The most stable solution is obtained when using Cockburn limiter. This limiter can be implemented on unstructured grids with cells of various shapes, but the use of this limiter does not guarantee the declared accuracy of the discontinuous Galerkin method. In contrast to Cockburn limiter, moment limiter retains the increased accuracy of the method, but is currently designed only for structured grids. The most promising in our opinion are slope limiter, WENO limiter,

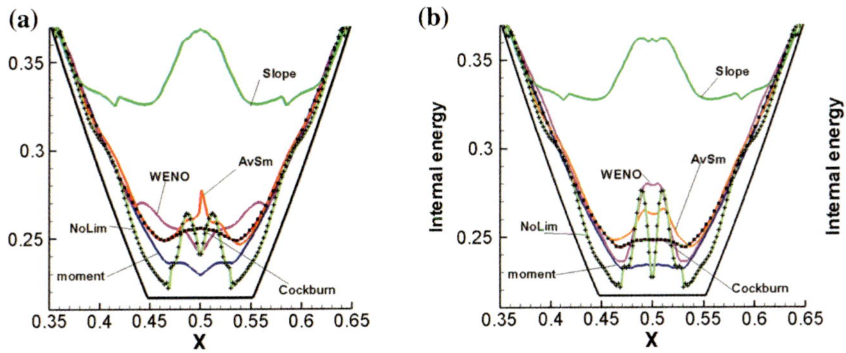

Fig. 6.6 Internal energy. Test 5. 500 points: **a** without pseudo-heat conductivity, and **b** with pseudo-heat conductivity

and limiter based on averaging, which are fairly simple to implement and generalize to multidimensional unstructured grids.

6.4 Conclusions

Numerical results show that when solving problems using the discontinuous Galerkin method, the application of moment limiter, slope limiter, WENO limiters, and limiter based on averaging allows to obtain a high order of accuracy on smooth solutions, as well as, the clear, non-oscillating profiles on shock waves provided with the appropriate constants for the correct definition of defective cells. In addition, slope limiter, WENO limiter, and averaging limiter are simple enough to implement and generalize to multidimensional unstructured grids.

References

1. Reed, W., Hill, T.: Triangular mesh methods for the neutron transport equation. Los Alamos, Scientific Laboratory Report LA-UR-73-79, United States (1973)
2. Cockburn, B.: An introduction to the discontinuous Galerkin method for convection—dominated problems. In: Quarteroni, A. (ed.) Advanced Numerical Approximation of Nonlinear Hyperbolic Equations, LNM, vol. 1697, pp. 150–268 (1998)
3. Nastase, C., Mavriplis, D.: High-order discontinuous Galerkin methods using an hp-multigrid approach. Comput. Phys. **213**, 330–357 (2006)
4. Volkov, A.: Features of the use of the Galerkin method for the solution of the spatial Navier-Stokes an-structured hexahedral equations. TsAGI Sci. J. **XL**(6), 695–718 (2009)
5. Bassi, F., Rebay, S.: Numerical evaluation of two discontinuous Galerkin methods for the compressible Navier-Stokes equations. Int. J. Numer. Meth. Fluids **40**, 197–207 (2002)

6. Krasnov, M., Kuchugov, P., Ladonkina, M., Lutsky, A., Tishkin, V.: Numerical solution of the Navier-Stokes equations by discontinuous Galerkin method. J. Phys. Conf. Ser. **815**(1), 012015.1–012015.9 (2017)
7. Bosnyakov, S., Mikhailov, S., Podaruev, V., Troshin, A.: Unsteady high order accuracy DG method for turbulent flow modeling. Math. Model. **30**(5), 37–56 (2018)
8. Yasue, K., Furudate, M., Ohnishi, N., Sawada, K.: Implicit discontinuous Galerkin method for RANS simulation utilizing pointwise relaxation algorithm. Commun. Comput. Phys. **7**(3), 510–533 (2010)
9. Ladonkina, M., Nekliudova, O., Ostapenko, V., Tishkin, V.: On the accuracy of the discontinuous Galerkin method in the calculation of shock waves. Comput. Math. Math. Phys. **58**(8), 1344–1353 (2018)
10. Krivodonova, L.: Limiters for high-order discontinuous Galerkin methods. J. Comput. Phys. **226**(1), 276–296 (2007)
11. Zhong, X., Shu, C.-W.: A simple weighted essentially non oscillatory limiter for Runge-Kutta discontinuous Galerkin methods. J. Comput. Phys. **232**, 397–415 (2013)
12. Volkov, A., Lyapunov, S.: Monotonization of the finite element method in gas dynamics problems. TsAGI Sci. J. **XL**(4), 15–27 (2009)
13. Ladonkina, M., Nekliudova, O., Tishkin, V.: Construction of the limiter based on averaging of solutions for discontinuous Galerkin method. Math. Model. **30**(5), 99–116 (2018)
14. Persson, P.-O., Peraire, J.: Sub-cell shock capturing for discontinuous Galerkin method. AIAA-2006-112.1–AIAA-2006-112.13 (2012)
15. Kovyrkina, O., Ostapenko, V.: On the practical accuracy of shock-capturing schemes. Math. Model. **25**(9), 63–74 (2013)
16. Rusanov, V.: Calculation of the interaction of non-stationary shock waves with obstacles. USSR Comput. Math. Math. Phys. **2**, 267–279 (1961)
17. Lax, P.: Weak solutions of nonlinear hyperbolic equations and their numerical computation. Commun. Pure Appl. Math. **7**(1), 159–193 (1954)
18. Godunov, S.: Difference method for the numerical calculation of discontinuous solutions of the equations of hydrodynamics. Math. Notes **47**(89)(3), 271–306 (in Russian) (1959)
19. Ladonkina, M., Neklyudova, O., Tishkin, V.: Application of averaging to smooth the solution in DG method. Preprint KIAM 89, Moscow (in Russian) (2017)

Part II
Numerical Simulation of Multiphase Flows, Combustion, and Detonation

Chapter 7
Numerical Simulation of Detonation Initiation: The Quest of Grid Resolution

Alexander I. Lopato, Artem G. Eremenko, Pavel S. Utkin and Dmitry A. Gavrilov

Abstract The chapter is dedicated to the numerical investigation of the grid resolution influence on the detonation initiation process in the multifocused system with the profiled end-wall. Two-dimensional system of Euler equations coupled with the single-step Arrhenius kinetic reaction mechanism was solved on completely unstructured triangular grid using the numerical scheme of second approximation order. The important technical features of interaction with SALOME software used to build an unstructured triangular computational grid, including differences in the results of the triangulation algorithms, are discussed. The content of the structure elements of the output file format CGNS of SALOME is considered. The mechanisms of detonation initiation in the multifocused system are investigated. The grid convergence problem and the influence of the resolution on flow structures are considered.

7.1 Introduction

Mathematical modeling of detonation processes is a significant tool for clarifying the mechanisms, reducing the risks and costs of providing experiments, and supplementing the existing knowledge about the processes occurring during detonation. As noted in the review [1], the majority of the works on numerical studies of gas

A. I. Lopato (✉) · P. S. Utkin
Institute for Computer Aided Design of the RAS, 19/18, Vtoraya Brestskaya ul., Moscow 123056, Russian Federation
e-mail: lopato2008@mail.ru

P. S. Utkin
e-mail: pavel_utk@mail.ru

A. I. Lopato · A. G. Eremenko · P. S. Utkin · D. A. Gavrilov
Moscow Institute of Physics and Technology (National Research University), 9, Institutskiy per., Dolgoprudny, Moscow Region 141701, Russian Federation
e-mail: artem-temoff@mail.ru

D. A. Gavrilov
e-mail: gavrilov.da@mipt.ru

L. C. Jain et al. (eds.), *Advances in Theory and Practice of Computational Mechanics*, Smart Innovation, Systems and Technologies 173,
https://doi.org/10.1007/978-981-15-2600-8_7

detonation problems are related to the use of mainly structured computational grids. At the same time, the problem of Detonation Wave (DW) initiation in complex structures, which include the curvilinear boundaries, protrusions, and bodies of various forms in the geometry of the problem to be solved, occupies a special place in this branch of science. The presence of such entities increases the role of unstructured grids allowing not only to cover the entire computational domain under consideration but also to adapt the computational grid in the areas, where it turns out to be necessary. In addition, effective adaptation of the grid can also reduce the total CPU cost in comparison with a uniform fine grid [1]. Finally, programming of adaptive computational grids for unstructured grids in some cases is relatively easier than for structured grids, as discussed in [1–3]. Although earlier works with calculations on unstructured computational grids use computational methods of low approximation order, gradually, methods and tools of the order increase (limiters, Weighted Essentially Non-Oscillatory (WENO)-schemes [4], spectral volume method [5]) appear and develop in literature. On the other hand, the unstructured grid method is more diffusive than the structured one (see [1], for example). The accuracy of the gradient calculation on the unstructured grid is less than on the structured one, especially in the area of radically changing physical values. As a result, the solution of problems with strong Shock Waves (SWs) including detonation requires a finer resolution, which means a greater number of cells and computational sources and times in comparison with structured grids.

Let us consider some works devoted to the study of DW on the unstructured grids. In [1], attention was paid to the numerical study of the formation of cellular detonation in H_2/air mixture in the tube in the two-dimensional case. The question of the required resolution of the unstructured grid and comparison with the results obtained using a structured grid in the simulation of detonation in the mixture was considered. The series of calculations on unstructured grids with an average cell size of 1, 2.5, 3, 5 μm and the calculation on a structured grid with a cell size of 5 μm were carried out. The detailed model of Petersen and Hanson was used to model the chemical reactions occurring in the mixture. The results show that the unstructured grid simulation carried out with the identical grid resolution of structured grid study could not capture the key DW features.

It was shown that in contrast to the calculation on the structured grid, some key elements of the solution, such as triple point, are not fully and clearly captured using the unstructured grid with cell size 5 μm due to the coarse grid resolution. The effect of the grid resolution of the transverse waves depends not only on the resolution around the detonation front but also on the resolution around the shear layer behind the detonation front. The simulations that were carried out using cell size of 2.5 μm and more could not capture the vortices in the area of shear layer. On the other hand, the calculation with cell size 1 μm captures the vortices like 5 μm structured grid. Thus, insufficient grid resolution leads to under-resolution of the key elements of DW structure (shear layers, vortices), which leads to the serious qualitative differences from the results obtained with sufficient resolution. The significant contribution to this feature is the instability of the key elements of the solution. The authors noted that the most important advantage of unstructured grids is easy application of adaptive grid refinement that allows adding/removing nodal points without modifying the code drastically.

The work [6] focused on the numerical studies of DW structure in high and low activation energy model mixtures characterized by their irregular and regular detonation structure, respectively. The mathematical model was represented by 2D Euler equations with the single-step Arrhenius kinetic reaction mechanism.

The ignition mechanism in irregular structure that corresponds to the high activation energy is due to the both shock compression behind the main front and by turbulent mixing for unburned gas pockets of hot and cold gases at shear layers associated with vortices characterized by hydrodynamic instabilities, mainly Richtmyer–Meshkov instability, while the mechanism in regular structure that corresponds to the low activation energy is completely defined by the shock compression. It was also shown in [6] that DW structure depends on the grid resolution. The low resolution (50 and 125 cells per half-reaction length) gives very poor insight about the structure and instabilities behind DW front and due to this fact demonstrates very regular structure, while for finer resolution (300 and 600 cells per half-reaction length) the structure is irregular. Moreover, there are no unburned gas pockets in the case of low resolution of the grid. The lack of the pockets in case of coarse grid is associated with insufficient resolution in numerical diffusion which artificially accelerated the reaction rate.

The regular structure in low activation energy mixtures totally differs from the irregular structure in considered problem. The number of elements including large vortices of the detonation structure is much less than it was in irregular structure. The whole gases which have passed through DW front are burned completely, and no unburned gas pockets are apparent in the detonation structure. The results justify that the grid resolution does not change qualitatively the detonation structure of the mixture in low activation energy. This fact is associated with the absence of the elements required for changes, mainly secondary triple points and hydrodynamic instability. Thus, the work demonstrates that some differential problems may be unstable depending on the dimensionless activation energy and may not have a grid convergence in terms of classical theory.

Mentioned above researches demonstrate the possibility of productive study of DW flows using unstructured computational grids. Among many problems devoted to DW, an important place is occupied by the problem of optimization of detonation initiation. The optimization can be reached with the help of SW reflection from the focusing system. The system of elliptical reflectors was considered in works of Vasil'ev and named "multifocused" system [7].

The motivation of this work is related to the study of detonation issues in multifocused systems and clarification of the basic mechanisms accompanying the detonation process using unstructured computational grids, as well as, the study of the manifested features using the unstructured grids approach for numerical studies in this branch.

The chapter is organized as follows. The statement of the problem with the geometry of the considered domain is represented in Sect. 7.2. The governing equations for two-component fluid flows are introduced and approximated by the numerical scheme of second approximation order on completely unstructured computational grids in Sect. 7.3. Section 7.4 discusses the features of software used for grid generation and visualization of obtained results. Section 7.5 gives numerical experiments with a discussion of the results, in particular, grid convergence problem. Finally, we conclude this work in Sect. 7.6.

7.2 Problem Statement

Consider 2D channel with the elliptical end-wall shape filled with stoichiometric hydrogen–oxygen mixture. The multifocused system with elliptical curves is represented schematically in Fig. 7.1. The values of semi-axis of considered elliptical curves are equal to 5 and 3.5 mm. At the initial time moment, the incident SW is at the distance of 28 mm from the left end of the channel. The considered Mach number of the incident SW is 2.5. The pressure and temperature of the mixture in the area $x > 28$ mm are 0.04 atm and 298 K. For the sake of computation cost diminishing, the computational domain corresponds to the one half of the channel. The bottom half part of the geometry is used in computations. As a result, the symmetry conditions are set at the upper boundary, the slip-conditions at the right and bottom boundaries and the inflow conditions with parameters of the incident SW at the left boundary.

Note that the sizes of the geometries of the considered computational domain correspond to the geometry from the experimental work of Vasil'ev [7]. Elliptical reflectors were fabricated by means of milling with milling cutter angle 45°.

Fig. 7.1 Schematic statement of the problem

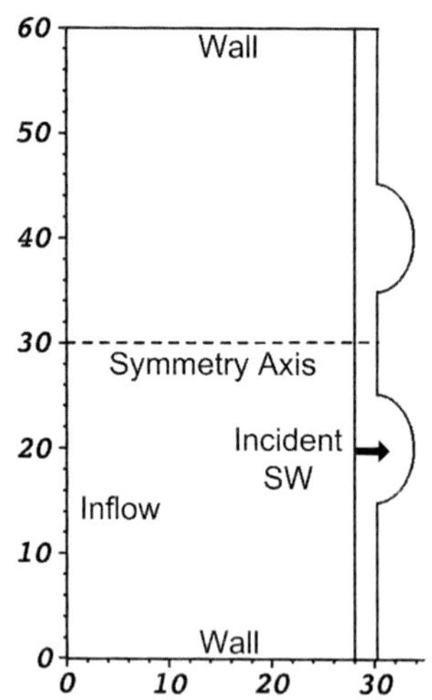

7.3 Mathematical Model and Numerical Method

The mathematical model that produces the results discussed below includes 2D Euler equations supplemented by one-step chemical reaction global model of hydrogen–oxygen combustion described by the first-order Arrhenius kinetics. The governing equations for considered two-component flows may be written in Cartesian frame using Eq. 7.1.

$$\frac{\partial \mathbf{U}}{\partial t} + \frac{\partial \mathbf{F}}{\partial x} + \frac{\partial \mathbf{G}}{\partial y} = \mathbf{S}$$
$$\mathbf{U} = \begin{bmatrix} \rho \\ \rho u \\ \rho v \\ e \\ \rho Z \end{bmatrix} \mathbf{F} = \begin{bmatrix} \rho u \\ \rho u^2 + p \\ \rho u v \\ (e+p)u \\ \rho Z u \end{bmatrix} \mathbf{G} = \begin{bmatrix} \rho v \\ \rho v u \\ \rho v^2 + p \\ (e+p)v \\ \rho Z v \end{bmatrix} \mathbf{S} = \begin{bmatrix} 0 \\ 0 \\ 0 \\ 0 \\ \rho\omega \end{bmatrix}$$
$$e = \tfrac{\rho}{2}\left(u^2+v^2\right) + \rho\varepsilon,\ \varepsilon = \tfrac{p}{\rho(\gamma-1)} + ZQ,\ p = \tfrac{\rho}{\mu}RT,\ \omega = -A\rho Z\exp\left(-E\big/RT\right). \tag{7.1}$$

The nomenclature used here is the standard one. The specific heat ratio γ, molar mass μ, heat release Q, and activation energy E are taken from the database [8]:

$$\gamma = 1.23,\ \mu = 12\,\frac{\mathrm{g}}{\mathrm{mole}},\ Q = 7.37\,\frac{\mathrm{MJ}}{\mathrm{kg}},\ E = 76.2\,\frac{\mathrm{kJ}}{\mathrm{mole}},\ A = 9.16\cdot 10^8\,\frac{\mathrm{m}^3}{\mathrm{kg\,s}}.$$

Parameter A is calculated with the use of the reaction zone time, corresponded to Konnov reaction mechanism:

$$A = \frac{1}{\tau\rho_{\mathrm{vN}}}\exp\left(E\big/RT_{\mathrm{vN}}\right),$$

where $\tau = 0.47\ \mu\mathrm{s}$, $\rho_{\mathrm{vN}} = 0.541\ \mathrm{kg/m^3}$, and $T_{\mathrm{vN}} = 1682$ K in accordance with [8].

The main feature of the computational technique is the numerical solution of the governing equations on completely unstructured computational grids with triangular cells. The computational algorithm is based on Strang splitting principle in terms of physical processes called as the convection and chemical reactions. The spatial part of Eq. 7.1 is discretized using the finite-volume method. The flux is calculated using Advection Upstream Splitting Method (AUSM) [9] extended for the case of a two-component mixture. Note that the chosen AUSM scheme of flux calculation is not a necessary requirement. Thus, in [10], the calculation of DW flows was successfully carried out with Courant–Isaacson–Rees flux scheme. For the approximation order increase, the special reconstruction of the grid functions and minmod limiter are applied [11]. The numerical flux in the local frame is computed using the parameters from the left and the right sides of the edge. Time integration is advanced by the second-order Runge–Kutta method [12]. The time step is chosen dynamically from the stability condition. On the second stage of the algorithm, the system of ordinary differential equations of chemical kinetics for Z variable and temperature in each computational cell of the grid is solved.

The detail description of the numerical method with the verification of the algorithm can be found in [13].

7.4 Technical Features

The realization of the computational algorithm includes three main steps called as the filling of a given area with a set of computational cells with given properties or the construction of a computation grid, carrying out computations in accordance with the chosen computational algorithm, and visualizing and processing the obtained results. In order to build a computational grid in this work, we use software SALOME [14], which has ample opportunities for the construction of the geometry of the computational domain and filling it with cells of various types. Software can be downloaded and installed not for Linux platforms but also has experimental build for Windows. SALOME module of grid construction provides a wide range of algorithms particularly suited for finite-element and finite-volume methods. Group naming provides the identification of local boundaries and contains various output formats of files with results that can facilitate the visualization or other post-processing operations.

In grid construction module, it is possible to create a group of grid entities, in particular, edges. Create a group with the name "in" for the edges with the inflow boundary condition. For our computational domain, these are the edges that form the left vertical boundary of the channel of the computational domain. The edges that form the rest of the boundary will be combined into another group called "wall". For these edges, the slip-conditions are realized. Groups are exported along with grid objects to a small number of file formats, one of which is Computational Fluid Dynamics (CFD) General Notation System (CGNS) [15].

CGNS provides a standard for recording and recovering computer data associated with the numerical solution of equations of fluid dynamics [15]. The intent is to facilitate the exchange of CFD data between sites, applications codes, and across computing platforms, and to stabilize the archiving of CFD data. Application Program Interface (API) is a platform-independent and includes the realization in C/C++ application. The data are stored in a compact, binary format and are accessible through a complete and extensible library of functions.

CGNS file containing a grid is a hierarchical data type having the following structure. This entity is organized into a set of "nodes" in a tree structure according to the certain rules that allow users to easily access the necessary information. The rules are described in Standard Interface Data Structures (SIDS) [15]. The topmost node is called the "root node". Each node can be a "parent" for one or more "child" nodes. A node can also have a child node reference to a node elsewhere in the file or to a single node. The links are transparent to the user: the user "sees" the associated child nodes as if they really exist in the current tree. The structure of the file with the studied grid is shown in Fig. 7.2.

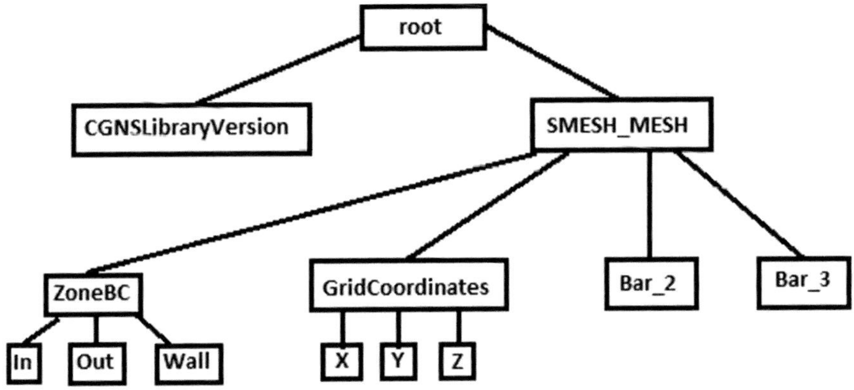

Fig. 7.2 Hierarchical data type of CGNS file format

The first level of the hierarchical structure of CGNS file determines the version of CGNS file—section CGNSLibraryVersion and the dimension of the task—section SMESH_MESH. The CGNS version of the file defines API for interacting with the file structure and possible types of nodes. Version 3.4.0 [16] is used in this work. The next level characterizes the geometric model, number of the boundary vertices (section Bar_2), number of cells (section Bar_3), coordinates of points in space (section GridCoordinates), and boundary conditions (section ZoneBC). For various problems and geometries, three types of boundary conditions are introduced: "in" corresponds to the inflow conditions, "out" corresponds to the extrapolation conditions, and "wall" corresponds to the slip-conditions. Each node contains numbers of vertices that are located on one or another boundary. Section GridCoordinates describes directly the position of nodes in space on each of the axes: sections CoordinateX, CoordinateY, and CoordinateZ.

A triangular calculation grid may be generated in several ways using SALOME. In particular, there are three main algorithms of triangulation called as "Projection 1D-2D", "Delaunay", and "Frontal". The computations in this work were carried out using the Projection 1D-2D grid generation algorithm.

The algorithm is based on the advancing front method. The generation of the computational grid starts from the boundary of the computational domain. On the boundaries of the domain, nodes are searched for by recursively splitting the region in half to the fineness limited by the parameters of the algorithm (the reference size of the triangle edge). Further, the iterative process of searching for nodes is carried out, which are connected to the edges of the boundary and form a new layer of triangles and a new border separating the area with the generated grid from the area without it. As a result, the state of the algorithm is always represented by the advancing boundary front. The coordinates of the nodes of the new boundary are determined by the area in front of the advancing boundary, in accordance with the rules and criteria of [17].

The result of using this algorithm, as well as, two other algorithms is shown in Fig. 7.3. As a region of triangulation, we consider a rectangle 0.1×0.02 m^2. We specify in SALOME the edge length of 50 μm as the reference length of the edges of triangles. Delaunay triangulation shows the distribution of the maximum lengths of the triangles edges, which is close in shape to the normal Gauss distribution. For the Frontal algorithm, all the constructed triangular cells have the same value of maximum edge length. In the case of Projection 1D-2D triangulation, a part of the cells has a maximum edge length that is equal to the selected reference value in SALOME, the length of the remaining cells is higher and obeys a certain distribution, and the overwhelming number of triangles (about 75%) has a maximum edge length of about 87 μm. Thus, the selected algorithms give qualitatively different patterns of distribution of triangles in terms of geometrical parameters. The question of the optimal algorithm in the computations is quite complicated and determined by many factors including the time of triangulation, quality of the computational grid, geometry of the computational domain, and features of the computational algorithm when dealing with unstructured grids.

To visualize the results obtained, Visit software [18] is used. Earlier versions of this software had problems with the display of CGNS files, but since version 2.13.3 these problems have disappeared.

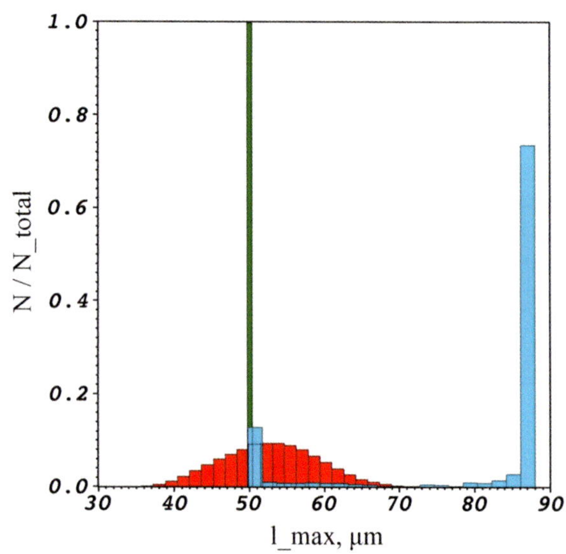

Fig. 7.3 The percentage distribution of the maximum lengths of triangles in the case of Delaunay algorithm (red color), Frontal (green color), and Projection 1D-2D (blue color). The horizontal axis corresponds to the maximum length of the edges of the triangles in μm. The vertical axis corresponds to the percentage of triangles with the selected maximum edge length

7.5 Numerical Experiments

A series of unstructured grids is prepared for resolution study. The average triangular element sizes equal 12.5, 20, and 35 μm, while the approximate numbers of cells are 2, 4, and 8 million cells. Consider the process of detonation initiation for the grid with the average cell size of 12.5 μm.

Shock waves that are formed by the reflection of the incident SW from the plane wall parts amplify the parameters inside the reflector cavity. The main reflected SW emerging from the elliptical cavity is created by the reflected incident wave and Mach stem. The combustion area gradually occupies the cavity of the reflector and spreads outward primarily in the area between the shear layers, which separate completely the burned gas of the jet flow with the compressed gases that have passed through the reflected wave. Detonation initiates at about 41 μs in several points at the symmetry axis of the channel outside the cavity of the reflector. First, the initiation of DW occurs when the combustion front is reflected from the symmetry axis. Second, the numerous collisions of the generated waves with the plane wall between the reflectors leads to the increase in the gas pressure and temperature near the plane wall between the reflectors and the appearance of the second place of DW initiation.

The main large-scale structures are successfully resolved using the relatively coarse grid while the fine grid better resolves shear layers and vortices. Figure 7.4 confirms this fact. Some secondary modes of obtained hydrodynamic instabilities are not apparent in case of low grid resolution. As a result, the present grid study demonstrates less irregular structure for low considered resolution and more irregular one for high grid resolution. In addition, better resolution of the grid leads to a smaller smearing of shocks. In the computations, these facts lead to the fact that grid resolution affects the combustion process and moment of DW initiation. For the coarsest grid, the detonation initiation occurs in the time range 40–41 μs. With the refinement of the grid, the same moment occurs closer to 40 μs. Thus, the noted peculiarities of computations can explain the weak change in the time moment of initiation of detonation waves during the refinement of the grid.

7.6 Conclusions

The numerical investigation of mechanisms of detonation wave initiation is carried out. The results of the computations show that the focusing of reflected waves, interaction with the generated waves, and generated hydrodynamic instabilities play an important role in DW initiation in the considered problem. Mathematical model is based on two-dimensional Euler equations written in Cartesian frame and supplemented by one-stage chemical reaction model. The numerical method of second approximation order for integration of the governing equations on fully unstructured triangular grid is proposed. The features of the technical component of the work

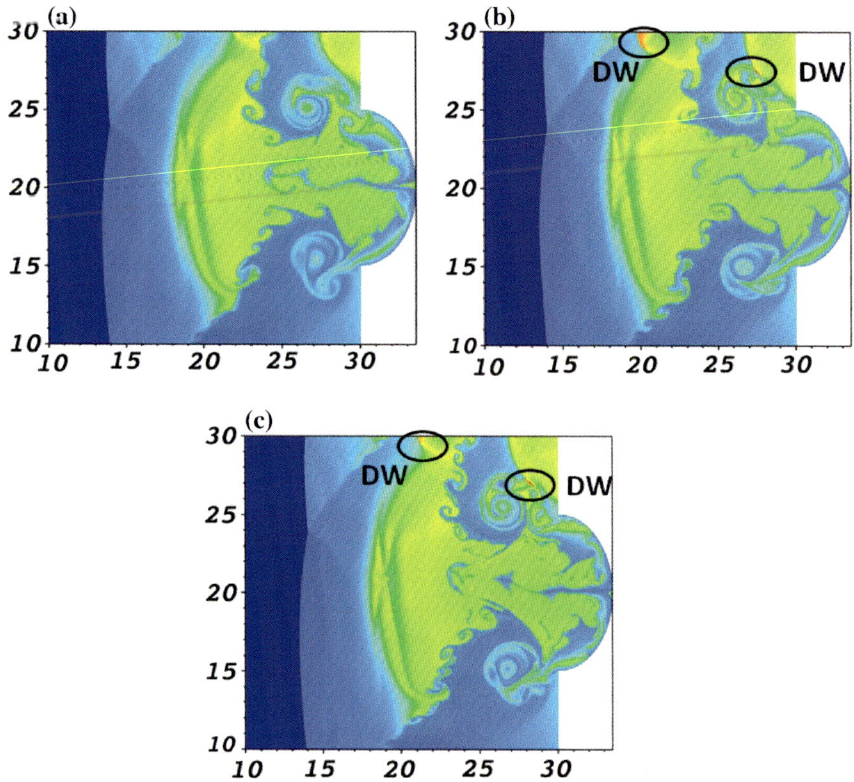

Fig. 7.4 Predicted temperature distributions at the time moment 40 μs for different grid resolution. The considered average triangular element sizes are: **a** 35 μm, **b** 20 μm, and **c** 12.5 μm

responsible for the triangulation of the computational domain are considered. Construction of the computational grid is carried out using the software SALOME. A comparison of three algorithms for triangulation using the example of a test problem was made. The structure of the output file in CGNS format relevant for our computations is described.

The series of unstructured grid is applied for resolution study. The computations show that the initiation of detonation occurs later on the considered coarse grid than on the detailed ones. The results obtained are in qualitative agreement with the results of numerical studies by other authors. In particular, our calculations are confirmed by the observations in [1]. The computations on unstructured grids require the use of sufficiently detailed grids since the unstructured grid method is more diffusive than the structured one. Insufficient grid resolution in the calculations causes the smearing of shocks and under-resolution of hydrodynamic instabilities that lead to certain differences in the flow patterns of the gas mixture and the weak change in the time moment of detonation initiation. However, the use of unstructured grids allows us to explore the areas of arbitrary geometries, which is relevant in solving a number of problems and expands the range of applicability of the computational method.

Acknowledgements The work of A. I. Lopato and P. S. Utkin is carried out under the state task of the ICAD of the RAS.

References

1. Togashi, F., Lohner, R., Tsuboi, N.: Numerical simulation of H_2/air detonation using unstructured mesh. Shock Waves **19**, 151–162 (2009)
2. Loth, E., Sivier, S., Baum, J.: Adaptive unstructured finite element method for two-dimensional detonation simulations. Shock Waves **8**(1), 47–53 (1998)
3. Pimentel, C., Azevedo, J., Silva, L.: Numerical study of wedge supported oblique shock wave-oblique detonation wave transitions. J. Braz. Soc. Mech. Sci. Eng. **24**, 149–157 (2002)
4. Luo, H., Baum, J., Lohner, R.: A Hermite: WENO-based limiter for discontinuous Galerkin method on unstructured grids. J. Comput. Phys. **225**, 686–713 (2007)
5. Wang, Z.J.: Spectral (finite) volume method for conservation laws on unstructured grids: basic formulation. J. Comput. Phys. **178**, 210–251 (2002)
6. Mahmoudi, K., Mazaheri, K.: High resolution numerical simulation of the structure of 2-D gaseous detonations. Proc. Combust. Inst. **33**, 2187–2194 (2011)
7. Vasil'ev, A.A.: Cellular structures of a multifront detonation wave and initiation (Review). Combust. Explos. Shock Waves **51**(1), 1–20 (2015)
8. Schultz, E., Shepherd, J.: Validation of detailed reaction mechanisms for detonation simulation. CalTech Explosion Dynamics Lab, Report No. FM99-5 (2000)
9. Liou, M.-S., Steffen Jr., C.J.: A new flux splitting scheme. J. Comput. Phys. **107**, 23–39 (1993)
10. Lopato, A.I., Utkin, P.S.: The usage of grid-characteristic method for the simulation of flows with detonation waves. In: Petrov, I.B., Favorskaya, A.V., Favorskaya, M.N., Simakov, S.S., Jain, L.C. (eds.) Proceedings of the Conference on 50 Years of the Development of Grid-characteristic Method, SIST, vol. 133, pp. 281–290. Springer (2019)
11. Chen, G., Tang, H., Zhang, P.: Second-order accurate Godunov scheme for multicomponent flows on moving triangular meshes. J. Sci. Comput. **34**, 64–86 (2008)
12. Shu, C.W., Osher, S.: Efficient implementation of essentially non-oscillatory shock-capturing schemes. J. Comput. Phys. **77**, 439–471 (1988)
13. Lopato, A.I., Utkin, P.S.: Numerical study of detonation wave propagation in the variable cross-section channel using unstructured computational grids. J. Combust. **3635797**, 1–8 (2018)
14. Salome. The open source integration platform for numerical simulation. https://www.salome-platform.org. Accessed 20 June 2019
15. CGNS. CFD data standard. http://cgns.github.io. Accessed 20 June 2019
16. CGNS Version 3.4.0. https://github.com/CGNS/CGNS/releases/tag/v3.4.0. Accessed 20 June 2019
17. Schoberl, J.: An advancing front 2D/3D-mesh generator based on abstract rules. Comput. Vis. Sci. **1**, 41–52 (1997)
18. Visit. https://wci.llnl.gov/simulation/computer-codes/visit. Accessed 20 June 2019

Chapter 8
On the Stability of a Detonation Wave in a Channel of Variable Cross Section with Supersonic Input and Output Flows

Vladimir Yu. Gidaspov and Dmitry S. Kononov

Abstract The possibility of the formation of the stationary Shock Wave (SW) and Detonation Wave (DW) in a variable cross section channel with the hydrogen–air and hydrogen–oxygen mixtures in a quasi-one-dimensional non-stationary formulation is investigated. The channel has a form of two successively arranged Laval nozzles. A comparative analysis of the solutions in stationary equilibrium, stationary frozen, non-stationary frozen, and non-stationary non-equilibrium formulations is presented. Equilibrium initial approximation was proposed for non-equilibrium flows modeling. Configurations of variable cross section channel with a stationary (detonation) wave in the first expanded area are obtained. It is shown that non-equilibrium stationary solutions in the first narrowing part of a dual Laval nozzle channel are unstable, and non-equilibrium stationary solutions in the second expanded part are unstable too, but they stabilize in the first one. The range of flow rates, at which a stationary detonation wave exists, can be predicted with a high degree of accuracy by the equilibrium stationary theory.

8.1 Introduction

Mathematical modeling of high-speed flows of combustible mixtures has great practical importance in the studying of processes of combustion and detonation in channels. This approach substantially complementing the natural experiment makes possible to increase a reliability of the designed power plants with lower material costs and helps to understand the physical/chemical phenomena observed in them.

Possibility of the formation of the stationary SW and DW in a variable cross section channel in a quasi-one-dimensional non-stationary formulation is investigated in this

V. Yu. Gidaspov · D. S. Kononov (✉)
Moscow Aviation Institute (National Research University), 4, Volokolamskoe Shosse, Moscow 125993, Russian Federation
e-mail: dr.kononoff@yandex.ru

V. Yu. Gidaspov
e-mail: gidaspov@mai.ru

L. C. Jain et al. (eds.), *Advances in Theory and Practice of Computational Mechanics*, Smart Innovation, Systems and Technologies 173,
https://doi.org/10.1007/978-981-15-2600-8_8

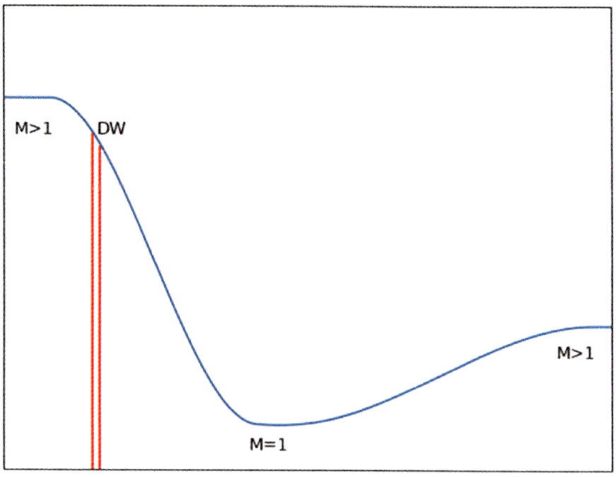

Fig. 8.1 The location of the stationary DW in the channel

chapter. A comparative analysis of the obtained solutions and the results from [1], in which the case of fuel combustion in Laval nozzle in the stationary DW is considered with subsequent acceleration of combustion products to supersonic rates (Fig. 8.1) in a quasi-one-dimensional stationary approximation, is presented. The gas flow before DW is considered frozen, whereas after DW is equilibrium. In [1], the stationary solution depends exclusively on the ratio of the channel areas in the current, inlet, and minimum cross sections.

In [1–5], it is noted that in a frozen flow the stationary SW is stable in an expanding area of channel and unstable in a narrowing one. To study DW stability in a variable cross section channel, a numerical simulation of a frozen and chemically non-equilibrium quasi-one-dimensional flow in a variable-section channel having the form of two successively arranged Laval nozzles in series (Fig. 8.2) was carried out. The contour is given by the analytical way. The radii of the critical sections of the channel are R_1 and R_2, respectively, with $R_1 > R_2$. The radius of the inlet cross section is R_0. The problem is studied in the non-stationary and stationary quasi-one-dimensional formulations. Mathematical model does not take into account the effects of viscosity, thermal conductivity, and diffusion. The flow at the channel inlet was supersonic, and SW/DW was realized inside the channel, in which the flow was braked to subsonic speeds. Approximately, at the critical section of the second Laval nozzle, the flow rate passed through the sound speed, and in the expanding part of the second nozzle the flow accelerated to supersonic rates. Hereinafter, the narrowing part of the first Laval nozzle will be referred to as the first narrowing part, the expanding part of the first Laval nozzle will be referred to as the first expanding part, the narrowing part of the second Laval nozzle will be referred to as the second narrowing part, and the expanding part of the second Laval nozzle will be referred to as the second expanding part.

In [3], the supersonic inlet flow with the constant heat implementation was investigated using the analytical and numeric methods. According to [3], DW stabilization

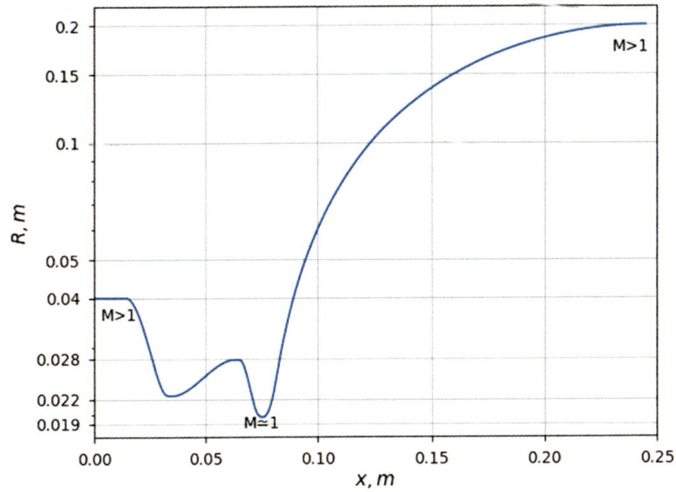

Fig. 8.2 Dependence of the radius of the investigated channel of variable cross section on the longitudinal coordinate

is possible with non-constant heat release because of variable composition of the combustion mixture.

In the equilibrium quasi-one-dimensional stationary formulation of the problem according to the method from [1] for the above-described two-contour channel, an analysis of possible flow variants with the stationary SW/DW was carried out (Figs. 8.3 and 8.4). The relative radii of the location of SW/DW in the channel (Fig. 8.3) and the temperature behind SW/DW (Fig. 8.4) are determined depending on the flow rate at the channel inlet. It is shown (Fig. 8.3) that, in the absence of chemical reactions in a hydrogen–air mixture, SW can be in the second tapering part at flow rates of ~1300–1500 m/s, in a hydrogen–oxygen mixture at ~1800–2100 m/s, and the first tapering and the first expanding parts at flow rates exceeding 1500 m/s and 2100 m/s, respectively. In the case of a chemically reacting mixture (Fig. 8.3), DW may be in the second tapering part for the hydrogen–air mixture at flow rates of ~2200–2600 m/s, in the hydrogen–oxygen mixture at ~3100–3700 m/s, and in the first tapering and first expanding parts at flow rates exceeds 2600 m/s and 3700 m/s, respectively. It should be noted that in the investigated range of parameters a chemically reacting flow is always realized since the temperature behind SW/DW exceeds the auto-ignition temperature of the considered combustible mixtures (Fig. 8.4). Further, the stable positions of SW/DW in the investigated channel (Fig. 8.2) were studied by simulating a quasi-one-dimensional non-stationary flow of a multicomponent gas.

The chapter is organized as follows. Section 8.2 presents a mathematical model. Testing methodology is described in Sect. 8.3. The results of mathematical modeling are reported in Sect. 8.4. Section 8.5 concludes the chapter.

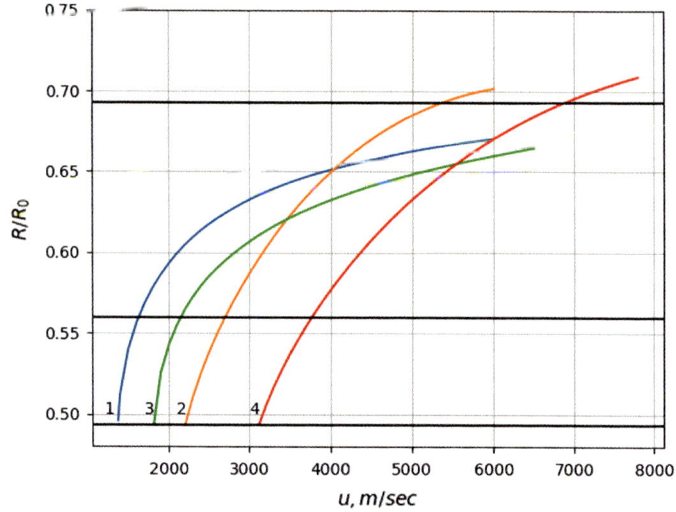

Fig. 8.3 Dependence of the relative cross-sectional radii of channel from the flow rate at the channel inlet, in which the stationary SW/DW is implemented. Solid horizontal lines indicate the radius values of the cross section of the channel respect to the local extrema of the profile, where curve 1—the frozen hydrogen–air mixture, curve 2—reacting hydrogen–air mixture, curve 3—frozen hydrogen–oxygen mixture, and curve 4—reacting hydrogen–oxygen mixture

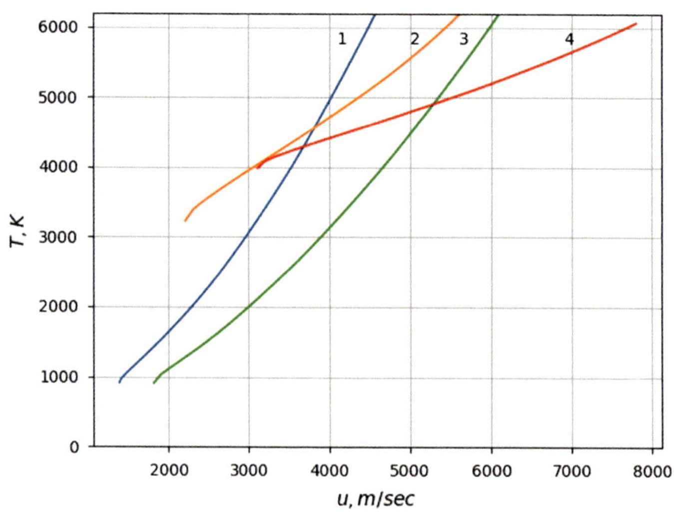

Fig. 8.4 Dependence of the temperature of the mixture beyond SW/DW from the flow rate at the channel inlet, where curve 1—the frozen hydrogen–air mixture, curve 2—the reacting hydrogen–air mixture, curve 3—the frozen hydrogen–oxygen mixture, and curve 4—the reacting hydrogen–oxygen mixture

8.2 Mathematical Model

Non-stationary chemically non-equilibrium gas flow in the channel is described by the system of Euler equations supplemented by thermal and caloric ones of state and N equations of chemical kinetics [6] is provided by Eq. 8.1.

$$\begin{cases} \frac{\partial \rho F}{\partial t} + \frac{\partial \rho u F}{\partial x} = 0 \\ \frac{\partial \rho u F}{\partial t} + \frac{\partial (\rho u^2 + P) F}{\partial x} = P \frac{dF}{dx} \\ \frac{\partial \rho \left(E \frac{u^2}{2} \right) F}{\partial t} + \frac{\partial \rho u F \left(E \frac{P}{\rho} + \frac{u^2}{2} \right)}{\partial x} = 0 \\ \frac{\partial \rho F \gamma_i}{\partial t} + \frac{\partial \rho u F \gamma_i}{\partial x} = F W_i \;\; i = 1 \ldots N \\ P = \rho R T \sum_{i=1}^{N} \gamma_i \\ E = \sum_{i=1}^{N} \gamma_i E_i(T) \end{cases} \tag{8.1}$$

Here, x is the longitudinal coordinate, t is the time, ρ is the density, u is the mixture rate, P is the pressure, T is the temperature, H is the enthalpy, $E_i(T)$ and γ_i are the functions for internal energy from catalogs and concentration of the ith component in the mixture, respectively, $\bar{\gamma}$ is the vector of molar mass concentrations of the components of the mixture, $F = F(x)$ is the dependence of the cross-sectional area of the channel on the longitudinal coordinate. When considering frozen flows, it is assumed that $W_i = 0, i = 1 \ldots N$.

The quasi-one-dimensional stationary chemically non-equilibrium flow is described by a similar system of equations, in which all the complexes $\frac{\partial (\cdot)}{\partial t}$ are set equal to zero.

A mixture of perfect gases is considered, whose thermodynamic properties are described by defining the expression for Gibbs potential [7, 8]:

$$G(P, T, \bar{\gamma}) = \sum_{i=1}^{N} \gamma_i \left(G_i^0(T) + RT \ln \frac{P_i}{P_0} \right), \tag{8.2}$$

where R is the universal gas constant, P_0 is the normal pressure, $G_i^0(T)$ are the known dependences [7] of the temperature part of Gibbs molar potential of a separate component of the mixture. The internal energy and density of the mixture are expressed in terms of Gibbs potential and its partial derivatives provided by Eqs. 8.3–8.4.

$$E(T, \gamma_i) = G - T \left(\frac{\partial G}{\partial T} \right)_P - P \left(\frac{\partial G}{\partial T} \right)_T = \sum_{i=1}^{N} \gamma_i E_i(T) \tag{8.3}$$

$$\rho(P, T, \bar{\gamma}) = 1/\left(\frac{\partial G}{\partial P}\right)_T = \frac{P}{RT\sum_{i=1}^{N}\gamma_i} \tag{8.4}$$

SW holds the conservation laws (Rankine–Hugoniot equations):

$$\begin{cases} \rho_1 v_1 = \rho_2 v_2, \\ P_1 + \rho_1 v_1^2 = P_2 + \rho_2 v_2^2, \\ E_1 + \frac{P_1}{\rho_1} + \frac{v_1^2}{2} = E_2 + \frac{P_2}{\rho_2} + \frac{v_2^2}{2}, \\ \bar{\gamma}_1 = \bar{\gamma}_2, \end{cases} \tag{8.5}$$

where the index “2” marks the values after SW, the index “1” is before $v = D - u$, where D is the rate of SW.

To obtain a numerical solution of the system of Euler equations, the method of Godunov [9] was used. For integrating the system of equations of the stationary model, the difference approximation was taken from [1]. To integrate the system of ordinary differential equations of chemical kinetics, the method of Pirumov from [6] was used.

8.3 Testing

To test the numerical method of integrating the equations of chemical kinetics, a test model of the adiabatic reaction at constant density provided by Eq. 8.6 is used.

$$\begin{aligned} &\sum_{i=1}^{N}\gamma_i E_i(T) = E \\ &\rho\frac{d\gamma_i}{dt} = W_i(\rho, T, \gamma_1, \ldots, \gamma_N) \quad \gamma_i(0) = \gamma_i^0, i = 1, \ldots, N \\ &\rho = \text{const} \quad E = \text{const} \end{aligned} \tag{8.6}$$

The kinetic mechanism for this test consisted of 19 reversible stages involving 8 components [10]. The initial temperature was set at $T_0 = 1000$ K, the initial density was $\rho_0 = 0.1$ kg/m^3, and a stoichiometric mixture of hydrogen and oxygen was taken as the reacting gas. A characteristic ignition delay was obtained at the time $t \simeq 50$ μs, the temperature of the combustion products was set at ≈ 3300 K (Fig. 8.5), and the change in the concentrations of the mixture components during ignition (Fig. 8.6) was similar to the results in [9, 10].

As a test for checking the numerical simulation of non-stationary flows, we took the problem of the interaction of an isolated shock wave propagating in a channel filled with a non-reacting ideal gas with an adiabatic parameter $k = 1.4$. It considers the breakup of an arbitrary discontinuity at the point $x = 0.2$ with the specification

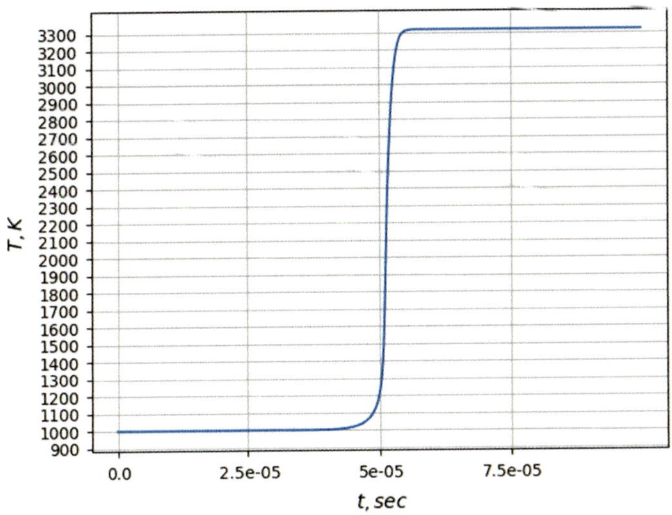

Fig. 8.5 Temperature distribution in the adiabatic reaction model at constant density

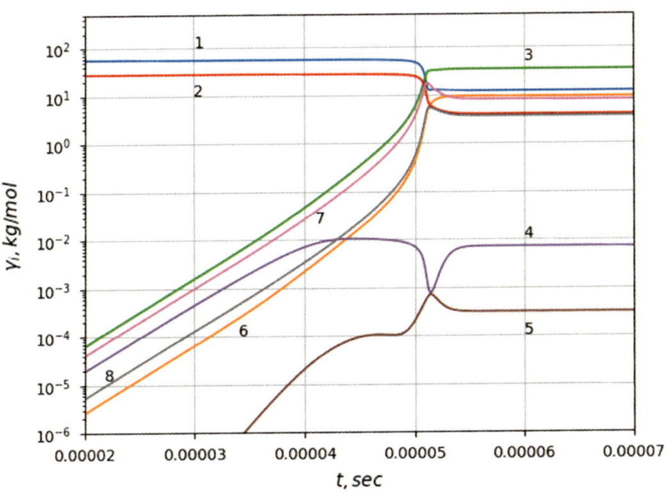

Fig. 8.6 The distribution of the concentrations of the components of the mixture in the model of the adiabatic reaction at constant density, where curve 1—H_2, curve 2—O_2, curve 3—H_2O, curve 4—HO_2, curve 5—H_2O_2, curve 6—OH, curve 7—H, and curve 8—O

of the dimensionless initial parameters provided by Eq. 8.7.

$$\rho = \begin{cases} 3.857143 & \text{if } x < 0.2 \\ 1 + 0.2\sin(50(0.2 - x)) & \text{if } x > 0.2 \end{cases}$$

$$u = \begin{cases} 3.857143 & \text{if } x < 0.2 \\ 1 + 0.2\sin(50(0.2 - x)) & \text{if } x > 0.2 \end{cases}$$

$$P = \begin{cases} 10.33333 & \text{if} x < 0.2 \\ 1 & \text{if} x > 0.2 \end{cases} \quad (8.7)$$

The calculation was carried out up to the time point $t = 0.18$ s. In the absence of an analytical solution, the numerical solution obtained on such a detailed grid was taken as an exact one, so that as the number of points increases, it remains unchanged. In this chapter, this number of points is equal to 15,000 (Fig. 8.7). The area behind SW is divided into subareas of high-frequency, low-frequency oscillations, and unperturbed parameters, which are in good agreement with the results given in [11].

Testing of the simulation method of stationary quasi-one-dimensional non-equilibrium flow was carried out by simulating the flow of a stoichiometric mixture of hydrogen with oxygen ($H_2 + 0.5O_2$) in a constant section channel with the presence of SW in the channel inlet section and without it with an initial temperature of 1000 K and an initial pressure of 1 atm. Chemical transformations here and hereinafter were modeled by 8 reversible stages (Table 8.1) [10].

The stoichiometric mixture of hydrogen and oxygen $H_2 + 0.5O_2$ with 6 reacting components H, O, H_2, O_2, H_2O, and OH and the hydrogen–air mixture $H_2 + 0.5O_2 + 1.881N_2$ with 7 reacting components H, O, H_2, O_2, N_2, H_2O, and OH were investigated.

The obtained results are presented on Pressure–Volume (PV)-diagram in the form of Rayleigh–Michelson Lines (RMLs) (Fig. 8.8). Each of RMLs is limited by the

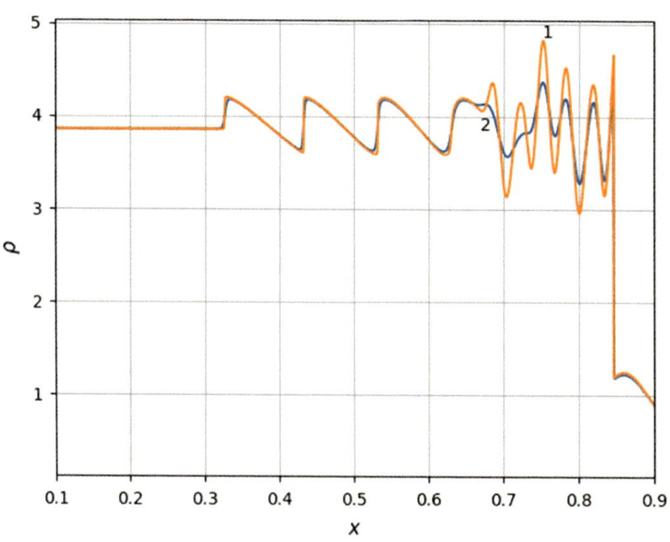

Fig. 8.7 Density distribution at time $t = 0.18$ s, where curve 1—the solution obtained by dividing the computational domain into 15,000 points and curve 2—the solution obtained by dividing the computational domain into 3,000 points

Table 8.1 Kinetic mechanism

Reaction	A, mol, m^3, s, K	n	E, K
$H_2 + M = H + H + M$	5.5 E18	−1.0	51.987
$O_2 + M = O + O + M$	7.2 E18	−1.0	59.340
$H_2O + M = OH + H + M$	5.2 E21	−1.5	59.386
$OH + M = O + H + M$	8.5 E18	−1.0	50.830
$H_2O + O = OH + OH$	5.8 E13	0	9.059
$H_2O + H = OH + H_2$	8.4 E13	0	10.116
$O_2 + H = OH + O$	2.2 E13	0	8.455
$H_2 + O = OH + H$	7.5 E13	0	5.586

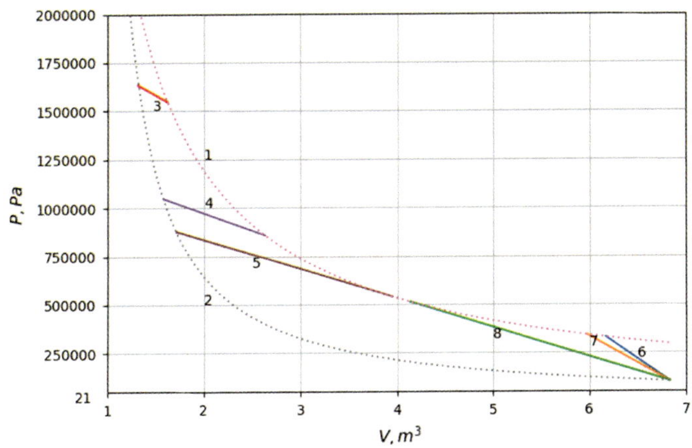

Fig. 8.8 PV-diagram of stationary non-equilibrium flow in a constant shape channel, where curve 1—equilibrium DA, curve 2—SA, curve 3—RML for flow with $u_0 = 3600$ m/s behind SW, curve 4—RML for flow with $u_0 = 2900$ m/s behind SW, curve 5—RML for flow with $u_0 = 2661.3$ m/s behind SW, curve 6—RML for a flow with $u_0 = 4000$ m/s, curve 7—RML for a flow with $u_0 = 3600$ m/s, and curve 8—RML for a flow with $u_0 = 2664.7$ m/s

Shock Adiabats (SA) and Detonation Adiabats (DA), which indicates the fulfillment of conservation laws. Chapman–Jouguet mode ($u_0 = 2661.3$ m/s) obtained by calculation, in which RML DA touch (curve 5) is observed, with a decrease of rate, has no solution. Unstressed flow regimes (curves 6–8) are received. It should be noted that the minimum flow rate, at which there is a solution with $u_0 = = 2664.7$ m/s, is slightly greater than Chapman–Jouget rate. The calculation results are in good agreement with the data of [9].

8.4 The Results of Mathematical Modeling

At first, the problem of modeling a frozen flow in a contour has been solved (Fig. 8.9). As an initial approximation of the solution, it was assumed that in the whole computational domain $u_0 = 4297$ m/s, $T_0 = 400$ K, $P_0 = 1$ atm, a stoichiometric mixture of hydrogen and oxygen flows. The problem of obtaining a stable state flow was solved, after which the emergence of SW was simulated in the expanding part, and the search for the stable state solution was carried out again. The method of establishing a solution was obtained with a stationary SW in the first expanding part of the investigated contour. It should be noted that the temperature behind the steady SW $\approx$ 3500 K is consistent with the dependence of the temperature behind a stationary SW on the input flow rate shown in Fig. 8.4.

By varying the inlet flow rate in the channel, it was possible to obtain stable state solutions with SW in the range of the initial rate from 2172 to 4297 m/s (Fig. 8.10).

Then, the hydrogen–air reacting gas was considered as an investigated mixture. Based on the method given in [1] for a chemically equilibrium hydrogen–air mixture in the investigated channel ($R_1/R_0 = 0.56$, $R_2/R_0 = 0.493$), an RR-diagram can be constructed, according to which DW in this problem can be in the second narrowing part of the channel in question at an input flow rate of more than 2200 m/s, in the first tapering and first expanding parts of the channel in question at an input flow rate of more than 2700 m/s (Fig. 8.11).

It should be noted that the equilibrium solution with a stationary SW depends exclusively on the ratio of the channel radii in the current and inlet sections. Thus,

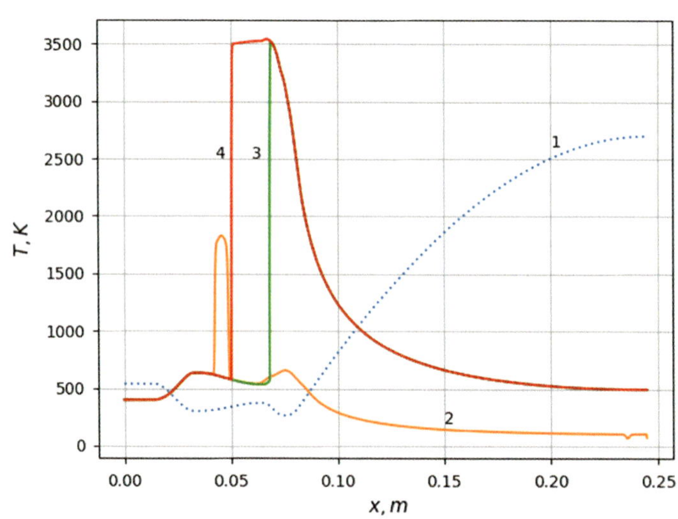

Fig. 8.9 Temperature distribution in time layers, frozen stoichiometric mixture of hydrogen with oxygen, $u_0 = 4297$ m/s, where curve 1—channel contour shape, curve 2—start of propagation of the implemented shock wave, $t = 0.000255$ s, curve 3—position of the shock wave at time $t = 0.000375$ s, and curve 4—position of the shock wave at time $t = 0.02$ s (stable state)

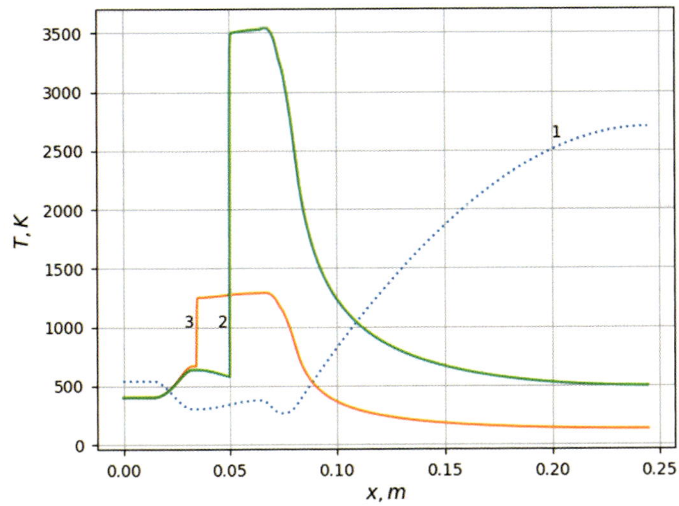

Fig. 8.10 The range of inlet rates in the channel and the corresponding stable state solutions, a frozen stoichiometric mixture of hydrogen with oxygen, where curve 1—channel contour form, curve 2—stable state solution, $u_0 = 4297$ m/s, and curve 3—stable state solution, $u_0 = 2172$ m/s

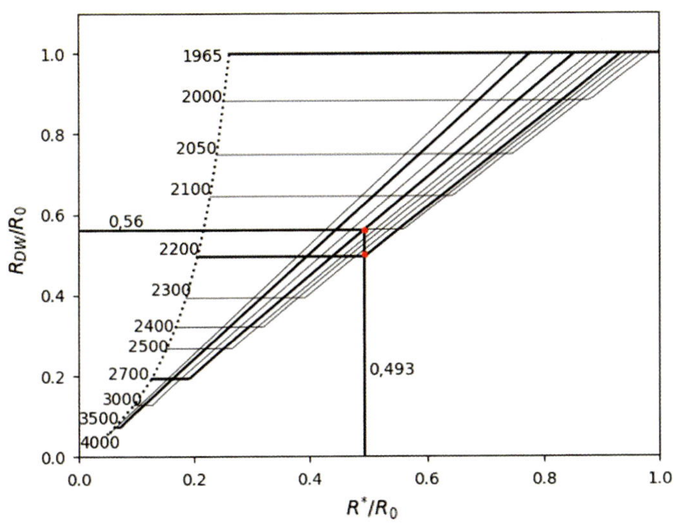

Fig. 8.11 RR-diagram for a chemically equilibrium hydrogen–air mixture at $T_0 = 400$ K

from the equilibrium stationary simulation it follows that there are three possible positions of the stationary SW for the considered channel configuration (Fig. 8.12).

The stability of the position of SW/DW was investigated by the method of relaxation [3]. When setting the initial approximation with SW in the first narrowing part of the channel with an input rate of 2750 m/s, DW turned out to be unstable

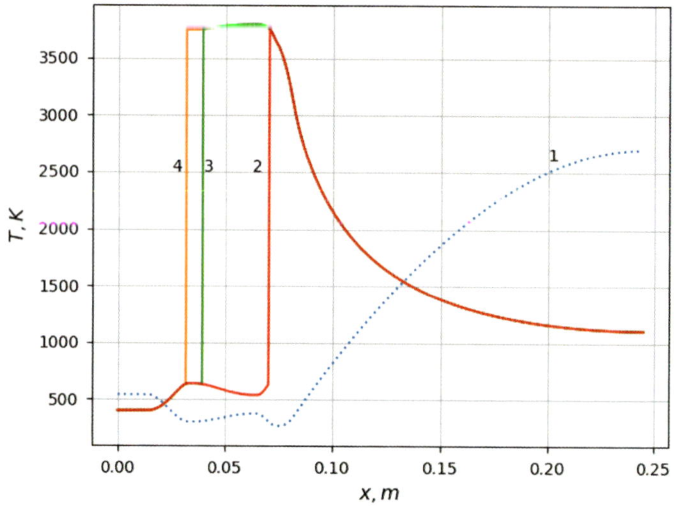

Fig. 8.12 Solutions with the stationary SW for a chemically equilibrium hydrogen–air mixture with an inlet flow rate of 2750 m/s at $R/R_0 = 0.57$, R is the section radius, in which SW is realized, where curve 1—contour form, curve 2—solution with stationary SW in the second narrowing part, curve 3—solution with stationary SW in the first expanding part, and curve 4—solution with stationary SW in the first narrowing part

and moved to the left, against the direction of flow leaving the computational area (Fig. 8.13). This fact is consistent in RR-diagram depicted in Fig. 8.11.

When setting the initial approximation with SW in the first expanding part of the channel with an inlet rate of 2750 m/s, DW was stable and only slightly shifted to the left relative to the initial approximation (Fig. 8.14).

When setting the initial approximation with SW in the second narrowing part of the channel in question with an input rate of the mixture of 2750 m/s, DW was unstable in this position. It began to move to the left against the direction of flow and stabilized with the same cross-sectional area of the channel as in the above case (Fig. 8.15).

When specifying an initial approximation with SW in the second narrowing part of the channel with an inlet rate of a mixture of 2650 m/s, DW is also unstable, but, at the same time, unlike in the previous case it completely went beyond the limits of the computational domain (Fig. 8.16).

The solutions obtained by the relaxation method were compared with the solutions obtained by solving the direct problem of nozzle theory (quasi-one-dimensional stationary case) with passing a singular point in the vicinity of the critical section (Fig. 8.17) using the modified algorithm from [12]. The obtained results are in good agreement in all computational areas with the exception of DW vicinity. This difference can be explained by the fact that to simulate the fine structure of DW, very small discretization of the computational domain is required, which was implemented for solving the problem with the stationary formulation only. When solving the problem

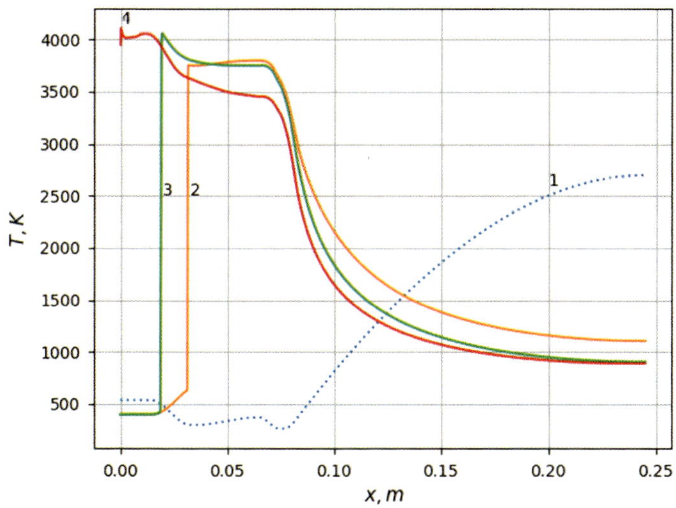

Fig. 8.13 Non-equilibrium non-stationary calculation of the flow of a hydrogen–air mixture with DW in the first narrowing area, where curve 1—contour form, curve 2—initial approximation, curve 3—temperature distribution at time $t = 0.000684$ s, and curve 4—temperature distribution at time $t = 0.000776$ s

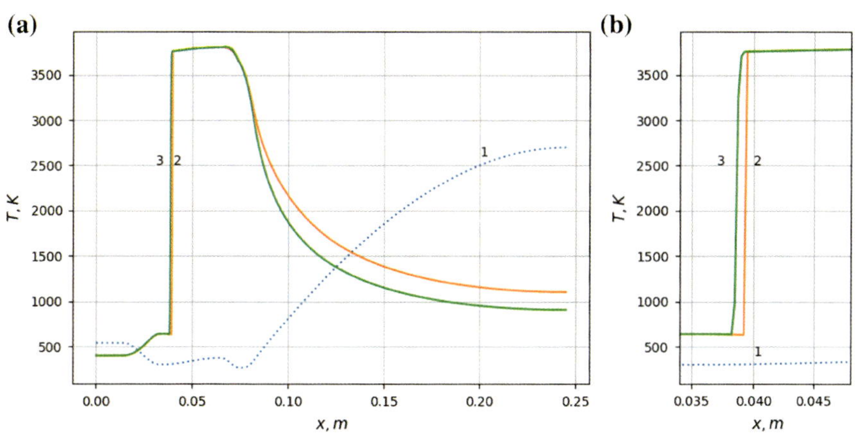

Fig. 8.14 Non-equilibrium non-stationary calculation of the flow of a hydrogen–air mixture with DW in the first expanding area, where curve 1—contour form, curve 2—initial approximation, and curve 3—stable state solution: **a** temperature distribution in the channel, **b** temperature distribution in the vicinity of the relaxation of DW

in the non-stationary formulation, the fine discretization in the vicinity of DW was not done.

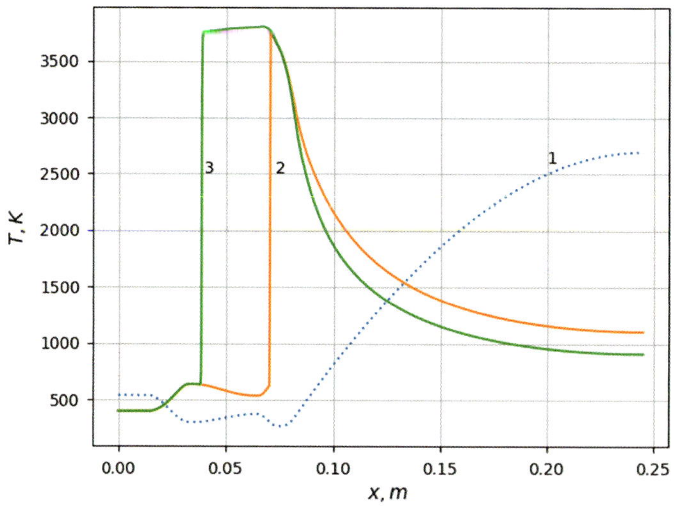

Fig. 8.15 Non-equilibrium non-stationary calculation of the flow of a hydrogen–air mixture with DW in the second narrowing area, where curve 1—form contour, curve 2—initial approximation, and curve 3—stable state temperature distribution

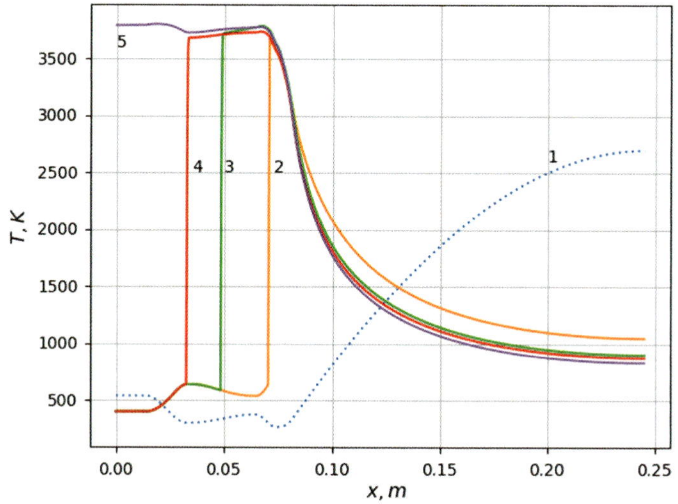

Fig. 8.16 Non-equilibrium nonstationary calculation of the flow of a hydrogen–air mixture with DW in the second expanding area where curve 1—contour form, curve 2—initial approximation, curve 3—temperature distribution at time $t = 0.000811$ s, curve 4—temperature distribution at time $t = 0.004928$ s, and curve 5—temperature distribution at time $t = 0.007188$ s

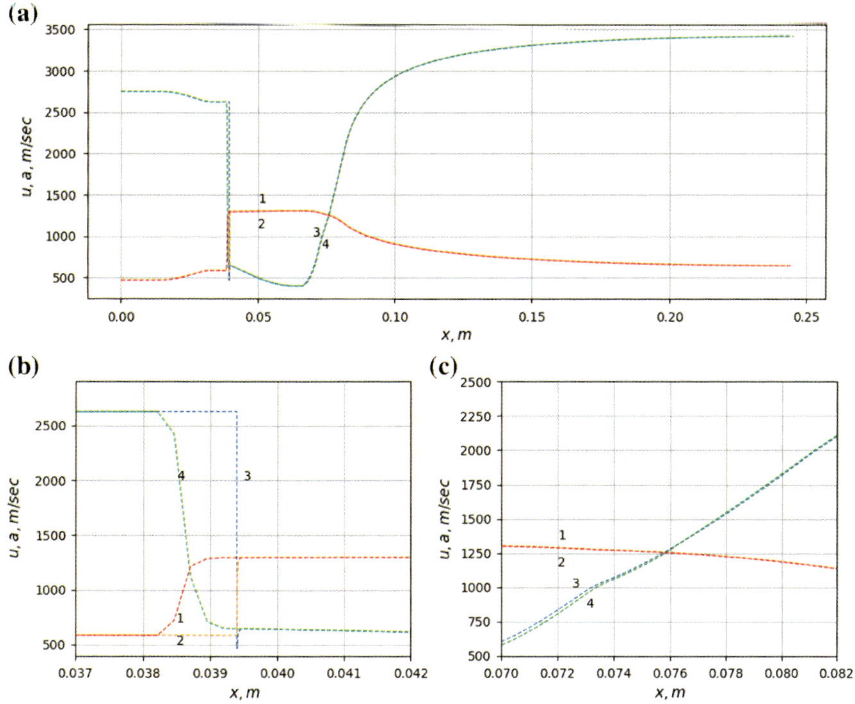

Fig. 8.17 Chemically non-equilibrium mixture of hydrogen–air, inlet flow rate of 2750 m/s: **a** velocity distribution (curves 3 and 4) and sound velocity (curves 1 and 2) in the channel, **b** in the vicinity of DW, **c** in the vicinity of the critical section (curves 1 and 4—method of relaxation) and (curves 2 and 3—direct problem of nozzle theory)

8.5 Conclusions

A quasi-one-dimensional non-stationary formulation of a chemically non-equilibrium flow of a combustible hydrogen–air mixture in a channel consisting of two consecutive Laval nozzles with fuel combustion in a stationary detonation wave with supersonic flow and at the channel inlet and outlet was investigated. By calculation, it was obtained that a stationary detonation wave is stable in the first expanding part of the channel and unstable in narrowing parts. The range of the flow rates at the channel inlet, at which the formation of a stationary detonation wave is possible, is obtained. It is shown that for a hydrogen–air mixture in the investigated channel, the range of flow rates, at which a stationary detonation wave exists, can be predicted with a high degree of accuracy by the equilibrium stationary theory.

Acknowledgements This work was carried out within the state task no. 9.7555.2017/BCh.

References

1. Gidaspov, V.Yu.: Numerical simulation of one-dimensional stationary equilibrium flow in engine detonation. Trudy MAI 83, 1–20 (in Russian) (2015)
2. Cherniy, G.G.: Gas dynamics. Moscow, Nauka Publ., (in Russian) (1988)
3. Levin, V.A., Manuilovich, I.S., Markov, V.V.: Stabilization of detonation waves in a supersonic flow. Moscow Univ. Mech. Bull. **66**(4), 77–82 (2011)
4. Kraiko, A.N., Shironosov, V.A.: Investigation of flow stability in a channel with a closing shock at transonic flow velocity. J. Appl. Math. Mech. **40**, 533–541 (1976)
5. Grin, V.T., Kraiko, A.N., Tillyaeva, N.I., Shironosov, V.A.: Analysis of one-dimensional flow stability in a channel with arbitrary variation of stationary flow parameters between the closing shock cross section and channel outlet. J. Appl. Math. Mech. **41**, 651–659 (1977)
6. Pirumov, U.G., Roslyakov, G.S.: Gas Flow in Nozzles. Springer, Berlin, Heidelberg (1986)
7. Gurvich, L.V., Veyts, I.V., Alcock, C.B.: Thermodynamic Properties of Individual Sub-stances, 4th edn, vol. 1. Elements O, H(D, T), F, Cl, Br, I, He, Ne, Ar, Kr, Xe, Rn, S, N, P, and Their Compounds. Part 2. New York, Hemisphere Pub. Corp. (1989)
8. Gidaspov, V.Yu., Severina, N.S.: Numerical simulation of the detonation of a propane-air mixture, taking irreversible chemical reactions into account. High Temp. **55**(5), 777–781 (2017)
9. Godunov, S.K.: Finite difference methods for numerical computations of discontinuous solutions of the equations of fluid dynamics. Matematicheskii Sbornik **47**, 271–306 (1959)
10. Gardiner, W.C. (ed.): Combustion Chemistry. Springer, New York (1984)
11. Shu, C.-W., Osher, S.: Efficient implementation of essentially non-oscillatory shock-capturing schemes. J Comp. Phys. **83**, 32–78 (1989)
12. Gidaspov, V.Yu.: Numerical simulation of chemically non-equilibrium flow in the nozzle of the liquid-propellant rocket engine. Aerosp. MAI J. **20**(2), 90–97 (2013)

Chapter 9
Physical and Kinematic Processes Associated with Meteoroid When Falling in the Earth's Atmosphere

Viktor A. Andrushchenko, Vasiliy A. Goloveshkin and Nina G. Syzranova

Abstract We consider the motion of a number of known meteor bodies in the Earth's atmosphere and their fall out in the form of meteorites on the Earth's surface. Various mechanisms of their destruction in the atmosphere are explored within the framework of the extended theory of meteor physics. For some of them, the proposed hypothetical model of the formation of the surface relief of falling meteoroids and, accordingly, their fallen remains was tested, which turned out to correspond to the real structure of the surfaces of the meteorites found—smooth in some cases and dotted with irregularities in the form of rhegmaglypts in others.

9.1 Introduction

Meteor bodies are divided into three main classes: stone, iron, and ironstone. In the percentage terms it is as follows: stone meteorites—92.5%, iron meteorites—5.7%, ironstone meteorites—1.3%, and anomalous—0.5% [1]. The average number of falling fireballs per year recorded by satellites and infrasound stations is more than 4 times higher than the average number of meteorites, which confirms the fact that most of them "burns" in the atmosphere before reaching the Earth's surface. Here, the term "burns" refers to either an air explosion or progressive fragmentation with already real combustion of small fragments left as a consequence of these phenomena.

V. A. Andrushchenko · N. G. Syzranova (✉)
Institute of Computer Aided Design of the RAS, 19/18, Vtoraya Brestskaya ul., Moscow 123056, Russian Federation
e-mail: nina-syzranova@ya.ru

V. A. Andrushchenko
e-mail: andrusviktor@ya.ru

V. A. Goloveshkin
Institute of Applied Mechanics of the RAS, 7, Leningradsky prt, Moscow 125040, Russian Federation
e-mail: nikshevolog@ya.ru

L. C. Jain et al. (eds.), *Advances in Theory and Practice of Computational Mechanics*, Smart Innovation, Systems and Technologies 173,
https://doi.org/10.1007/978-981-15-2600-8_9

Falling meteorites on the Earth is accompanied by a number of physical phenomena: light, sound, and mechanical. The meteor body begins to glow at altitudes of about 130–80 km, and at altitudes of 20–10 km its movement is significantly slowed down. On this segment of the path called the region of delay, the heating and evaporation of the meteor body (or fragments) stop, the luminosity disappears, and a thin molten layer on the surface of the fragments quickly solidifies, forming a melting crust on the surface. Under the influence of the temperature field depending on the nature of the movement by this time, the surface takes a smooth or rhegmaglypts-strewn structure of the relief. Having overcome the region of delay, fragments of meteor bodies fall almost vertically under the influence of gravity and fall out in the form of fragments—meteorites.

In this chapter, using the methods of mathematical modeling, we study the processes that occur when moving at hypersonic speeds of meteoroids in the Earth's atmosphere, and assess their impact on the nature of the destruction of bodies. The study of the movement and destruction of each particular meteoroid is an independent task since each of them is significantly different: its form is arbitrary, and the structure is heterogeneous. As an example, the motion and destruction of several specific meteorites significantly different in size, properties, composition of the material, and trajectory parameters are analyzed.

The chapter has a following structure. Section 9.2 discusses a problem of determining heat fluxes to the surface of meteoroids. In Sect. 9.3, the model of meteoroid's fragmentation is described, and in Sect. 9.4 the mechanisms of destruction of the bodies due to thermal stresses are presented. Section 9.5 concludes the chapter.

9.2 Heat Transfer to the Surface of Meteoroids

Problems of heat transfer of bodies moving at hypersonic speeds within the atmosphere are quite fully studied. This is primarily due to the research of spacecraft flights and the providing of their thermal protection. In meteor physics, the knowledge of the heat transfer coefficient is necessary for the most realistic assessment of the behavior of cosmic bodies at the entrance to the atmosphere.

Equations 9.1 of the extended physical theory of meteors in a form of their motion in the exponential atmosphere [2] are used for the numerical study of the problem, where V, M, and θ are the speed of the body, its mass, and angle of incidence of the trajectory to the horizon, respectively, R_E is the radius of the Earth, C_D, C_H, and H_{eff} are the coefficients of resistance, heat transfer to the surface of the body, and effective enthalpy of vaporization of the meteoroid, respectively, S_{mid} is the cross-sectional area of the body, z is the height of the meteor body above the Earth's surface, ρ_0 is the density of the atmosphere at $z = 0$, h is the characteristic scale of height.

$$\begin{aligned} &M\frac{\mathrm{d}V}{\mathrm{d}t} = Mg\sin\theta - C_D S_{\mathrm{mid}}\frac{\rho V^2}{2} \quad V\frac{\mathrm{d}\theta}{\mathrm{d}t} = g\cos\theta - \frac{V^2\cos\theta}{R_E+z} \\ &H_{\mathrm{eff}}\frac{\mathrm{d}M}{\mathrm{d}t} = -C_H S_{\mathrm{mid}}\frac{\rho V^3}{2} \quad \frac{\mathrm{d}z}{\mathrm{d}t} = -V\sin\theta \quad \rho = \rho_0\exp(-z/h) \end{aligned} \tag{9.1}$$

In the high-temperature gas stream, there are two heat transfer mechanisms from the gas to the surface of the body: the convective and radiative heat transfers.

The following formula [3] is used for the convective heat flux at the critical point of the spherical surface of the meteorite:

$$q_{c0} \approx 3.3 \times 10^{-5} \left(\frac{\rho_\infty}{R} \right)^{1/2} V_\infty^{3.2}, \text{ W/m}^2.$$

Here, R is given in m, ρ_∞ is given in kg/m^3, and V_∞ is given in m/s. The index ∞ marks the parameters of the incident flux. ReVelle formula, the parameters of which are presented in [3], is used for the radiation heat transfer coefficient at the critical point provided by Eq. 9.2.

$$C_{Hr} = f \cdot e^{A_1} \rho^{A_2 + A_3 V - 1} R^{A_4 + A_5 V + A_6 V^2} V^{A_7 + A_8 V + A_9 V^2 - 3} \quad (9.2)$$

Accordingly, the heat flow at the critical point is written as $q_{r0} = 0.5 \rho_\infty V_\infty^3 C_{Hr}$.

The heat flux distribution along the spherical surface for the convective heat flux is approximated by Eq. 9.3 [4], where β is the meridional cross-sectional angle measured from the direction to the critical point, and for radiative flux we have [5]: $q_r = q_{r0} \cos^n \beta$, $n = 1/(0.051V - 0.43) + 1.811$.

$$q_c = q_{c0}(0.55 + 0.45 \cos 2\beta) \quad (9.3)$$

The total heat flux to the body surface is defined as $q = q_c + q_r$.

The calculated ratios of the finite mass M (near the Earth's surface) to the initial mass of the atmospheric entry M_e for four meteoroids (Tunguska [6], Sikhote-Alin [7], Kunya-Urgench [8], and Chelyabinsk [9]) are presented in Table 9.1. The calculations took into account only the ablation of bodies, and did not take into account the process of fragmentation meteoroids.

The calculated data on the heat transfer to the surface of meteor bodies show that for large meteoroids (Tunguska or Chelyabinsk), the radiative heat flux was several orders of magnitude higher than the convective one along the entire trajectory of the fall (high speed of flight and large body size). At the same time, for a relatively small meteorite Kunya-Urgench (lower flight speed and small body size), the values of radiation heat fluxes on almost the entire trajectory were significantly less than convective. This fact was manifested in the process of mass loss.

Table 9.1 The parameters of meteoroids

Meteoroid	M_e (t)	V_e (km/s)	Material, density (g/sm^3)	M/M_e (%)
Tunguska (1908)	1,000,000	30	Ice, 1.0	1
Sikhote-Alin (1947)	500	15	Iron, 7.8	7
Kunya-Urgench (1998)	3	13	Stone chondrite, 3.3	84
Chelyabinsk (2013)	13,000	19	Stone chondrite, 3.3	5

9.3 Fragmentation of Meteoroids

Calculation of the meteoric-body ablation requires us to take into account its fragmentation. Statistics of falls of meteoroids shows that most of them fell onto the Earth as fragmented pieces.

In [2], one of the models of successive fragmentation that uses the statistical theory of strength is considered. It is known that the structure of meteoric bodies that penetrate the atmosphere has strength thin homogeneity. The statistical theory of strength studies structurally inhomogeneous bodies [10]. In the context of this theory, fragmentation occurs along with defects and cracks, which are available in space bodies, i.e., the structural inhomogeneity of meteoric bodies influences the process of fragmentation. As a result, the fragmentation appears as a process of successive elimination of defects with increase in the load by means of body disruption along these defects, so that the fragments created have a greater strength than the initial body. In this connection, the fragmentation process will complete as soon as the velocity of head begins to diminish.

According to this model, the fragment strength can be written as

$$\sigma_f^* = \sigma_e (M_e / M_f)^\alpha,$$

where σ_e and M_e are the strength and mass of the meteoroid before entering the atmosphere, σ_f^* and M_f are the same characteristics for the fragment, and α is the exponent of the material nonhomogeneity (for larger α, the nonhomogeneity is higher). The body fragmentation in the atmosphere takes place under condition:

$$\rho_* V_*^2 = \sigma^*,$$

where the air pressure is equal to σ^* that is one of the strength characteristics of the body material (compression, tensile, and shear strength). The fragmentation altitude z_* in the exponential atmosphere is determined from the condition $\rho_* = \rho_0 \exp(-z_*/h)$. From this altitude, instead of a single body a swarm of cleaving fragments with increasing number N falls that is accounted in the set of characteristic variables. Assume that the formed fragments are spheres of equal mass their number depending on the current values of the air pressure and the total mass of all fragments, and the midsection area is obtained (determined on the assumption that the fragments do not overlap) [2]. The problem of the fragmenting meteor body motion is solved in three stages. At the first stage, the movement of a single body from the entry altitude to the fragmentation altitude is considered. At the second stage, from the fragmentation altitude to the maximum air pressure altitude the motion of swarm of fragments is considered. At the third stage, the motion of only one fragment is tracked since it is believed that the fragments are of the same size.

The strength of meteoroids varies widely. For a typical iron Sikhote-Alin meteorite (Primorsky Krai, 1947) as a result of processing of observations [7], it was found that the first fragmentation occurred at an altitude of ~58 km. According to these data, the critical value of the strength parameter, at which the fragmentation process began, is estimated by the value $\sigma_e = 10^5$ N/m^2. The variations in the total mass of the meteor body depending on the altitude of the flight for the single body model and for the fragmenting meteoroid at different values of parameter α are given in Fig. 9.1. Parameter α depends on the degree of the inhomogeneity of the material and size of the body. It should be noted that during the fragmentation of meteor bodies, the removal of mass grows rapidly at first due to the increasing surface area in the flow. However, this leads to an increase in the deceleration of the fragmenting body, i.e., to a decrease in the velocity of the meteor body and its fragments, which, in turn, reduces the heat flow to the surface and slows the process of ablation. For this reason, as the calculations show, the final mass of the fragmenting meteoroid, as a rule, turns out to be greater than in the case of the single body model. For example, the final mass for the fragmenting bolide with parameter $\alpha = 0.25$ is ~90 t, while the final mass for a single body is ~36 t. However, as the calculations show, at $\alpha = 0.125$ and altitude of 40 km, the meteoroid splits into a multitude $N = 2 \times 10^7$ of small fragments, which leads to intensive removal of mass even before their deceleration. In this case, the final mass of the meteoroid will be approximately 10 t. It is well known that for Sikhote-Alin meteorite, in total, several tens of thousands of fragments with a total mass of 27 t were collected.

The cosmic body can collapse into several large fragments, which then fly autonomously or split into a cloud of small fragments united by a common shock wave and flying as a whole. This cloud usually expands rapidly and slows down during flight causing a bright flash of radiation. When a large meteor body is destroyed, both fragmentation scenarios can occur simultaneously. In both cases, the fragments of the body acquire speeds in a direction perpendicular to the trajectory, which can lead to transverse scattering of fragments.

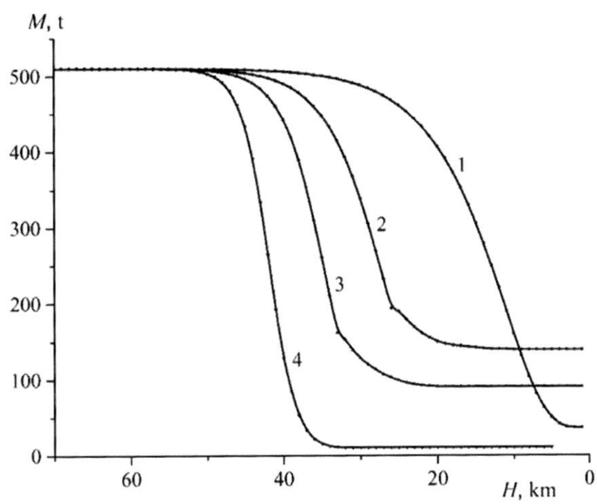

Fig. 9.1 Variation in mass of meteoroid along flight trajectory for different values of parameter α, where curve 1—single body (without fragmentation), curve 2—$\alpha = 0.5$, $N = 2.4\times10^3$, curve 3—$\alpha = 0.25$, $N = 2.0\times10^5$, and curve 4—$\alpha = 0.125$, $N = 2.1\times10^7$

In 1991, in the Czech Republic fell one of the largest bolide registered by the European network—Benešov bolide [11]. It belonged to the intermediate and relatively poorly studied class of cosmic bodies, which are almost completely "burned" (or rather evaporated) in the atmosphere. Unique observational data including radiation spectra were obtained for this bolide. The fragmentation of the Benešov bolide was recorded starting from the height of 42 km. Observations [11] showed that relatively small fragments were separated from the main body in the height range of 24 < H < 42 km. At an altitude of 24 km was the final crushing of the bolide, and its complete extinction took place at an altitude of about 19 km. The dynamic model of the gross-fragmentation [12] taking into account the crushing of the main body gave a value for the mass entrance of 80–300 kg. Photometric mass of the bolide [13] made up of 13,000 kg. The significant discrepancy between estimates of the mass of the meteoroid calculated using the above model was explained in the following way. Abnormally, high glow of the bolide was due to the movement of not a single body, but a compact swarm of fragments with a large total area of the glowing surface. Moreover, it was assumed that the process of fragmentation of the bolide began at altitudes $H = 60-50$ km at pressures $(1-5) \times 10^5$ N/m^2. Analysis of the dynamics of the bolide and its radiation suggested that this meteoroid began to split up at high altitude, about 60 km. Due to the relatively low speed in the direction perpendicular to the main trajectory, the fragments at high altitude diverged at a distance no more than the accuracy of photographic measurements. However, this leads to an increase in the deceleration of the fragmenting body, i.e., to a decrease in the velocity of the meteor body and its fragments, which, in turn, reduces the heat flow to the surface and slows the process of ablation.

In work [14], the assessment of the possible expansion of fragments at altitudes of ~60–40 km showed that due to the low density of air, the fragments did not fly over long distances, but flew compactly that confirms the established fact that the observed braking of the bolide belongs to the leading fragment. With a decrease in the altitude of the flight, the scattering of individual fragments can be significant—up to 300 m. The comparison of the trajectory characteristics of the bolide obtained during the computational experiment with the observational data showed good corrective agreement.

A special place among the meteorites is occupied by carbonaceous chondrites, fragments of which are very rare. Interesting is the study of a meteorite that fell with an explosion on January 18, 2000 near Tagish lake in northwestern Canada [15]. The mass of the meteoroid, officially called Tagish lake, reached 200 t before entering the Earth's atmosphere. It was found that the rate of entry of the meteoroid into the atmosphere at an angle of ~16.5° to the horizon was ~15.8 km/s, and the transverse body size to destruction—from 4 to 6 m. Unlike stone meteorites, the substance, of which is similar to terrestrial rocks, such bodies are very fragile and resemble "dried silt". Entering the dense Earth atmosphere, they just crumble. The found fragments of Tagish lake are rare specimens of one of the carbonaceous chondrites rich in volatile substances. Scientists from the Southwest research Institute of the United States made a sensational statement—they suggested that this meteorite came to us from Kuiper belt. This is the conclusion they came to after careful consideration of

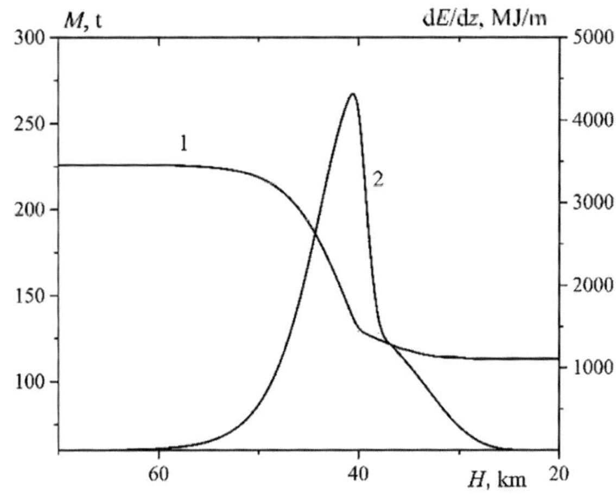

Fig. 9.2 The total mass (curve 1) and kinetic energy loss per unit length (curve 2) with respect to altitude of flight of Tagish lake meteoroid

the composition of the meteorite. In most cases, meteoroids fall on our planet from the main asteroid belt. The objects there are mostly composed of rocks and metals, while the objects from Kuiper belt consist of volatile substances (ammonia, methane, and water). If the theory is confirmed, the meteorite from Tagish lake will be the most distant alien to our planet.

In Fig. 9.2, the calculation data showing the change in mass (curve 1) and the loss of kinetic energy per unit length (curve 2) depending on the altitude of Tagish lake meteoroid are presented. The model of progressive crushing with the degree of heterogeneity of the material $\alpha = 0.25$ was used. As calculations have shown, the maximum loss of energy, the so-called explosion, happened at a height of $H = 40$ km with the number of fragments formed is ~10^5.

It should be noted that when moving along the trajectory, the mass of the fracturing Tagish lake meteoroid decreased by about 2 times that is the total mass of the fragments of the meteoroids considered was much higher than the observations. This fact is also confirmed in the study of motion and destruction and other meteoroids. Thus, it is necessary to consider other mechanisms of destruction of meteoroids.

9.4 Thermal Stress

It should be noted that at the final stage of motion of meteor bodies, the process of destruction may continue due to the temperature extension, which does not play a significant role for large meteor bodies. However, if the sizes of fragments reach several centimeters, the emerging temperature gradients may further destroy them to the size of coarse dust, which rapidly melts and evaporates in air at a high temperature. The estimates performed in [16] show that for a body with a radius of 10 cm, the time needed to reach critical tension is ~4 s. Thus, over the time of passing through the atmosphere, such a body may be repeatedly destroyed due to the emerging temperature tension. The calculations show that at the final stage of fragmentation at $\alpha = 0.25$, the number of fragments of Sikhote-Alin meteoroid reached 2×10^5, and the radius of an individual fragment was 3 cm. Thus, the process initiated by the temperature extension leads to the additional removal of mass of the fragmenting meteoroid.

This work analyzes also the influence of nonuniform temperature field on the meteoroids stress-deformable state during their falling in the Earth atmosphere. As a result of atmospheric interaction, the warming-up of the meteoroid thin near-surface layer is formed. At this, if the object is rapidly rotated, its warming-up along the surface will be naturally uniform. If the object is not rotated, then essential nonuniformity of temperature field will occur, both in depth and over the surface.

To analyze the influence of induced temperature field on the iron meteoroid stress-deformable state, the following problem is considered. The meteoroid is simulated by the elastic infinite isotropic cylinder of radius $R+h$ with the warmed-up near-surface layer of thickness h ($h \gg R$). Suppose its velocity is perpendicular to the axis, and in the first case the cylinder is rapidly rotated around the axis, and in the second case the cylinder is moving without rotation. The initial equilibrium equations in the fixed polar coordinates related to the body in partial derivatives for the stresses have a view of Eq. 9.4, where σ_r, σ_φ, $\sigma_{r\varphi}$ are the components of the stress tensor.

$$\frac{\partial \sigma_r}{\partial r} + \frac{1}{r}\frac{\partial \sigma_{r\varphi}}{\partial \varphi} + \frac{\sigma_r - \sigma_\varphi}{r} = 0 \quad \frac{\partial \sigma_{r\varphi}}{\partial r} + \frac{2}{r}\sigma_{r\varphi} + \frac{1}{r}\frac{\partial \sigma_\varphi}{\partial \varphi} = 0 \tag{9.4}$$

According to theory of elasticity, the stress tensor components depend on temperature and gradients of radial $U(r, \varphi)$ and azimuthal $W(r, \varphi)$ displacements:

$$\begin{aligned}
\sigma_r &= \lambda\left(\frac{\partial U}{\partial r} + \frac{U}{r} + \frac{1}{r}\frac{\partial W}{\partial \varphi}\right) + 2\mu\frac{\partial U}{\partial r} - \beta T, \\
\sigma_\varphi &= \lambda\left(\frac{\partial U}{\partial r} + \frac{U}{r} + \frac{1}{r}\frac{\partial W}{\partial \varphi}\right) + 2\mu\left(\frac{U}{r} + \frac{1}{r}\frac{\partial W}{\partial \varphi}\right) - \beta T, \\
\sigma_{r\varphi} &= \mu\left(\frac{1}{r}\frac{\partial U}{\partial \varphi} + \frac{\partial W}{\partial r} - \frac{1}{r}W\right),
\end{aligned} \tag{9.5}$$

where λ, μ are Lame constants, $\beta = (2\lambda + 3\mu)\alpha_1$, where α_1 is the material thermal-expansion coefficient, T is the temperature set by a known coordinate function:

$$\begin{aligned}
0 < r < R &: T = 0, \\
R < r < R + h &: T = T(\varphi),
\end{aligned}$$

where $T = T(\varphi)$ is an even function φ.

The system of equations (Eqs. 9.4–9.5) is solved at the boundary conditions $r = R$ (condition of continuity of normal and tangential stresses) and displacements provided by Eq. 9.6.

$$r = R + h : \sigma_r(R + h, \varphi) = \sigma_{r\varphi}(R + h, \varphi) = 0 \tag{9.6}$$

The system of equations (Eqs. 9.4–9.5) with boundary conditions (Eq. 9.6) is solved by decomposition of the required functions into Fourier series by azimuthal angle φ.

Decomposing the function $T = T(\varphi)$ into a Fourier series, we obtain:

$$T(\varphi) = T_0 + \sum_{n=1}^{\infty} T_n(\varphi) \cos n\varphi.$$

The solutions for the functions $U(r, \varphi)$, $W(r, \varphi)$ are sought in the form:

$$U(r, \varphi) = U_0(r) + \sum_{n=1}^{\infty} U_n(r) \cos n\varphi \quad W(r, \varphi) = \sum_{n=1}^{\infty} W_n(r) \sin n\varphi.$$

After substitution of Fourier series system of equations (Eqs. 9.4–9.5) is transformed to a system of ordinary differential equations along radial coordinate. Solving this system and determining arbitrary constants using the boundary conditions, we determine resulting stresses. Due to the bulkiness of the solution, only the finite formulas for the maximum shear stresses required for the analysis of the solution results are given below.

Two cases are considered: rapidly rotating cylinder and nonrotating cylinder. In the first case, at the high speed of the object rotation, the thin near-surface layer will be naturally heated uniformly in azimuth angle φ, viz., function $T(\varphi) = \theta$, where θ is constant. This means the maximum shear stress along the layer is equal to:

$$\tau_{\max}^{(1)} \approx \frac{\mu}{(\lambda + 2\mu)} \beta\theta. \tag{9.7}$$

In the second case, with no rotation, the temperature field in the near-surface layer corresponds to the function $T(\varphi) = \theta + \theta \cos\varphi$, viz., the maximum temperature is realized in the head point of the falling object and is equal to 2θ, in the back is equal to 0, and the average temperature over the surface is equal to θ. This means that the maximum shear stresses are, respectively, equal to:

$$\tau_{\max} = \frac{\mu}{(\lambda + 2\mu)} \beta\theta + \frac{\mu}{(\lambda + 2\mu)} \beta\theta \cos\varphi.$$

The largest value of the maximum shear stress is reached in this case at $\varphi = 0°$ and is equal to:

$$\tau_{\max}^{(2)} \approx \frac{2\mu}{(\lambda + 2\mu)} \beta\theta. \tag{9.8}$$

Suppose, to be definite, the meteoroid is an isotropic stone cylinder, for which Lame constants are equal to: $\lambda \approx 22.5 \times 10^9\,\text{N/m}^2$, $\mu \approx 28.7 \times 10^9\,\text{N/m}^2$, the temperature expansion coefficient $\alpha_1 \approx 0.74 \times 10^{-5}\,(\text{grad})^{-1}$ and the strength critical value $\sigma^* \approx 1.6 \times 10^8\,\text{N/m}^2$ [17].

In the case of the exceeding by any of the stresses of the strength critical value in some points of the moving meteoroid, its destruction occurs in them. At the warm-up of the cylinder near-surface layer, simulating the stone meteoroid to the temperature $\theta = 500\,°\text{C}$ reached at the most moderate values of the air density in the atmosphere at its initial falling velocity (hypervelocity) that means at sufficiently high altitudes, the value of the maximum shear stress, in case of the cylinder rapid rotation (Eq. 9.7), is equal $\tau_{\text{max}}^{(1)} \approx 1.8 \times 10^8\,\text{N/m}^2$. Thus, the exceeding by the maximum shear stress a critical value of the material strength parameter of the stone meteoroid $\tau_{\text{max}}^{(1)} > \sigma^*$ is on the whole warmed-up near-surface layer h of the cylinder, and, so, the destruction, delamination, and peeling near the surface are realized [18]. This effect is repeated multiply; therefore, the external surface of the falling meteoroid takes a relatively smooth form.

Now consider the case of the cylinder meteoroid falling without rotation (Eq. 9.8). In the stone meteoroid, the maximum shear stress exceeds the strength critical value $\tau_{\text{max}}^{(2)} \approx 3.6 \times 10^8\,\text{N/m}^2 > \sigma^*$, firstly, in the head point of the warmed-up near-surface layer h at $\varphi = 0°$ and in this point, the local cavity appears. This leads to the local erosion of the cylinder flown by the hypervelocity air that creates Taylor-Görtler flow instability [19] even for the single rough element. Following this instability in the boundary layer of the incoming flow to the roughness areas, Görtler vortices are formed [20], which are tornado-type vortices rotating at a great speed [13]. These vortices increase substantially the pressure and heat exchange on surface of cylinder that leads to intensification of local destructions of the falling meteoroid with the rhegmaglypts formation on its surface [21]. During the rhegmaglypts formation (new cavities), the tornado genesis process of Görtler vortices is intensified incrementally, distributing from the head upstream part of the meteoroid-cylinder over its upper and lower side parts to the back—leeward part.

Therefore, the calculations and analysis of the received results confirm the fact that according to the type of the surface relief of the fall meteorites it is possible to estimate the kinematics nature of the relative falling meteoroids. Their relatively smooth surface corresponds to the fall of the rapidly rotating objects, and the surface specked with rhegmaglypts corresponds to nonrotating ones.

In Fig. 9.3, there are photos of the corresponding stone meteoroids illustrating these assertions.

Fig. 9.3 Reliefs of meteorite fragments: **a** Kunashak (1949) [22], **b** Chelyabinsk (2013) [23]

9.5 Conclusions

Thus, the main processes occurring during the motion from meteor bodies in the Earth's atmosphere have been considered. We have studied the mechanisms of destruction of the bodies due to thermal stresses. The results obtained qualitatively correctly reflect the process of destruction of bodies in the atmosphere. We have theoretically studied the mechanisms of destruction of the bodies due to thermal stresses. The results obtained qualitatively correctly reflect the observed processes of destruction of bodies in the atmosphere.

References

1. Hughes, D.W.: Meteorite falls and finds: some statistics. Meteorites **19**, 269–281 (1981)
2. Syzranova, N.G., Andrushchenko, V.A.: Simulation of the motion and destruction of bolides in the Earth's atmosphere. High Temp. **54**(3), 308–315 (2016)
3. Andrushchenko, V.A., Syzranova, N.G., Shevelev, Yu.D.: An estimate of heat transfer to blunt bodies moving with hypersonic velocity in the atmosphere. J. Appl. Math. Mech. **71**(5), 747–754 (2007)
4. Murzinov, I.N.: Laminar boundary layer on a sphere in hypersonic flow on a equilibrium dissociating air. Fluid Dyn. **1**(2), 131–132 (1966)
5. Apshtein, E.Z., Vartanyan, N.V., Sakharov, V.I.: Distribution of radiant heat flux over the surface of three-dimensional and axisymmetric bodies in a supersonic ideal-gas flow. Fluid Dyn. **21**(1), 78–83 (1986)
6. Korobeinikov, V.P., Shurshalov, L.V., Vlasov, V.I., Semenov, I.V.: Complex modeling of the Tunguska catastrophe. Planet. Space Sci. **46**(2/3), 231–244 (1998)
7. Fisher, D.E.: "Ages" of the Sikhote Alin iron meteorite. Science **139**(3556), 752–753 (1963)
8. Mukhamednazarov, S.: Observation of a fireball and the fall of the first large meteorite in Turkmenistan. J. Astron. Lett. **25**(2), 117–118 (1999)
9. Emel'yanenko, V.V., Popova, O.P., Chugai, N.N., Shelyakov, M.A., Pakhomov, Yu.V., Shustov, B.M., Shuvalov, V.V., Biryukov, E.E., Rybnov, Yu.S., Marov, M.Ya., Rykhlova, L.V., Naroenkov, S.A., Kartashova, A.P., Kharlamov, V.A., Trubetskaya, I.A.: Astronomical and physical aspects of the Chelyabinsk event. Solar Syst. Res. **47**(4), 240–254 (2013)

10. Weibull, W.: A statistical theory of the strength of materials. Proc. Roy Swedish Inst. Eng. Res. **151**, 1–45 (1939)
11. Spurny, P.: Recent fireballs photographed in central Europe. Planet. Space Sci. **42**(2), 157–162 (1994)
12. Ceplecha, Z., Spurny, P., Borovička, J., Keclikovd, J.: Atmospheric fragmentation of meteoroids. Astron. Astrophys. **279**, 615–626 (1993)
13. Borovička, J., Spurny, P.: Radiation study of two very bright terrestrial bolides and an application to the Comet S-L 9 collision with Jupiter. Icarus **121**, 484–510 (1996)
14. Andrushchenko, V.A., Maksimov, F.A., Syzranova, N.G.: Simulation of flight and destruction of the Benešov bolid. Comput. Res. Model. **10**(5), 605–618 (2018)
15. Grossman, J.N.: A meteorite falls on ice. Science **290**, 283–285 (2000)
16. Andrushchenko, V.A., Syzranova, N.G., Shevelev, Yu.D., Goloveshkin, V.A.: Destruction mechanisms of meteoroids and heat transfer to their surfaces. Math. Model. Comput. Simul. **8**(5), 506–512 (2016)
17. Slyuta, E.N.: Physical and mechanical properties of stony meteorites. Sol. Syst. Res. **51**(1), 64–85 (2017)
18. Kholin, N.N., Goloveshkin, B.A., Andruschenko, V.A.: Mathematic Simulation of the Wave Activity in Condensed Medium and Meteoroids Dynamics. Lenand Publ., Moscow (in Russian) (2016)
19. Chuvakhov, P.V., Borovoy, V.Y., Egorov, I.V., Radchenko, V.N., Olivier, H., Roghelia, A.: Effect of small bluntness on formation of Görtler vortices in a supersonic compression corner flow. J. Appl. Mech. Tech. Phys. **58**(6), 975–989 (2017)
20. de la Chevalerie, D.A., Fontenean, A., de Luca, L., Cardone, G.: Görtler-type vortices in hypersonic flows: the ramp problem. Exp. Therm. Fluid Sci. **15**(2), 69–81 (1997)
21. Laganelli, A.L., Nestler, D.E.: Surface ablation patterns: a phenomenology study. AIAA J. **7**, 1319–1325 (1969)
22. Vernadsky Geological Museum of the RAS. https://ok.ru/sgm_ran/topic/69255458122647. Accessed 05 Oct 2019 (in Russian)
23. Vernadsky Geological Museum of the RAS. https://i.ytimg.com/vi/fKbt9Uxb8hI/maxresdefault.jpg. Accessed 05 Oct 2019 (in Russian)

Chapter 10
Computational Modeling of Rarefied Plasma and Neutral Gas Effusion into Vacuum

Vadim A. Kotelnikov and Mikhail V. Kotelnikov

Abstract Physical, mathematical, and computational models of neutral gas and rarefied plasma effusion into vacuum space have been considered in this chapter. Gas or plasma effusion is viewed as the process in which they escape from a container through a relatively small hole shaped like a narrow long slot. Results of computational experiments have been provided: the distribution functions of charged particles and neutral gas, their velocity field and concentration field. The impact of latitudinal magnetic field on the distribution functions and momentums of charged particles effusion has been studied. Parameters of evolutionary processes have been considered, from the moment, when the slot is formed through the transition of the effusion process into stationary mode.

10.1 Introduction

Effusion is an outflow of neutral or ionized gases through relatively small holes and is commonly found in various fields of modern technology. As spacecraft travel through the Earth's ionosphere or the space, they are at a risk of depressurization due to the accidents, defects, welded seam deterioration, collisions with meteors or bits of space junk, and a number of other causes.

Designing innovative aerospace engines, e.g., pulsed plasma thrusters, involves designing vacuum benches, including the vacuum chambers, vacuum pumps, vacuum furnaces, electric vacuum devices, vacuum insulation, and a number of other devices. Any depressurization in those devices causes the effusion of neutral and ionized gases that leads to their malfunction.

V. A. Kotelnikov (✉) · M. V. Kotelnikov
The A.A. Blagonravov Institute of Machine Science of the RAS, 4 Maly Kharitonyevsky Per, Moscow 101990, Russian Federation
e-mail: mvk_home@mail.ru

Moscow Aviation Institute (National Research University), 4, Volokolamskoe Shosse, Moscow 125993, Russian Federation

L. C. Jain et al. (eds.), *Advances in Theory and Practice of Computational Mechanics*, Smart Innovation, Systems and Technologies 173,
https://doi.org/10.1007/978-981-15-2600-8_10

This chapter presents the results of research of the neutral gas and plasma effusion into vacuum space found by computational modeling using the kinetic theory. We assume that the dimensions of the slot, through which a gas or plasma effuse into vacuum space, are considerably shorter than free paths of particles effusing.

There exist multiple works dedicated to studies of gas effusion into vacuum through relatively small holes [1–8]. However, we failed to find any work based on the computational solutions referred to Vlasov kinetic equation [9, 10]. As we studied plasma flows, we supplemented the above kinetic equation with Poisson's equation to find the self-consistent electrical field included in Vlasov equation.

Next, the chapter will consider the physical, mathematical, and computational models of the problem, as well as, the results of computational experiments. Section 10.2 discusses the physical, mathematical, and computational models. Methodical calculations are described in Sect. 10.3. Results of computational experiments are presented in Sect. 10.4, while conclusions are given in Sect. 10.5.

10.2 Physical, Mathematical, and Computational Models of the Problem

Problems pertaining to the plasma and neutral gas effusion have a lot in common, as they both use Vlasov kinetic equation. Both of the problems prove to be four-dimensional problems in phase space. The geometrical shape of the effusion hole is a narrow rectangle. It is assumed that there exists vacuum space within the computational domain at the initial time moment. The Distribution Functions (DFs) on the cross-section of a hole are in both cases assumed to Maxwell–Bolztmann distribution functions. Computational models of both of the problems are similar. Nonetheless, there are substantial differences between them also, to name the principal ones:

- Neutral gas models use only one Vlasov equation, while plasma models use two Vlasov equations, each for ions and electrons, respectively.
- Mathematical models of plasma are supplemented with Poisson's equation for a self-consistent electrical field.
- Scale systems used for non-dimensionalization of gas models differ from those of plasma problems.

Here, the rarefied gas or rarefied plasma is considered that effuses from a tank having a volume V through a small hole and into vacuum space, the hole's dimensions being shorter than the lengths of free paths of particles effusing. The gas or the plasma contained in the tank is considered as being in a state of equilibrium. The tank wall thickness is neglected. No particle collisions are assumed to occur within the computational domain. The latter case can be often found in practice.

In a general case, the problem in question is a six-dimensional in phase space (x, y, z, v_x, v_y, v_z) and non-stationary problem [11]. We propose considering the hole as a rectangle having one side much longer than the other. Due to symmetry shear, this modeled shape allows considerable reduction of the problem's dimensions (in the

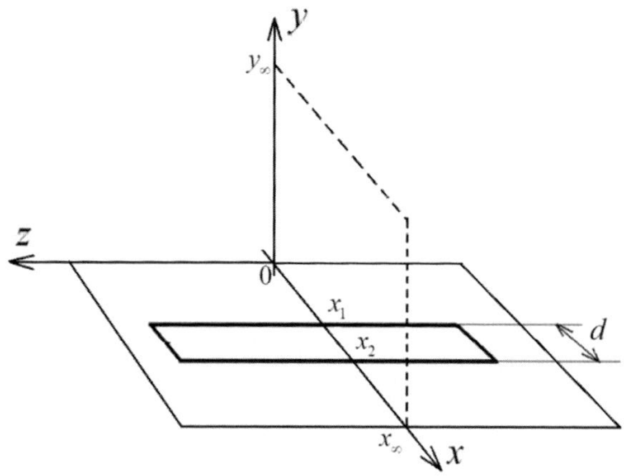

Fig. 10.1 Geometry of the problem

latter case, the problem depends on x, y, v_x, v_y in phase space) [10], and establishing all the principal parameters of effusion phenomenon. The slot-shaped hole used in our model is commonly found in real life and comes in the shape of cracks in aircraft shells or in housings of vacuum devices. Please refer to Fig. 10.1 to see the geometry of the problem, where d is the width of the hole, x_1 and x_2 are the respective coordinates of the edges of the slot, x_∞ and y_∞ are the coordinates of the outer edges of the computational domain.

First and foremost, let us formulate a mathematical model of plasma effusion into vacuum space. The system of equation is written down as shown below [10]:

$$\frac{\partial f_\alpha}{\partial t} + v_x \frac{\partial f_\alpha}{\partial x} + v_y \frac{\partial f_\alpha}{\partial y} + \frac{q_\alpha}{m_\alpha}\left(\left(E_x + v_y B\right)\frac{\partial f_\alpha}{\partial v_x} + \left(E_y - v_x B\right)\frac{\partial f_\alpha}{\partial v_y}\right) = 0, \quad (10.1)$$

$$\frac{\partial^2 \varphi}{\partial x^2} + \frac{\partial^2 \varphi}{\partial y^2} = -\frac{1}{\varepsilon_0}\sum q_\alpha n_\alpha, \ \mathbf{E} = -\nabla\varphi, \alpha = i, e, \quad (10.2)$$

$$n_\alpha = \left(\frac{2kT_\alpha}{m_\alpha}\right)^{\frac{1}{2}} \int_{-\infty}^{+\infty}\int_{-\infty}^{+\infty} f_\alpha\left(x, y, v_x, v_y, t\right) dv_x dv_y. \quad (10.3)$$

The mathematical model includes Vlasov equation for ions and electrons (Eq. 10.1), Poisson's equation for the self-consistent electrical field, a formula to link strength and potential (Eq. 10.2), and a formula to link plasma concentrations with component distribution (Eq. 10.3). Here, t is the time, E and φ are the strength and potential of the electrical field, respectively, E_x and E_y are the strength components along axes OX and OY, and f_α, q_α, m_α, and n_α are the distribution function, the charge, the mass, and the concentration of charged particles, respectively. The index i denotes ions and e denotes electrons.

Let us consider the system of initial and boundary conditions. It is assumed that there exists vacuum space within the computational domain at the initial time moment. The distribution function at the hole's cross-section (at the "inflow" boundary) is written down using Eq. 10.4.

$$f_\alpha = (n_0/\pi)(m_\alpha/(2kT_\alpha))^{3/2}\exp[-m_\alpha\{v_{\mathbf{x}}^2 + v_y^2\}/(2kT_\alpha)] \qquad (10.4)$$

Equation 10.4 is Maxwell function for ions and electrons. Here, n_0 is the concentration of charged particles contained in plasma at the hole's cross-section. The plasma that effuses from the hole is assumed to be quasi-neutral, with the potential at the hole's cross-section assumed to be zero. "Softer" boundary conditions were stipulated for the rest of the boundaries of the computational domain (the "outflow" boundaries) and found through extrapolation of plasma parameters from adjacent computational layers.

System (Eqs. 10.1–10.4) was reduced to a dimensionless form by means of a system of scales below [9–11], where M_n is the concentration scale, $M_n = n_0$, M_L is the length scale, $M_L = r_D = \left(\varepsilon_0 kT_{i\infty}\big/ n_\infty e^2\right)^{1/2}$, M_φ is the potential scale, $M_\varphi = kT\big/|q_e|$, M_{V_α} is the velocity scale, $M_{V_\alpha} = \left(2kT_\alpha\big/m_\alpha\right)^{1/2}$, $\alpha = i, e$, M_E is the electrical field strength scale, $M_E = M_\varphi/M_L$, M_B is the electrical field induction scale, $M_B = 2M_E/M_{V_i}$.

Here, rD is Debye length, and ε_0 is the electrical constant. The rest of the scales are found from dimension formulas. Dimensionless parameters are obtained as the system is reduced to a dimensionless form, on which the solution to the problem hinges:

$$r_0 = r_p/M,\ \varphi_0 = \varphi_\mathrm{p}/M_\varphi,\ \varepsilon = T_i/T_e,\ B_0 = B/M_B. \qquad (10.5)$$

The computational model of the problem is based on the iterative method, whereby a transitional process from the initial to the final stationary state (finding the plasma parameter distribution within the computational domain) is modeled. The method of characteristics was used to solve Vlasov equations [12], while Poisson's equations were solved by means of spectral methods [11].

Now let us proceed to the mathematical model of neutral gas effusion into vacuum space. The collisionless gas is expressed by Vlasov equation [9, 10] provided by Eq. 10.6.

$$\frac{\partial f}{\partial t} + v_x\frac{\partial f}{\partial x} + v_y\frac{\partial f}{\partial y} = 0 \qquad (10.6)$$

The gas particle concentration is found from Eq. 10.3, where α determines the gas molecule concentration.

The boundary distribution functions at the hole's cross-section (the "inflow" boundary) are described by Eq. 10.7.

$$f_{zp} = \frac{n_0}{\pi}\left(\frac{m}{2kT}\right)^{\frac{3}{2}}\exp\left(-\frac{m}{2kT}\left(v_x^2 + v_y^2\right)\right) \qquad (10.7)$$

"Softer" boundary conditions were stipulated for the rest of the boundaries of the computational domain (the "outflow" boundaries) and found through extrapolation of plasma parameters from adjacent computational layers.

The calculation formulas for mean velocities of gas particles are as follows:

$$v_{x\ mean} = \frac{\int\limits_{-\infty}^{+\infty}\int\limits_{-\infty}^{+\infty} v_x f dv_x dv_y}{\int\limits_{-\infty}^{+\infty}\int\limits_{-\infty}^{+\infty} f dv_x dv_y}, \quad v_{y\ mean} = \frac{\int\limits_{-\infty}^{+\infty}\int\limits_{-\infty}^{+\infty} v_y f dv_x dv_y}{\int\limits_{-\infty}^{+\infty}\int\limits_{-\infty}^{+\infty} f dv_x dv_y}. \tag{10.8}$$

The formula for flow of particles from the hole is calculated as shown below:

$$J_{from\ orifice} = \left(\frac{2kT}{m}\right)^{\frac{1}{2}} \int\limits_{x_1}^{x_2}\int\limits_{-\infty}^{+\infty}\int\limits_{0}^{+\infty} v_y f_{boundary} dx dv_x dv_y. \tag{10.9}$$

In Eq. 10.9, x_1 and x_2 are the coordinates of the edges of the effusion hole.

The formula for a flow of particles over the outer boundary of the computational domain can be written in a view of Eq. 10.10.

$$\begin{aligned} J_{outer\ boundary} = & \left(\frac{2kT}{m}\right)^{\frac{1}{2}} \int\limits_{0}^{x_\infty}\int\limits_{-\infty}^{+\infty}\int\limits_{-\infty}^{0} v_y f\big(t, x, y_0, v_x, v_y\big) dx dv_x dv_y + \\ & + \left(\frac{2kT}{m}\right)^{\frac{1}{2}} \int\limits_{0}^{x_\infty}\int\limits_{-\infty}^{+\infty}\int\limits_{0}^{+\infty} v_y f\big(t, x, y_\infty, v_x, v_y\big) dx dv_x dv_y + \\ & + \left(\frac{2kT}{m}\right)^{\frac{1}{2}} \int\limits_{0}^{y_\infty}\int\limits_{-\infty}^{0}\int\limits_{-\infty}^{+\infty} v_y f\big(t, x_0, y, v_x, v_y\big) dy dv_x dv_y + \\ & + \left(\frac{2kT}{m}\right)^{\frac{1}{2}} \int\limits_{0}^{y_\infty}\int\limits_{0}^{+\infty}\int\limits_{-\infty}^{+\infty} v_y f\big(t, x_\infty, y, v_x, v_y\big) dy dv_x dv_y \end{aligned} \tag{10.10}$$

The mathematical model was reduced to a dimensionless form using a system of scales, where M_n is the concentration scale, $M_n = n_0$, M_L is the length scale, where d is the short side of the rectangle, see Fig. 10.1, $M_L = d$, M_V is the velocity scale, $M_V = (2RT/\mu)^{1/2}$, where $R = 8.314472$ J/(moles·K) is the gas constant, μ is the molar mass of gas, M_f is the distribution function scale, $M_f = M_n/(M_v)^3$, M_t is the time scale, $M_t = M_L/M_v$, M_N is the scale of flow of particles per unit of the slot length, $M_N = M_n M_v M_L$.

As with plasma effusion, the computational model of the problem is based on the iterative method, whereby a transitional process from the initial state to final stationary state, during which we obtain the plasma parameter distribution within the computational domain, is modeled. The method of characteristics algorithm was used to solve Vlasov equation [12].

10.3 Methodical Calculations

It is worth noting that the problem to be solved was multidimensional and non-stationary, and, in the case of plasma effusion, it also contained a considerable number of parameters, which resulted in heavy consumption of computational resource. The computational domain, in the case of plasma effusion, contained 3,075,200 cells of the computational mesh, the time step being 0.002 dimensionless units. The computation time was approximately 10 h for desktops PCs, each having a quad-core processor, with the clock rate of each core being 3.2 GHz, and the RAM being 3 GB.

The computation time was controlled visually, from a control page displayed throughout the computation, on which the diagram of ionic and electronic currents against time, inflowing from the hole and into the computational domain, and flowing beyond its limits through outer boundaries, was being plotted and replotted at each time step. The calculation was stopped whenever the ionic current flowing from the hole became practically equal to the ionic current flowing through the limits of the computational domain. Please refer to Fig. 10.2a, b for respective diagrams.

Those diagrams suggest that the calculation was stopped in 3.5 dimensionless time units after the ionic current flowing through the computational domain limits had become practically equal to the ionic current flowing from the nozzle, while the electronic current flowing through the limits of the computational domain had become practically equal to the electronic current flowing from the nozzle much earlier than that.

In the case of neutral gas effusion, the computational algorithm was implemented as a program in C++, with tools of OpenGL API used. The source code of the program was designed using Visual Studio 2017 IDE. The dimensions of the computational

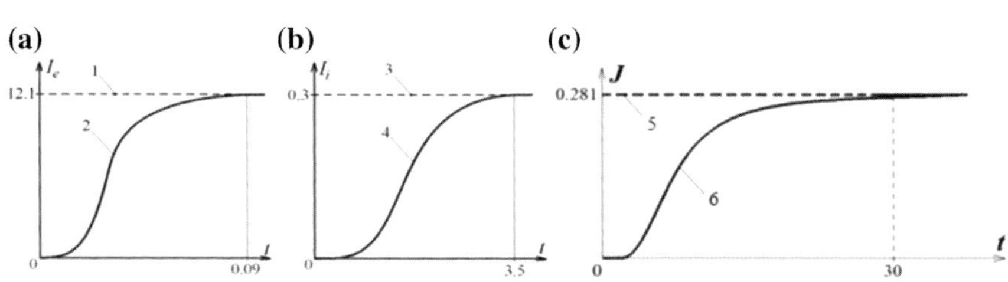

Fig. 10.2 Evolutions of ionic and electronic currents, as well as, of a flow of neutral gas particles through the limits of the computational domain: **a** curve 1—electronic current flowing from the hole and curve 2—the electronic current flowing from the computational domain, **b** curve 3—the ionic current flowing from the hole and curve 4—the ionic current flowing from the computational domain, **c** curve 5—a flow of gas particles flowing from the hole into the computational domain and curve 6—a flow of gas particles flowing from the computational domain

domain were 2 × 2 and contained 39,942,400 cells of the computational mesh, the time step being 0.01 dimensionless units. The computation time was several hours for a desktop PC having Intel Core i7-6700 K quad-core processor, the clock rate of each core being 4 GHz, and the PC's RAM being 32 Gb.

The computation stop time was controlled visually, from a control page displayed throughout the computation, on which the diagram of gas particles flow from the hole into the computational domain, and beyond its limits through outer boundaries, was being plotted and replotted at each time step. Please refer to Fig. 10.2c above for the diagram.

A moment, when a gas particle flow from the hole became practically equal to the gas flow through the limits of the computational domain, was spotted. Next, the following was visually observed and controlled on the monitor:

- The constant value of the gas particle flow crossing the computational domain, which was indicative of the establishment of a stationary solution to the problem.
- Practically, full similarity was in between the gas flow from the hole into the computational domain and the gas particle flow from the computational domain, which indicated that the law of conservation of the mass of the gas worked within the computational domain in an established stationary state.

Those two conditions having been met, the calculation was stopped with the results being subject to further analysis.

10.4 Results of Computational Experiments

Distribution functions of charged and neutral particles at various points of the region of interest, as well as, the momentums of those functions (the concentrations and velocities fields) were found through computational experiments.

First, let us present the results for plasma effusion. Figure 10.3a, b, and c shows isometric lines of the functions of distribution of ions recorded in the vicinity of the hole's cross-section at the time moment of 3.5 dimensionless units, which is

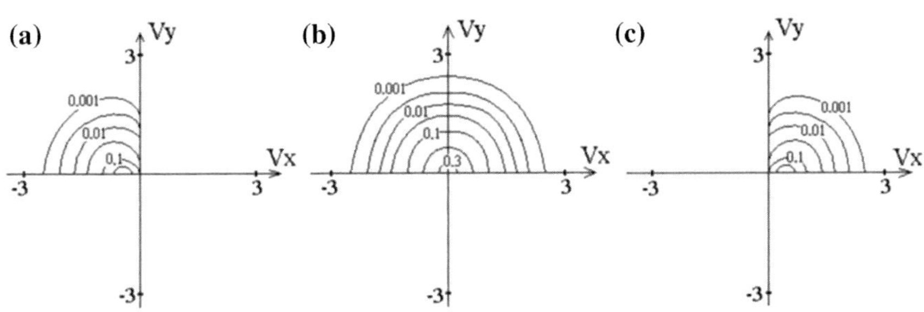

Fig. 10.3 Isometric lines of distribution functions of ions: **a** at the left-hand edge of the hole's cross-section, **b** in the middle of the hole, **c** at the right-hand edge of the hole's cross-section

in accord with the established solution to the problem. The distribution function shown in Fig. 10.3a was recorded at the left-hand edge of the hole, where there are no ions traveling at positive-value V_x or negative-value velocities V_y. Accordingly, the distribution function shown in Fig. 10.3c is determined by no ions traveling at positive-value V_x or negative-value velocities V_y at the left-hand edge of the hole. The distributions functions of ions in the vicinity of the middle of the cross-section of the hole are shaped like a half-dome of Maxwells' distribution function, and the whole of it lies within the positive-value velocities region V_y, while ions traveling at negative-valued velocities V_y are not present here either.

The calculations suggest that the distribution functions of electrons near the hole's cross-section have the same parameters as ions.

Figure 10.4a, b shows velocity fields of electrons and ions, respectively, their respective vectors of mean velocities are plotted from the centers of cells of computational meshes. The arrows of the vectors were omitted to avoid cluttering the diagram with lines. Vectors of mean velocities for their respective plasma components were only plotted wherever the condition $n_{i,e} > 0.1\ n_0$ was met, where n_0 is the concentration of charged particles at the time moment $t = 0$. The regions, where the condition in question was met, are highlighted in light-gray, while the hole's cross-section is highlighted in dark-gray. The thermal velocity of ions was selected as their velocity scale, while the thermal velocity of electrons was selected as their velocity scale.

The above diagrams suggest that the effusion velocity of the electronic component from the hole is much higher than that of the ionic component. As a result, a negative volumetric charge forms within the computational domain at the initial moment of evolution. The calculations suggest that the charge in question gradually decreases in the process of evolution.

Figure 10.5a, b shows the isometric lines of concentrations of electrons and ions, respectively. The parameters of the diagram in question are the same as those of Fig. 10.4 above. Moreover, Fig. 10.5a suggests that the concentration of electrons near the hole's cross-section exceeds 0.5 dimensionless units. The calculation suggests that the value exceeds that of the concentration of ions in the vicinity of the hole's

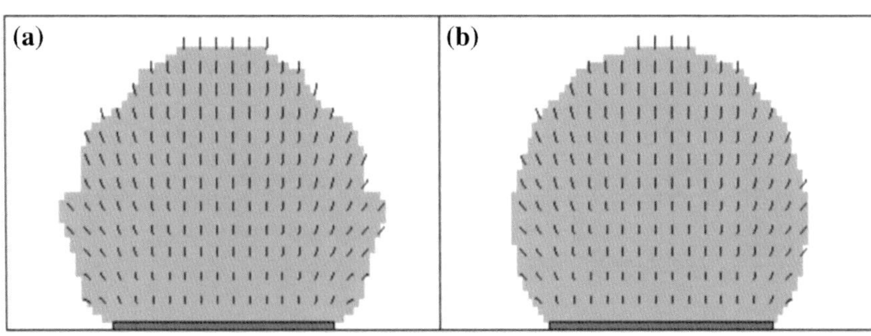

Fig. 10.4 Velocity fields of plasma components: **a** velocity field of electrons, $t = 0.032$, **b** velocity field of ions, $t = 1.23$

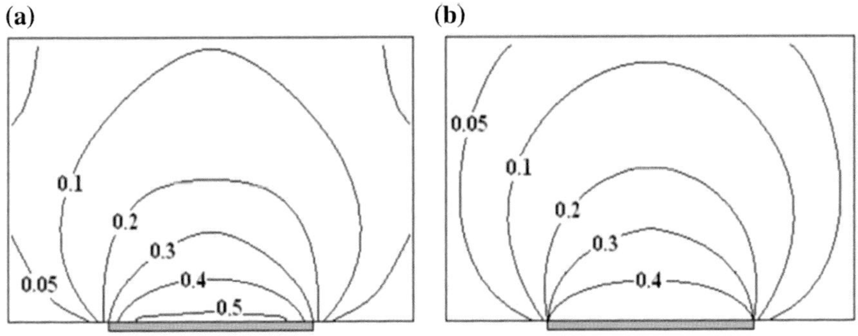

Fig. 10.5 Isometric lines of concentrations of plasma components: **a** electrons, $t = 0.032$, **b** ions, $t = 1.23$

cross-section at the same time moment. Thus, a low negative charge is present there, and it forms because electrons that flow from the hole repel from the volumetric negative charge located in the middle of the computational domain and concentrate in the vicinity of the hole's cross-section.

The impact of the axial magnetic field B_z on plasma effusing from the hole has also been studied in this chapter. Figure 10.6a shows the velocity field of the electronic component with the dimensionless magnetic induction value being $B_0 = 0.03$. The impact of Lorentz force on the electronic component can be clearly seen in Fig. 10.6a. The calculations suggest that the above value of magnetic induction has no decisive influence on the ionic component of plasma. Figure 10.6b shows the velocity field of the ionic components with the dimensionless magnetic induction value being $B_0 = 1.5$. Here, the impact of the magnetic field on the ionic component becomes substantial. Electrons, as they enter the computational domain, become magnetized, and a positive volumetric charge forms within the computational domain.

Now, let us present the results of modeled effusion of neutral gas. Figure 10.7

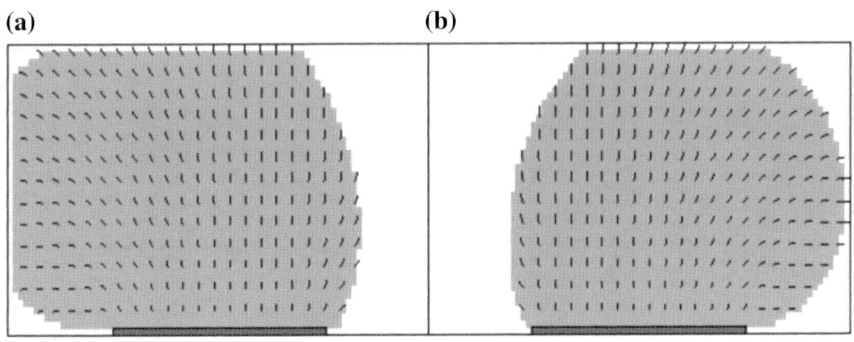

Fig. 10.6 Velocity field of plasma when an axial magnetic field is present: **a** electrons, $B = 0.03$, **b** ions, $B = 1.5$

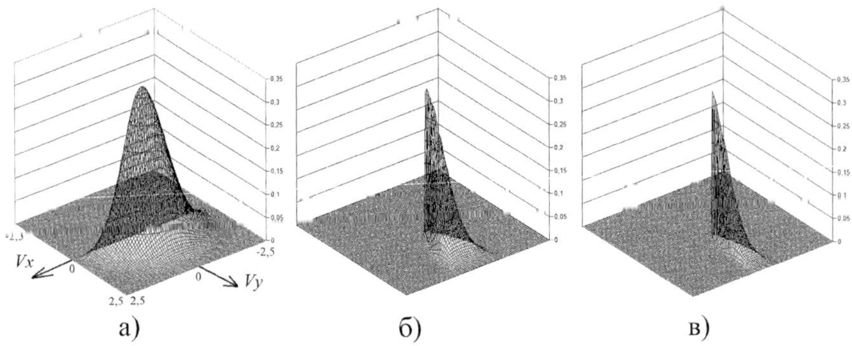

Fig. 10.7 Distribution function of gas particles versus coordinate y ($x = 5$, $t = 30$): **a** $y = 0.025$, **b** $y = 1.5$, **c** $y = 3$

shows the distribution function along the symmetry axis of the gas flow at the moment, when solution is established depends on the coordinate y.

The dependence in question has been visualized in Fig. 10.8 in the shape of isometric lines.

The above diagrams suggest that the distribution function changes its shape when sheared from the hole to the boundary of the computational domain along the symmetry axis of the flow.

The computational experiments suggest that any change in the shape of the distribution function due to its shift along axis OY causes a slight shift in its "center-of-gravity" toward an increase in the component. As the "center-of-gravity" of the distribution function corresponds to the vector of mean velocity of particles at a point under study, the mean velocity of particles within the flow increases with increasing

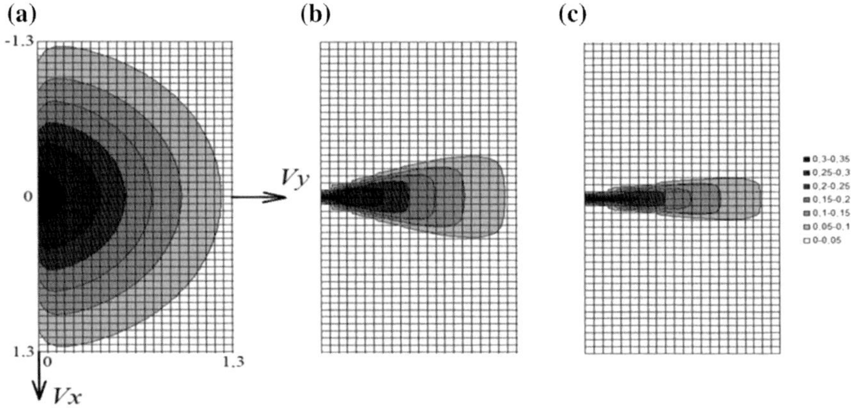

Fig. 10.8 Isometric lines of distribution functions of gas particles versus coordinate y ($x = 5$, $t = 30$): **a** $y = 0.025$, **b** $y = 1.5$, **c** $y = 3$

their distance from the hole, a phenomenon found through calculation, and is shown in Fig. 10.9.

Figure 10.10 shows the distribution function versus the coordinate x as it shears from the symmetry axis of the flow to the lateral boundary of the computational domain. Those distribution functions were obtained on the 20th computational layer from the hole.

Figure 10.11 visualizes the distribution functions from Fig. 10.10 in the shape of isometric lines.

Figure 10.12 shows the concentration field of particles within the computational domain, which was found through computational experiments. The diagram contains the computational mesh, and the boundaries between the gray regions are isometric lines. The concentration field has an axial symmetry.

The evolution of distribution of gas particles concentrations on axis OY along the symmetry axis of the flow at various time moments is shown in Fig. 10.13. Due to

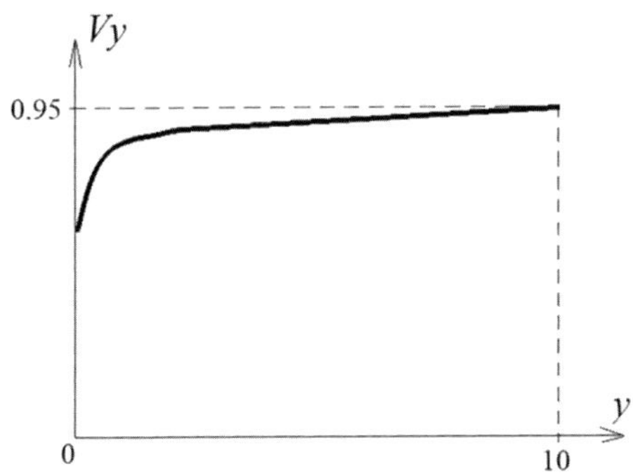

Fig. 10.9 Mean velocities of particles versus coordinate y along the symmetry axis of flow (t = 30)

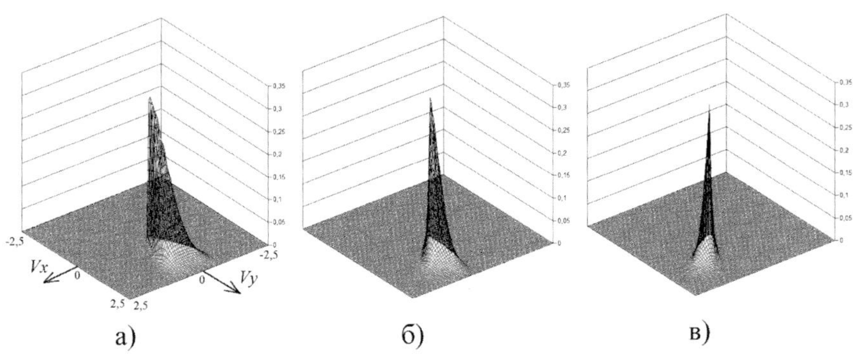

Fig. 10.10 Distribution function of gas particles versus coordinate x ($y = 1.5$, $t = 30$): **a** $x = 5$, **b** $x = 5.75$, **c** $x = 6.5$

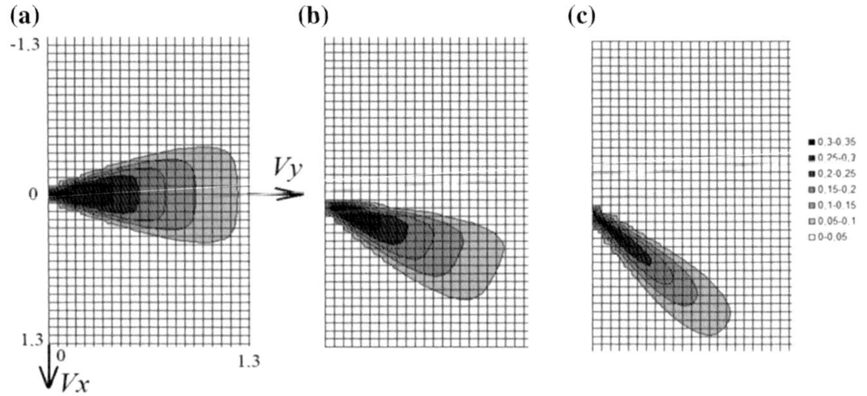

Fig. 10.11 Isometric lines of distribution of gas particles versus coordinate x ($y = 1.5$, $t = 30$): **a** $x = 5$, **b** $x = 5.75$, c $x = 6.5$

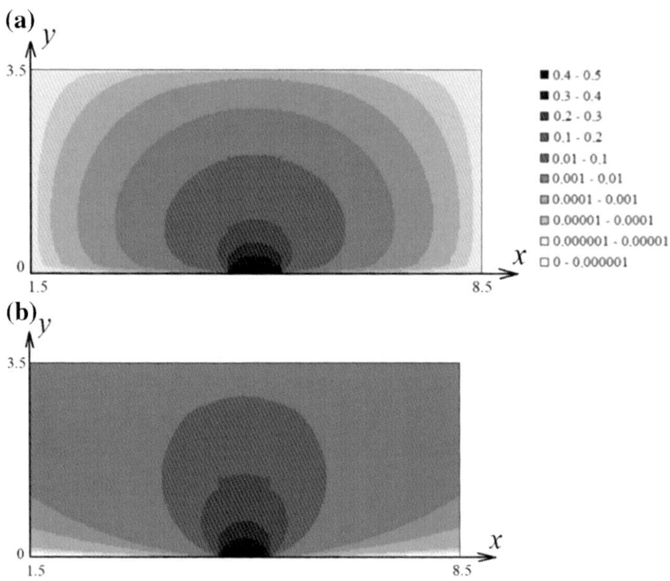

Fig. 10.12 Isometric lines of concentrations of gas particles within the computational domain: **a** $t = 1.5$, **b** $t = 30$

dissipation of the effusion flow, the concentration of gas particles decreases smoothly as the coordinate y increases.

Figure 10.14 shows a field of velocities of gas particles at the moment when the flow establishes. The arrows of the vectors of mean velocities of particles were omitted to avoid cluttering the diagram. The vectors of velocities were only plotted at the nodes of the computational mesh, where the concentration of particles was n

Fig. 10.13 Evolution of distribution of gas component concentration along axis OY ($x = 1$), curve 1—$t = 0.2$, curve 2—$t = 0.9$, and curve 3—$t = 10$

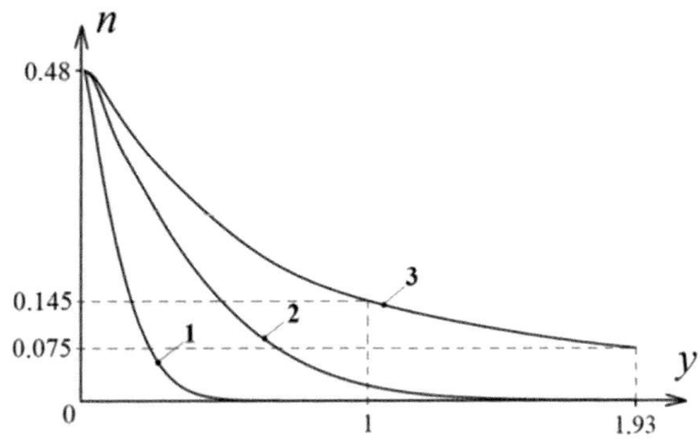

Fig. 10.14 Velocity field of gas particles ($t = 30$)

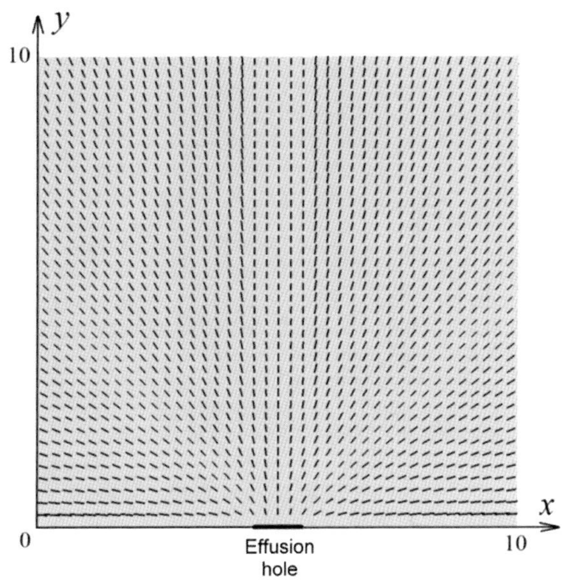

$> 0.1n_0$. The respective regions were highlighted in gray. The velocity field has an axial symmetry. The dissipation of the flow increases toward the edges of the hole.

10.5 Conclusions

The research into gas and plasma effusion through a hole shaped as a long rectangle provides a clear idea of the distribution functions, velocity fields, concentration fields, potential of the self-consistent electrical field, and other parameters of effusing flows of particles. The originally designed program codes allow further research into the area, while taking into account

- The shape of the effusion hole.
- The thickness of the walls and material of the volume, from which effusion occurs.
- The ratio between the inner and outer pressures.
- The impact of self-fields and outer fields on effusion.

The research results presented here may be of use to designers of portable small leak detectors for space stations and vacuum plants, in mass spectroscopy, and a number of other application areas.

References

1. Saksagansky, G.L.: Molecular Flows in Complex Vacuum Structures. Atomizdat Publishing House, Moscow (in Russian) (1980)
2. Rozanov, L.N.: Vacuum Technique. CRC Press, London, New York (2018)
3. Steckelmacher, W.: A review of the molecular flow conductance for systems of tubes and components and the measurement of pumping speed. Vacuum **16**(11), 561–584 (1966)
4. Semkin, N.D., Zanin, A.N., Piyakov, I.V., Voronov, K.E.: A time-of-flight mass spectrometer for detecting a point of air leakage from a spacecraft. Instrum. Exper. Techn. **50**(1), 108–112 (2007)
5. Holland, S.D., Roberts, R., Chimenti, D.E., Michael Strei, M.: Leak detection in manned spacecraft using structure-borne noise. Appl. Phys. Lett. **86**(17), 70–78 (2005)
6. Holland, S.D., Chimenti, D.E., Roberts, R., Michael Strei, M.: Locating air leaks in manned spacecraft using structure-borne noise. J. Acoust. Soc. Am. **121**(6), 3484–3492 (2007)
7. Nesterov, S.B., Astashina, M.A., Neznamova, L.O., Vasilyev, Y.K.: Problems and methods of studies of rarefied gas medium near spacecraft. Vacuum Dev. Technol. **18**(3), 183–186 (2007)
8. Rozanov, L.N., Skryabnev, A.Y.: Gas flow through circular duct at high pressure drops. Vacuum Dev. Technol. **20**(1), 3–8 (2010)
9. Kotelnikov, M.V.: The distribution functions of charged particles in the vicinity of a cylindrical body in a flow of collisionless plasma in magnetic field. High Temp. **46**(6), 757–762 (2008)
10. Kotelnikov, V.A., Kotelnikov, M.V.: Current–voltage characteristics of a flat probe in a rarified plasma flow. High Temp. **54**(1), 20–25 (2016)
11. Kotelnikov, V.A., Kotelnikov, M.V., Gidaspov, V.Y.: Computational modeling of flows of collisional and collisionless plasma around body. FizMatLit, Mosow (in Rissian) (2010)
12. Kotelnikov, V.A., Kotelnikov, M.V.: An advanced method of characteristics. Matem. Mod **29**(5), 85–95 (in Russian) (2017)

Chapter 11
Numerical Simulation of the Process of Phase Transitions in Gas-Dynamic Flows in Nozzles and Jets

Igor E. Ivanov, Vladislav S. Nazarov, Vladimir Yu. Gidaspov and Igor A. Kryukov

Abstract The chapter presents a development of condensation and evaporation in flows of two-phase gas-droplet mixture in the nozzles, jets, and external area in front of the nozzle. Condensation of pure water vapor and condensation vapor into wet stream mixture flow are considered. Two different models for modeling condensation process are used. One of them is a quasi-chemical method. Another method is Method Of Moments (MOM). Also, the task of gas mixture jump from metastable state to stationary state and the task of flow of superheated steam are reviewed.

11.1 Introduction

Condensation process contains a lot of natural phenomena and modern technical applications. Condensate can be formed in rarefaction areas around aerodynamic surfaces, when they maneuver in the engine blade or rocket nozzles, when gas intensely expands, and so on. Condensation can lead to a shock wave that influences on the airships aerodynamics, engine, and technical devices parameters. Sometimes in a

---

I. E. Ivanov · V. S. Nazarov (✉) · V. Yu. Gidaspov
Moscow Aviation Institute (National Research University), 4, Volokolamskoe Highway, Moscow 125993, Russian Federation
e-mail: naz.vladislav@yandex.ru

I. E. Ivanov
e-mail: ivanovmai@gmail.com

V. Yu. Gidaspov
e-mail: gidaspovvy@mai.ru

I. E. Ivanov
Lomonosov Moscow State University, 1, Leninskie Gory, Moscow 119991, Russian Federation

I. A. Kryukov
Ishlinsky Institute for Problems in Mechanics of the RAS, 101-1, Pr. Vernadskogo, Moscow 119526, Russian Federation
e-mail: ikryukov@gmail.com

L. C. Jain et al. (eds.), *Advances in Theory and Practice of Computational Mechanics*, Smart Innovation, Systems and Technologies 173,
https://doi.org/10.1007/978-981-15-2600-8_11

part of airship, where an expansion fan is located, the ice layer is formed. These phenomenas change the airships aerodynamics to negative level. In some applications, condensation is negative but for other applications it is positive. For example, there are the vacuum deposition surface technology [1], planarization of surfaces with the help of ion cluster beam [2], natural gas segregation, among others

Homogeneous condensation process of substance can be divided into two sequential stages. The first one is nucleation (self-generation droplets) and the second one is growing of droplets with the help of condensation coatings. Condensation process intensity is determined by the saturation parameter S. The saturation parameter represents the ratio of the vapor pressure of condensing substance to the saturation pressure at a determined temperature. At $S > 1$, the formed nucleuses begin to grow until they reach the chemical and thermodynamic equilibrium with the environment.

This study considers features of the condensation modeling methods. These methods are used to study a process of vapor condensation in Laval nozzles.

Condensation of gas mixtures components in nozzle have been researched since the middle of the last century [3, 4]. Calculation techniques of condensation vapor for stream from supersonic nozzles, discharging to vacuum from sonic nozzle stream, or discharging a stream to low-pressure ambient space mainly use the macroscopic modeling method in case of continuum equations, as well as, in modifications of classical nucleation theory [5–9]. Modern method, which fixes classical nucleation theory problems close to the formation to saturation parameters area of liquid phase, is considered in [10, 11].

Currently, microscopic (kinetic) approaches are widely used. In [12–14], a mathematical model of Monte-Carlo condensation process was considered. Clusters were formed as a result of particle collisions. The elastic collision of molecules, recombination of molecules, association of a cluster and a monomer, association of clusters, and evaporation of a monomer from a cluster were taken into account.

The quasi-chemical model of condensation is widely used [15, 16]. In the quasi-chemical cluster model, it is assumed that the pair consists of monomers and molecular aggregates—clusters formed from monomers connected by the forces of molecular interaction. It is assumed that the growth of clusters occurs through the addition of monomer to them, and their destruction happens through the loss of monomer. However, two-particle reaction for small clusters is unlikely because there are problems with the removal of excess heat of reaction. In this case, a cluster growth occurs at the expense of three-particle reaction. When a cluster becomes large enough to absorb the impact energy and excess heat, the cluster growth becomes dominant due to two-particle reactions. The simulation of the gas volume condensation process can be carried out using the kinetic equations for the droplet size distribution function [17, 18]. However, a direct solution of the kinetic equations is possible only for relatively simple model problems.

Method of moments is based on solving the equations for the moments of the droplet size distribution function, which have recently become widely used [19–21]. The method requires relatively small computational costs and combines organically with continuous modeling with Eulerian approach. The use of method of moments is limited to the case of small droplets, where speed and temperature are not much different from the corresponding parameters of the gas medium.

This study develops two ways for condensation modeling. The first one is a continual approach based on MOM. The second one is a kinetic approach based on a quasi-kinetic model.

The chapter is structured as follows: Section 11.2 presents two methods for condensation simulating: method of moments and quasi-chemical model of condensation. In Sect. 11.3, there is the growth dynamics of clusters test. Also, Sect. 11.3 presents the results of modeling a jet flowing into external area. Section 11.4 concludes the chapter.

11.2 Mathematical Models

In this section, two methods are under consideration: method of moments and quasi-chemical model of homogeneous condensation represented in Sects. 11.2.1 and 11.2.2, respectively.

11.2.1 Method of Moments

Two-phase substance is the multicomponent gas (carrier gas and vapor of condensate substance) and clusters (droplets) of condensing substance. To construct the mathematical model, following assumptions are introduced:

- The volume ratio of liquid phase is negligible.
- There is the mechanical and thermal equilibrium between the gas and liquid phase.
- There aren't collisions between droplets.

The system of Navier–Stokes equations written in a weakly divergent form can be adopted as a mathematical model of such a two-phase mixture:

$$\frac{\partial U}{\partial t} + \frac{\partial (F - F_v)}{\partial x} + \frac{\partial (G - G_v)}{\partial y} = S, \qquad (11.1)$$

where

$$U = \begin{bmatrix} \rho \\ \rho u \\ \rho v \\ \rho E \\ \rho Q_0 \\ \rho Q_1 \\ \rho Q_2 \\ \rho\alpha \\ \rho\alpha\text{max} \end{bmatrix}, \quad F = \begin{bmatrix} \rho u \\ \rho u^2 + p \\ \rho uv \\ (\rho E + p)u \\ \rho u Q_0 \\ \rho u Q_1 \\ \rho u Q_2 \\ \rho u\alpha \\ \rho u\alpha\text{max} \end{bmatrix}, \quad G = \begin{bmatrix} \rho v \\ \rho uv \\ \rho v^2 + p \\ (\rho E + p)v \\ \rho v Q_0 \\ \rho v Q_1 \\ \rho v Q_2 \\ \rho v\alpha \\ \rho v\alpha\text{max} \end{bmatrix},$$

$$F_v = \begin{bmatrix} 0 \\ \tau_{xx} \\ \tau_{xy} \\ u\tau_{xx} + v\tau_{xy} - q_x \\ 0 \\ 0 \\ 0 \\ 0 \\ 0 \end{bmatrix}, \quad G_v = \begin{bmatrix} 0 \\ \tau_{yx} \\ \tau_{yy} \\ u\tau_{xy} + v\tau_{yy} - q_y \\ 0 \\ 0 \\ 0 \\ 0 \\ 0 \end{bmatrix}, \quad S = \begin{bmatrix} 0 \\ 0 \\ 0 \\ 0 \\ J \\ r_* J + \dot{r}\rho Q_0 \\ r_*^2 J + 2\dot{r}\rho Q_1 \\ \frac{4}{3}\pi\rho_l\left(r_*^3 J + 3\dot{r}\rho Q_2\right) \\ 0 \end{bmatrix},$$

where

$$\tau_{xx} = \mu\left(2\frac{\partial u}{\partial x} - \frac{2}{3}\left(\frac{\partial u}{\partial x} + \frac{\partial v}{\partial y}\right)\right), \quad \tau_{yx} = \tau_{xy} = \mu\left(\frac{\partial u}{\partial x} + \frac{\partial v}{\partial y}\right),$$

$$\tau_{yy} = \mu\left(2\frac{\partial v}{\partial y} - \frac{2}{3}\left(\frac{\partial u}{\partial x} + \frac{\partial v}{\partial y}\right)\right), \quad q_x = -\lambda\frac{\partial T}{\partial x}, \quad q_y = -\lambda\frac{\partial T}{\partial y}.$$

Here, ρ is the density, p is the pressure, T is the static temperature, u is the velocity along the x-direction, v is the velocity along the y-direction, E is the total energy per unit volume, μ is the viscosity coefficient, λ is the thermal conductivity coefficient.

First four equations (in Eq. 11.1) describe the dynamics of a two-phase mixture in a two-dimensional volume, while other equations (the equations of moments) characterize the evolution of changes in the parameters of the liquid-droplet phase. Equations from the fifth to the seventh are obtained from the general equation of dynamics describing the process of nucleation and dynamics of clusters (droplets) with homogeneous condensation:

$$\frac{\partial f}{\partial t} + \frac{\partial}{\partial x}(uf) + \frac{\partial}{\partial r}(\dot{r}f) = \delta(r - r_*)J \tag{11.2}$$

Here, the function $f(x, t, r)$ is the size distribution function of the droplets, J is the rate of homogeneous nucleation [20, 22], $\delta(r - r_*)$ is the delta-function, r_* is the droplet critical radius, $\dot{r}$ is the droplet growth rate.

The derivation of the moment equations consists in the successive multiplication of Eq. 11.2 by r^k (k is the natural number) and integration over a radius in the range from 0 to ∞. The result is an infinite chain of moment equations, Hill chain [3]:

$$\frac{\partial}{\partial t}(\rho Q_k) + \frac{\partial}{\partial x_i}(\rho U_i Q_k) = (r_*)^k \rho J + k\rho Q_{k-1}\dot{r}, \quad k = 0, \infty, \tag{11.3}$$

where $\rho Q_n = \int_{x_*}^{\infty} r^n f(x, t, r)dr$ is the moments of the nth order.

In this case, for a unit mass of the mixture, the zero moment Q_0 equals to the number density of droplets per unit mass, Q_1 is the sum of radii of all clusters, Q_2 is the sum of the squares of the radii of all clusters (estimate of the surface area of all clusters), Q_3 is the sum of cubes of radii of all clusters (estimation of the volume of all clusters). Instead of the moment Q_3, it is convenient to use the mass fraction of the liquid phase α:

$$\alpha = \frac{4\pi}{3}\rho_l Q_3,$$

where ρ_l is the liquid phase density.

Thus, MOM describes the evolution of the liquid phase by a finite number of moment equations derived from the general equation of the dynamics of the distribution function $f(x, t, r)$ [3]. In this research, a modification of MOM is used, at which an additional equation is introduced that describes the dynamics of the mass fraction of the condensing phase $\alpha_{\max}$ (the sum of the mass fractions of droplets and vapor of the condensing substance). This allows to extend the class of tasks to be solved, for example, to consider problems with different contents of a condensable substance in different zones of the computational domain.

Thermodynamics model. The thermophysical properties of the mixture and equation of state for the mixture (caloric and thermal) are written using Eq. 11.4, where C_{Va}, C_{Pa} are the constant-volume and constant-pressure specific heats, respectively, C_{VV}, C_{PV} are the specific heats of the condensing medium (vapor). C_{Vmixt}, C_{Pmixt} are the specific heats of the two-phase mixture, C_l is the specific heat of the liquid water, R_a, R_V, R_{mixt} are the individual gas constants of the carrier gas condensing the medium and two-phase mixture, respectively, κ_f is the adiabatic exponent of the mixture [19].

$$\begin{aligned}
C_{Vmixt} &= (1 - \alpha_{\max})C_{Va} + \alpha_{\max} C_{VV} + \alpha(C_l - C_{VV}) \\
C_{Pmixt} &= (1 - \alpha_{\max})C_{Pa} + \alpha_{\max} C_{PV} + \alpha(C_l - C_{PV}) \\
R_{mixt} &= (1 - \alpha_{\max})R_a + \alpha_{\max} R_V - \alpha R_V) \\
\kappa_f &= \frac{C_{Pmixt}}{C_{Vmixt}}
\end{aligned} \tag{11.4}$$

The caloric and thermal equations of state are mentioned below:

$$\begin{aligned}
&e = (1 - \alpha_{\max})C_{va}T + \alpha_{\max} C_{vv} + \alpha(C_l - C_{vv}) + \alpha L_0, \\
&p = \rho T R_{mixt}, a_f^2 = \kappa_f \frac{p}{\rho}, L = L_1 T + L_0, L_1 = C_{PV} - C_l,
\end{aligned} \tag{11.5}$$

where T is the temperature of the mixture, a_f is the frozen velocity of sound of the mixture, L is the latent heat of vaporization, $e = e(T)$ is the mixture internal energy.

The right sides of the moment equations in system (Eq. 11.1) are determined using the parameters of the classical nucleation theory $(J, \dot{r} = dr/dt, r_*, f(x, t, r))$,

$$J = \frac{q_c}{(1+\eta)} \sqrt{\frac{2\sigma}{\pi m^3}} \frac{\rho_v^2}{\rho_l} \exp\left(-g \frac{4\pi}{3} \frac{r_*^2 \sigma}{R_V m T}\right),$$

$1/(1+\eta)$ is the corrective factor taking into account the non-stationarity of the process [23], q_c is the condensation coefficient ($q_c \approx 1$),

$$\eta = 2\frac{\kappa_f - 1}{\kappa_f + 1} \frac{L}{R_v T} \left(\frac{L}{R_v T} - \frac{1}{2}\right),$$

$\sigma = k_\sigma \sigma_\infty$, σ_∞ is the flat film surface tension, k_σ is the correction factor taking into account the curvature of the drop, g is the nucleation correction factor multiplier, $S = p_V / p_S$ is the saturation parameter.

To determine the magnitude of the growth rate of a drop, one of the following models is used for the free-molecular and continual regime of flow of a condensable substance around a cluster-drop.

One model is Hertz-Knuth model, where $\frac{dr}{dt}$ is defined as

$$\frac{dr}{dt} = \frac{\beta}{\rho_l} \frac{p_V - p_{S,r}}{\sqrt{2\pi R_V T}}, \tag{11.6}$$

where $p_{S,r}$ is the saturation pressure on the surface of a drop of average radius size,

$$p_{S,r} = p_S \exp \frac{2\sigma}{\rho_l R_V T r_{Hill}},$$

p_S is the surface saturation pressure, β is the evaporation coefficient [3],

$$r_{Yill} = \begin{cases} \sqrt{\frac{Q_2}{Q_0}} & \text{if } \alpha > 10^{-6} \\ 0 & \text{if } \alpha \le 10^{-6} \end{cases}, \quad r_* = \begin{cases} \frac{2\sigma}{\rho_l R_V T \ln S} & \text{if S} > 1 \\ \infty & \text{if S} \le 1 \end{cases}. \tag{11.7}$$

Another model is Hill–Young model [3, 24], where $\frac{dr}{dt}$ is defined as

$$\frac{dr}{dt} = \frac{p_v}{\rho_l L \sqrt{2\pi R_v T}} \frac{C_{p_v} + C_{v_v}}{2} (T_s(p_v) - T),$$

T_s is the saturation temperature.

One more model is Gyarmathy model [22, 25, 26], where $\frac{dr}{dt}$ is defined as

$$\frac{dr}{dt} = \frac{\lambda_v (T_s(p_v) - T)\left(1 - \frac{r^*}{r_{hill}}\right)}{\rho_l L \cdot r_{hill} \cdot (1 + 3.18K_n)}, \quad (11.8)$$

where λ_v is the thermal conductivity of vapor to active substance condensate, K_n is the Knudsen number characterizing the flow around a drop of steam, $K_n = \frac{l}{2r_{hill}}$, l is the free length of the vapor molecule:

$$l = \frac{1.5\mu_v\sqrt{R_v T}}{p_v},$$

where μ_v is the dynamic viscosity of vapor of condensable substance.

The system of Eqs. 11.1–11.8 is solved by an explicit method of control volume of the second order of accuracy in space, while for approximating inviscid flows Godunov scheme is used in conjunction with the approximate Advection Upstream Splitting Method (AUSM) and method for solving Riemann problem, and central-difference approximation is used for approximating viscous flows.

11.2.2 Quasi-Chemical Model of Homogeneous Condensation

The clustering process is described by the following, generally speaking, infinite-dimensional kinetic model. It is assumed that the number of monomers per unit volume is much more than dimers, trimers, etc. When this condition is fulfilled in the kinetic mechanism, which determines the course of the whole condensation process, the addition reactions of monomers to the cluster of the following type prevail:

$$M_i + M_1 + \underset{K_i^{-(1)}}{\overset{K_i^{+(1)}}{\longleftrightarrow}} M_{i+1}, \quad (11.9)$$

$$M_i + M_1 + M_1 \underset{K_i^{-(2)}}{\overset{K_i^{+(2)}}{\longleftrightarrow}} M_{i+1} + M_1, \quad i = 1, 2, 3, \ldots, \quad (11.10)$$

$$M_i + M_1 + M_A \underset{K_i^{-(3)}}{\overset{K_i^{+(3)}}{\longleftrightarrow}} M_{i+1} + M_A, \quad (11.11)$$

where M_1, M_i, M_A are the symbolic designation of the monomer, i-measure, and inert gas molecules, $K_i^{+(r)}$, $K_i^{-(r)}$ $(r = 1, 2, 3)$ are the rate constants of the corresponding requirements. The plus and minus markers correspond to cases of monomer addition and detachment.

In accordance with the kinetic mechanism, the change in concentration over time is described by a system of ordinary differential equations γ_i:

$$\rho\frac{d\gamma_i}{dt} = (K_{i-1}^{+(1)} + K_{i-1}^{+(2)}\rho\gamma_1 + K_{i-1}^{+(3)}\rho\gamma_A)\rho\gamma_1\rho\gamma_{i-1}$$
$$- (K_i^{+(1)} + K_i^{+(2)}\rho\gamma_1 + K_i^{+(3)}\rho\gamma_A)\rho\gamma_1\rho\gamma_i - (K_{i-1}^{-(1)} + K_{i-1}^{-(2)}\rho\gamma_1 + K_{i-1}^{-(3)}\rho\gamma_A)\rho\gamma_i$$
$$+ (K_i^{-(1)} + K_i^{-(2)}\rho\gamma_1 + K_i^{-(3)}\rho\gamma_A)\rho\gamma_{i+1}.$$

By analogy with chemical kinetics, we can get a connection between the rate constants of the forward and reverse reactions.

From the equilibrium condition of reactions (Eqs. 11.9–11.11), it follows:

$$(K_i^{-(1)} + K_i^{-(2)}\rho\gamma_1 + K_i^{-(3)}\rho\gamma_A) = (K_i^{+(1)} + K_i^{+(2)}\rho\gamma_1 + K_i^{+(3)}\rho\gamma_A)\rho\gamma_1\frac{\rho\gamma_i}{\rho\gamma_{i+1}}.$$

From the conditions of thermodynamic equilibrium follows the equality of chemical potentials:

$$G_i(p,T) + RT\ln x_i + G_1(p,T) + RT\ln x_1 = G_{i+1}(p,T) + RT\ln x_{i+1}.$$

It follows that

$$\frac{x_i}{x_{i+1}} = \frac{\rho\gamma_i}{\rho\gamma_{i+1}} = \exp\left(\frac{G_{i+1}(p,T) - G_i(p,T) - G_1(p,T) - RT\ln(\gamma_1 m_\Sigma)}{RT}\right)$$

or if we enter the value

$$\varepsilon_i = \exp\left(i\ln x_1 - \frac{G_i(p,T) - iG_1(p,T)}{RT}\right),$$

then

$$\frac{\rho\gamma_i}{\rho\gamma_{i+1}} = \frac{\varepsilon_i}{\varepsilon_{i+1}}.$$

Denoted by

$$\nu_{\Sigma i} = (K_i^{+(1)} + K_i^{+(2)}\rho\gamma_1 + K_i^{+(3)}\rho\gamma_A)\rho\gamma_1,$$

we get

$$\frac{d\gamma_i}{dt} = \nu_{\Sigma i-1}\gamma_{i-1} - \nu_{\Sigma i}\gamma_i - \nu_{\Sigma i-1}\frac{\varepsilon_{i-1}}{\varepsilon_i}\gamma_i + \nu_{\Sigma i}\frac{\varepsilon_i}{\varepsilon_{i+1}}\gamma_{i+1} = \nu_{\Sigma i-1}\varepsilon_{i-1}\left(\frac{\gamma_{i-1}}{\varepsilon_{i-1}} - \frac{\gamma_i}{\varepsilon_i}\right) - \nu_{\Sigma i}\varepsilon_i\left(\frac{\gamma_i}{\varepsilon_i} - \frac{\gamma_{i+1}}{\varepsilon_{i+1}}\right),$$

$$i = 2, 3, \ldots, \infty.$$

Expressions of condensation reaction rates can be calculated by the formulas of the liquid-drop theory [4, 15]. It should be noted that in practice the systems of finite dimensionality N obtained from the source system are used by truncating it. In this case, the system removes the equations for $i > N$, and in the amounts used, the corresponding terms are removed. It is additionally assumed that clusters of a size larger than N are not formed. In accordance with [4, 15], this system is approximated by a finite system with dimension N (in real calculations $N = 10{,}000$–$10{,}000{,}000$):

$$\frac{d\gamma_i}{dt} = \nu_{\Sigma i-1}\gamma_{i-1} - (\nu_{\Sigma i} + \nu_{\Sigma i-1}\frac{\varepsilon_{i-1}}{\varepsilon_i})\gamma_i + \nu_{\Sigma i}\frac{\varepsilon_i}{\varepsilon_{i+1}}\gamma_{i+1}, \quad i = 2, 3, \ldots, N.$$

The concentration of monomers in this case is obtained from the normalization condition:

$$\sum_{i=1}^{N} i\gamma_i = \text{const} = \gamma_0, \quad \gamma_A = \text{const}.$$

The system of condensation kinetics equations is approximated by a semi-implicit difference principle, the sweep method can be solved for each time step, and the nonlinearity of γ_1 is eliminated by the iteration method.

Comments. Also, in numerical modeling, an approximation of the source system by a suitable finite-dimensional system of a similar type with effective values of its coefficients is used. The latter are found using a transformation $i = n(j)$ that translates an infinite interval $i = 1, 2, \ldots$ into a finite $j = 1, 2, \ldots, N, N + 1$. In this case, the normalization condition is

$$\sum_{j=1}^{N} n_j n_j' \gamma_j = \gamma_0, \quad n_j' = \frac{dn(j)}{dj}.$$

Thermodynamic parameters of the mixture are found by formulas, in which it is assumed that infinite sums with the participation of some thermodynamic parameter Ai are approximated as follows:

$$\sum_{i=1}^{\infty} \gamma_i A_i \approx \sum_{j=1}^{N} n_j' \gamma_j A_j.$$

All parameters depending on the index j are calculated using the appropriate formulas for $i = n(j)$.

Thermodynamics model. The mixture of perfect gases and condensable component described in terms of the model of perfect gases is considered. The thermodynamic properties of such a mixture can be described by Gibbs specific potential G:

$$G(p, T, \gamma) = \gamma_A(G_A(p, T) + RT\ln x_A) + \sum_{i=1}^{\infty} \gamma_i(G_i(p, T) + RT\ln x_i),$$

$$x_i = \gamma_i m_\Sigma,\ x_A = \gamma_A m_\Sigma,\ m_\Sigma = (\gamma_A + \sum_{i=1}^{\infty} \gamma_i)^{-1},$$

$$\gamma_A = \sum_{i=1}^{N} \tilde{\gamma}_i,\ G_A(p,T) = \sum_{i=1}^{N} \tilde{\gamma}_i \tilde{G}_i(p,T)/\gamma_A.$$

The parts of the Gibbs molar potentials that are independent of the concentrations for i parameters $G_i(p,T)$ ($I = 1, 2, 3 \ldots$) and non-condensable molecules are determined by the formulas:

$$G_j(p,T) = RT\ln(p/p_0) + G_j^\circ(T), \quad j = A,\ 1,\ 2,\ \ldots$$

The thermodynamic properties of the clusters were calculated within the framework of the liquid-drop model in its standard form. In particular, in order to write $G_i^\circ(T)$ for clusters, i.e., at $i = 2, 3, \ldots$ in the standard reference system of enthalpies, one can use the expression:

$$\Delta G_i^\circ(T) = i(G_L^\circ(T) - G_1^\circ(T)) + 4\pi r_i^2 N_A \sigma_i(T).$$

Here, r is the radius of the droplet which contains n molecules, $\sigma_i(T)$ is the surface tension of the i-measure. A cluster containing i gas molecules with molecular mass m_1 occupies the volume $V_i = \frac{im_1}{N_A\rho_L(T)} = \frac{4}{3}\pi r^3$. Therefore,

$$r = \left(\frac{3}{4\pi}\frac{im_1}{N_A\rho_L(T)}\right)^{\frac{1}{3}},$$

$$\Delta G_i^\circ(T) = i(G_L^\circ(T) - G_1^\circ(T)) + B_i(T)i^{2/3},$$

$$B_i(T) = \sigma_i(T)(36\pi)^{1/3}N_A^{1/3}(m_1/\rho_L(T))^{2/3},\ i = 2, 3\ldots$$

According to Gibbs potential of the mixture, all other thermodynamic parameters are determined.

Equilibrium distribution function. Consider the state of thermodynamic equilibrium between the gaseous and liquid phase consisting of drops, including a specified number of molecules, with a known number of molecules γ_0 condensing and number of inert phase in a kilogram of the mixture molecules γ_A. We write the equilibrium conditions between the gas and liquid phases at the phase transition, which is the equality of the corresponding chemical potentials μ_i:

$$\mu_i = \left(\frac{\partial G}{\partial \gamma_i}\right)_{P,T,\gamma_{j,j\neq i}} = G_i(p,T) + RT\ln x_i,$$

$$\mu_L = G_L(p,T)$$

on the saturation curve

$$G_L(p_H, T) = G_1(p_H, T) + RT \ln x_1.$$

This equation contains three unknown parameters: the pressure, temperature, and mole fraction of monomers in the mixture. The cluster size distribution function in the state of thermodynamic equilibrium at the phase transition, in accordance with the used model of thermodynamics and condensation kinetics, should be defined from the following conditions:

$$G_2(p_H, T) + RT \ln x_2 = 2(G_1(p_H, T) + RT \ln x_1) = 2G_L(p_H, T),$$
$$G_i(p_H, T) + RT \ln x_i = i(G_1(p_H, T) + RT \ln x_1) = iG_L(p_H, T),$$

where

$$x_i = \gamma_i \left(\sum_{i=1}^{\infty} \gamma_i + \gamma_A \right)^{-1}.$$

Parameter x_i is the unknown mole fraction of the ith component of the mixture, $i = 1, 2, \ldots, A$ is the unknown saturation pressure.

This system of equations can be supplemented by the normalization condition:

$$\sum_{i=1}^{\infty} x_i + x_A = 1.$$

This system is a system of nonlinear equations, from which the saturation curve, and corresponding to it, the cluster size distribution function can be found. From Eq. 11.2, we can get expressions for x_i:

$$x_i = \exp\left(\frac{iG_L(p_H, T) - G_i(p_H, T)}{RT} \right).$$

From the relationship between molar fractions and molar mass concentrations, the expression for the molar fraction of the inert component has a view:

$$x_A = \gamma_A \left(\sum_{i=1}^{\infty} \gamma_i + \gamma_A \right)^{-1},$$

$$\sum_{i=1}^{\infty} \gamma_i + \gamma_A = \frac{\gamma_A}{x_A}.$$

We write the law of conservation of the condensable component:

$$\sum_{i=1}^{\infty} i\gamma_i = \gamma_0 => \quad \sum_{i=1}^{\infty} ix_i = \gamma_0 \left(\sum_{i=1}^{\infty} \gamma_i + \gamma_A \right)^{-1} = \frac{\gamma_0}{\gamma_A} x_A.$$

The mole fraction of inert gas can be written in the following form:

$$x_A = \frac{\gamma_A}{\gamma_O} \sum_{i=1}^{\infty} ix_i.$$

Thus, the saturation curve can be found from the normalization conditions:

$$\sum_{i=1}^{\infty} x_i + x_A = 1.$$

After substitution of molar fractions into it, we obtain the following espression:

$$F(p, T) = \sum_{i=1}^{\infty} \left(1 + \frac{\gamma_A}{\gamma_0} i \right) \exp\left(\frac{iG_L(p, T) - G_i(p, T)}{RT} \right) - 1 = 0. \qquad (11.12)$$

The nonlinear Eq. 11.12 is the relationship between the pressure and temperature on the saturation curve. Accordingly, at the known temperature from Eq. 11.12, the saturation pressure p_H can be found. To solve Eq. 11.12, Newton method is used in the work, the initial pressure approximation (p_H^0) is calculated from the equality of the chemical potentials of the monomers and the liquid phase:

$$\frac{p_H^0}{p_0} = \exp\left(\frac{G_L^0(T) - G_1^0(T)}{RT} \right).$$

For practice, it is often of interest to find the equilibrium distribution function for given two thermodynamic parameters, for example, p and T. The chemical potential of the monomers (gas phase) is less than the chemical potential of the liquid phase:

$$G_1(p, T) + RT \ln x_1 < G_L(p, T).$$

In the equilibrium state, for a cluster of any size i, the following equality holds:

$$G_i(p, T) + RT \ln x_i = i(G_1(p_H, T) + RT \ln x_1).$$

Therefore,

$$x_i = \exp\left(i \ln x_1 + \frac{iG_1(p, T) - G_i(p, T)}{RT} \right).$$

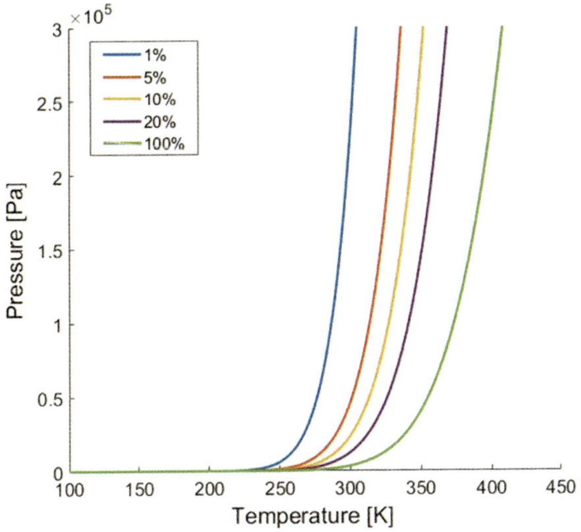

Fig. 11.1 Saturation curve of water vapor in nitrogen with different mass fractions of water (1, 5, 10, 20, and 100%)

Similar to the above, the concentration of monomers can be found from a nonlinear equation:

$$F(\ln x_1) = \sum_{i=1}^{\infty}\left(1 + \frac{\gamma_A}{\gamma_0} i\right)\exp\left(i \ln x_1 + \frac{iG_1(p,T) - G_i(p,T)}{RT}\right) - 1 = 0. \tag{11.13}$$

It should be noted that from this equation the cluster size distribution function can be found in the neighborhood of the saturation curve on the left (in the metastable region).

For water vapor in nitrogen, we present saturation curves on the P–T phase plane depending on the mass fraction of water vapor (Fig. 11.1). As expected, the saturation pressure increases significantly with decreasing mass fraction of water vapor.

11.3 Numerical Results

In this section, a special test with ideal constant volume adiabatic reactor is considered in Sect. 11.3.1, while a wet steam flow with spontaneous condensation in Laval nozzle is presented in Sect. 11.3.2.

11.3.1 Test: Ideal Constant Volume Adiabatic Reactor

A special test was used to test the dynamics of gas transition from a metastable state to an equilibrium state. At the initial moment of time, the parameters of the medium in the volume correspond to the strong nonequilibrium (metastable) state. Medium is motionless. In this chapter, we consider a mixture of nitrogen with water vapor. Mass fraction of water vapor is 0.02. The pressure in the mixture is 100,000 Pa, the temperature is 200 K. On the *P–T* diagram, this state corresponds to a point located in the region to the left of the saturation curve for water. Over time, the system will develop to an equilibrium state due to the formation of a liquid-droplet phase, the release of heat of condensation and an increase in temperature and pressure with a constant mixture density, but varying density of the gas phase. Thus, the simulation result depends only on the macrostate of the system, in which it will be after establishing equilibrium. This circumstance allows us to compare only the processes of gas transition to the equilibrium state separately from the gas-dynamic components of the system. In this test, the object of interest is both the final state of the system (pressure and temperature, the mass fraction of the condensed phase) and the dynamics of the transition from a nonequilibrium state to an equilibrium state (transition time and system trajectory).

Comparison of the calculating results of the transition process to the equilibrium state obtained by the quasi-chemical model and MOM is depicted in Fig. 11.2. The dynamics of temperature change and pressure in an ideal adiabatic reactor of constant volume is shown. For MOM, Eqs. 11.5–11.6 with different values of the parameters β и g were used. For all values of these parameters, the final state of the medium in the reactor coincides with the quasichemical model with an accuracy of 0.3 K and 40.0 Pa. The dynamics of condensation for $\beta = 1$ and $g = 1$ are very different in these two models (red dotted line and blue solid line in Fig. 11.2a, b). In a quasi-chemical model, the growth process of large clusters (droplets) proceeds

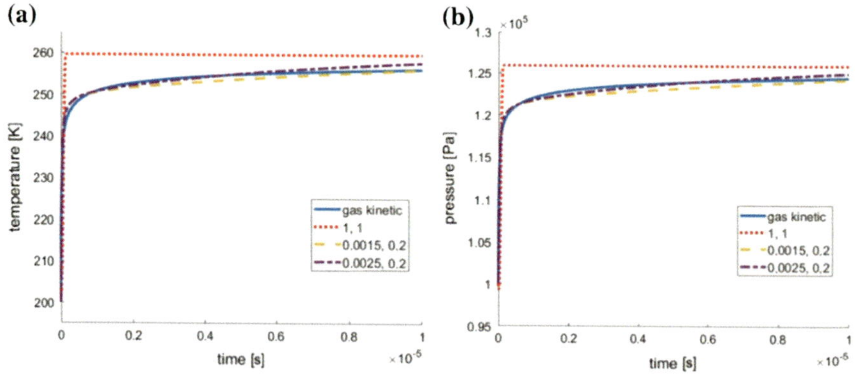

Fig. 11.2 The evolution of parameters in an ideal adiabatic constant volume reactor: **a** temperature, **b** pressure. Parameters of MOM ($\beta = 1, 0.0015, 0.0025$ and $g = 1, 0.2, 0.2$)

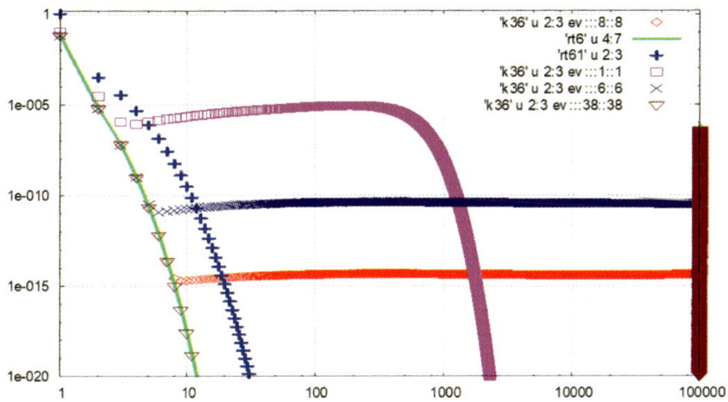

Fig. 11.3 Size distribution functions

more smoothly and nucleation occurs faster (Fig. 11.2). By accelerating the onset of nucleation in MOM ($g < 1$) and slowing the growth of droplets ($\beta < 1$), one can obtain a good correspondence of the dynamics for the two models under study.

Using a quasi-chemical model of condensation, the dynamics of the time variation of the cluster size distribution function were obtained (Fig. 11.3). The initial cluster size distribution function (blue cross) was taken from the saturation curve and contained clusters ranging in size from 1 to 10 with concentrations exceeding 10^{-10}. The green curve corresponds to the equilibrium distribution function for a given specific volume and internal energy and contains in significant concentrations clusters of sizes from 1 to 10, as well as, clusters with a size of 100,000, which in this model correspond to the liquid phase. The violet, blue, red, and brown curves show the dynamics of changes in the distribution function from the initial equilibrium state to the final one. Initially, a condensation wave is formed (purple graph), which spreads from small to large. After the condensation wave reaches the maximum of the cluster sizes taken into account, the concentrations of clusters grow from the vicinity of the maximum size, while the concentrations of clusters of intermediate sizes decrease (blue, red, and brown curves). Concentrations of clusters of small sizes gradually fall on the equilibrium distribution function (green curve).

11.3.2 Wet Steam Flow with Spontaneous Condensation in Laval Nozzle

The flow of superheated water vapor ($P_0 = 124{,}000$ Pa, $T_0 = 391.55$ K) in a flat nozzle and jet flowing out into outlet space ($P = 20{,}000$ Pa) is considered. The geometric parameters of the nozzle are taken from [27]. The height of the entrance section is 0.06947 m, the critical one is 0.04 m, the output one is 0.04733 m, the length of the supersonic part of the nozzle is 0.28625 m, the radius of the transonic part of

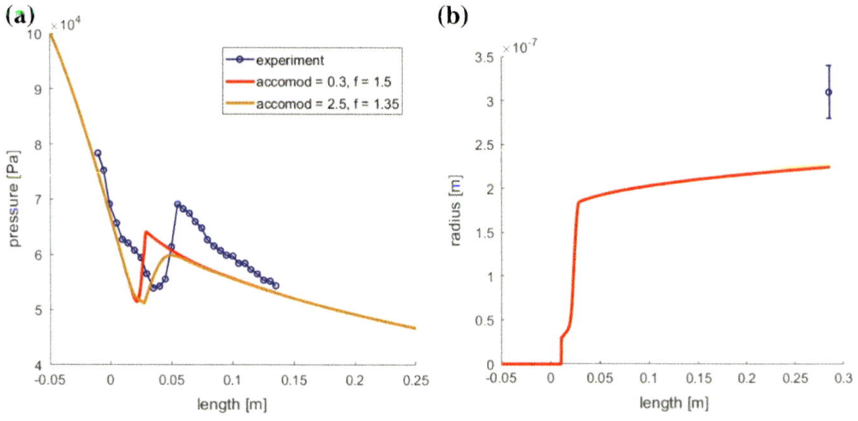

Fig. 11.4 The distribution of parameters along the axis of the nozzle: **a** pressure, **b** radius of droplets

the nozzle is 0.27145 m. The flow calculations were performed by MOM in 1D and 2D formulations using various models of the growth rate of a drop (Eqs. 11.6–11.8).

The results of 1D calculations in comparison with experimental data from [27, 28] are shown in Fig. 11.4a, b. Figure 11.4a shows the pressure distribution along OX axis in the plane of symmetry. The red curve corresponds to the calculations by MOM with the model of the growth rate of a drop (Eq. 11.6) with parameters $\beta = 0.3$ and the correction factor $g = 1.5$ in the ratio (Eq. 11.5). The orange curve corresponds to the calculation by MOM with the model of the growth rate of the drop (Eq. 11.10) with parameters $\beta = 0.3$ and $g = 1.5$. Figure 11.4b shows the distribution along OX axis of the drop radius in the calculations (line) and experiment (circle).

The results of 2D calculations for a jet flowing into external area are shown in Fig. 11.5. The mesh has about 100,000 cells. For the flow coming out of the nozzle, the second condensation jump occurs as a result of the additional expansion of the gas leaving the nozzle. The resulting shock wave behind Mach barrel causes a decrease in concentration α. The maximum concentration of water is reached in the inner area of the first barrel and is equal to 6.7%. Figure 11.6 shows Mach number contours.

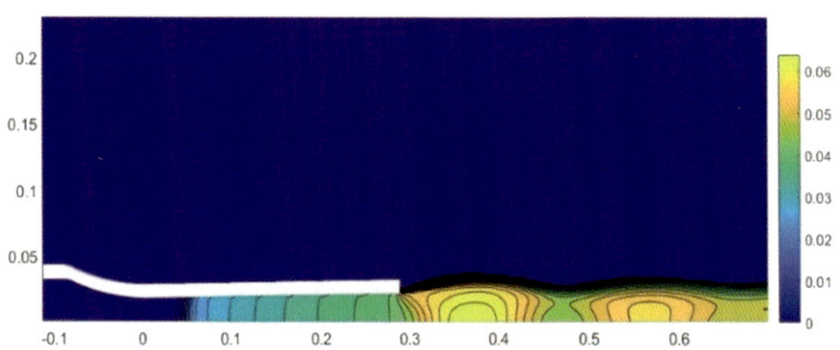

Fig. 11.5 The distribution of water concentration in the nozzle

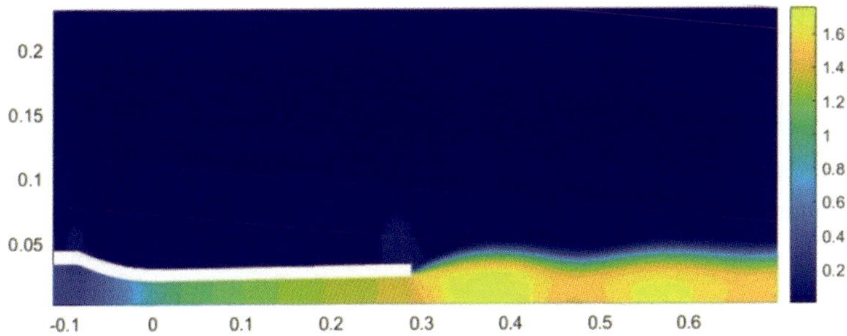

Fig. 11.6 Mach number contours

11.4 Conclusions

Two approaches to modeling the process of spontaneous condensation in gas-dynamic flows have been developed. The first approach is a kinetic approach, which is based on a quasi-chemical model. The second approach is a continual approach, which is based on MOM. The model of gas dynamics is Navier–Stokes equation in two-dimensional coordinate system and Euler equation in one-dimensional coordinate system. To solve the gas-dynamic system, a numerical method of enhanced order of accuracy based on Godunov method and AUSM method is used.

Using the two approaches developed, numerical simulation of condensation processes was carried out in an ideal adiabatic reactor of constant volume and with a flow of condensing matter in a nozzle using quasi-one-dimensional formulation and a jet flowing into outlet space. To simulate a jet flowing into flooded space, Navier–Stokes equations are used in two-dimensional formulation.

In the first case, spontaneous condensation is considered, when a mixture of nitrogen and water vapor (2% mass fraction) transfers from nonequilibrium (metastable) to equilibrium. Both approaches give similar results on the final state, the difference does not exceed 0.15% in temperature and 0.04% in pressure. However, the dynamics of the process of transition to the equilibrium state may differ significantly for the two approaches. Coherence high level of the dynamics of the process for the two studied models is achieved by a special selection of correction factors in MOM. Using a quasi-chemical model for water vapor in nitrogen, we obtained the saturation curves in the P–T phase plane depending on the mass fraction of water vapor.

In the second test case, the flow of superheated steam in a flat Laval nozzle is numerically investigated using MOM and the solution of Navier–Stokes equations. In the supersonic part of the nozzle, spontaneous condensation of water occurs with the formation of a region of increasing pressure and temperature (condensation jump) directly near the critical section. A qualitative agreement was obtained between the numerical and experimental pressure distributions on the plane of symmetry of the nozzle, and the accuracy of 23% of the values of the average radii of droplets on the nozzle section was obtained in the calculation and experiment.

Corrective factors in MOM selected for the considered problems (in the first problem due to match the quasi-chemical model and in the second one to match with the experiment) have significantly different sets of values. This indicates the need to improve MOM in terms of calculating the nucleation rate and rate of growth of the droplets' mass.

Acknowledgements This work was carried out within the state task no. 9.7555.2017/BCh.

References

1. Aksenov, I.I., Belous, V.A., Strel'nitskij, V.E., Aksyonov, D.S.: Vacuum-arc equipment and coating technologies in KIPT. Phys. Radio Technol. Ion Plasma Technol. 58–71 (2016)
2. Ieshkin, A.E., Shemukhin, A.A., Ermakov, Y.A., Chernysh, V.S.: The influence of the gas cluster ion beam composition on defect formation in targets. Mosc. Univ. Phys. Bull. **71**(1), 87–90 (2016)
3. Hill, P.G.: Condensation of water vapour during supersonic expansion in nozzles. J. Fluid Mech. 593–620 (1966)
4. Gorbunov, V.N., Ryzhov, YuA, Pirumov, U.G.: Noequilibrium condensation in high-speed gas flows. Gordon and Breach Science Publishers, Institute of Mechanics, Moscow University, USSR (1988)
5. Oswatitsch, K.: Kondensationserscheinungen in uberschallduzen. ZAMM **22**, 1–14 (1942)
6. Stiver, H.: A condensation phenomenon in high velocity flows. In: Emmons, W. (ed.) Fundamentals of Gas Dynamics. Chapter 3. Princeton University Press, New Jersey, U.S.A. (1958)
7. Saltanov, G.A., Seleznev, L.I., Tsiklauri, G.V.: Generation and growth of condensed phase in high-velocity flows. Int. J Heat Mass Transf. **16**, 1577–1587 (1973)
8. Kotake, S., Glass, I.I.: Flows with nucleation and condensation. Prog. Aerospace Sci. **19**, 129–196 (1979)
9. Bakhtar, F., Young, J.B., White, A.J., Simpson D.A.: Classical nucleation theory and its application to condensing steam flow calculations. In: Proceedings of the Institution of Mechanical Engineers, vol. 219, Part C. J. Mech. Eng. Sci., pp. 1315–1333 (2005)
10. Gorbunov, A.A., Igolkin, S.I.: Statistic simulation of crystal grids growing at vapor condensation. Matem. Mod. **17**(3), 15–22 (2005)
11. Bauer, S.Y., Frurip, D.J.: Homogeneous nucleation in metallic vapors. A self-consistent kinetic model. J. Chem. Phys. **81**(10), 1015–1024 (1977)
12. Bykov, N.Y., Gorbachev, Y.E.: Cluster formation in copper vapor jet expanding into vacuum: the direct simulation Monte Carlo. Vacuum **163**, 119–127 (2019)
13. Bykov, N.Y., Gorbachev, Y.E.: Mathematical models of water nucleation process for the direct simulation Monte Carlo method. Appl. Math. Comput. **296**, 215–232 (2017)
14. Bykov, N.Y., Safonov, A.I., Leshchev, D.V., Starinskiy, S.V., Bulgakov, A.V.: Gas-jet method of metal film deposition: direct simulation Monte-Carlo of He-Ag mixture flow. Mater. Phys. Mech. **38**, 119–130 (2018)
15. Volkov, V.A., Muslaev, A.V., Pirumov, U.G., Rozovskii, P.V.: Non Equilibrium condensation of metal vapors/inert gas mixture during expansion through the nozzles of cluster-beam generators. Fluid Dyn. **30**(3), 399–408 (1995)
16. Egorov, B.V., Markachev, Y.E., Plekhanov, E.A.: Correlation of the quasi-chemical cluster nucleation model when compared with the experimental data. Khimicheskaia Fizika **25**(4), 61–70 (in Russian) (2006)

17. Sternin, L.E.: Fundamentals of gas dynamics of two-phase nozzle flows. Moscow, Izdatel'stvo Mashinostroenie (in Russian) (1974)
18. Kortsenshteyn, N.M., Samuilov, E.V., Yastrebov, A.K.: Study of the volume condensation process in supersaturated vapor by the direct numerical solution of the kinetic equation for the droplet size distribution function. Colloid J. **69b**(4), 488–295 (2007)
19. Gidaspov, V.U., Ivanov, I.E., Kryukov, I.A., Nazarov, V.S., Malashin F.A.: Study of the condensation process in nozzles with a large degree of expansion. Phys. Chem. Kinetics Gas Dyn. **19**(2). Art. no. 737.1–737.17 (in Russian) (2018)
20. Luo, X., Cao, Y., Xie, H., Qin, F.: Moment method for unsteady flows with heterogeneous condensation. Comput. Fluids **146**, 51–58 (2017)
21. Wyslouzil, B.E., Heath, C.H., Cheung, J.L., Wilemski, G.: Binary condensation in a supersonic nozzle. J. Chem. Phys. **113**(17), 7317–7329 (2000)
22. Sova, L., Jun, G., Stastny, M.: Modifications of steam condensation model implemented in commercial solver. AIP Conference Proceedings 1889, 020039.1–020039.8 (2017)
23. Kantorowitz, A.: Nucleation in very rapid vapor expansions. J. Chem. Phys. **19**(9), 1097–1100 (1951)
24. Young, J.B.: The spontaneous condensation in supersonic nozzles. Phys. Chem. Hydrodyn. **3**(1), 57–82 (1982)
25. Gyarmathy, G.: Grundlageiner Theorie der Nassdampfturbine. Dissertation, Juris Verlad, Zurich (1960)
26. Choi, B., Shim, J., Kim, Ch., Park, J., You, D., Beak, J.: Numerical simulation of homogeneous condensing wet-steam flow using an Eulerian–Lagrangian method. In: Proceedings of Shanghai 2017 Global Power and Propulsion Forum, GPPS AME, pp. 17–91 (2017)
27. Dykas, S., Majkut, M., Smolka, K., Strozik, M.: Experimental research on wet steam flow with shock wave. Exp. Heat Transf. **28**(5), 417–429 (2014)
28. Dykas, S., Majkut, M., Smolka, K., Strozik, M.: An attempt to make, a reliable assessment of the wet steam flow field in the de Laval nozzle. Heat Mass Transf. **54**(9), 2675–2681 (2018)

Chapter 12
Numerical Study of the Injection Parameters Impact on the Efficiency of a Liquid Rocket Engine

Yulia S. Chudina, Evgenij A. Strokach, Igor N. Borovik and Vladimir Yu. Gidaspov

Abstract This chapter analyzes the effect of various parameters (such as droplet injection velocity components by a swirl injector in a cylindrical coordinate system and droplet size distribution parameters) of fuel injection in an oxygen-kerosene rocket engine on the efficiency of the workflow. The study was conducted for two cases of application of the wall film cooling of the combustion chamber and without it. It is shown that the parameters for fuel injection in the case of using a wall film cooling effect in an unobvious way. A description of the object of study, the main features of the numerical experiment, and models used in the course of the study are given, as well as, the results are analyzed and recommendations on their use and further research in this area are formulated. The use of the obtained results allows to form a technical task for the design of a mixing head that implements optimal combustion conditions in the combustion chamber. Verification of the calculated data was carried out using the results of experimental studies conducted at Moscow Aviation Institute (National Research University) at the Department of Rocket Engines. This experimental work was carried out using a specially designed DMTMAI-200OK rocket engine.

Y. S. Chudina · E. A. Strokach · I. N. Borovik (✉) · V. Yu. Gidaspov
Moscow Aviation Institute (National Research University), 4, Volokolamskoe Shosse, Moscow 125993, Russian Federation
e-mail: borovik.igor@mai.ru

Y. S. Chudina
e-mail: y.chudina@gmail.com

E. A. Strokach
e-mail: evgenij.strokatsch@mai.ru

V. Yu. Gidaspov
e-mail: gidaspov@mai.ru

L. C. Jain et al. (eds.), *Advances in Theory and Practice of Computational Mechanics*, Smart Innovation, Systems and Technologies 173,
https://doi.org/10.1007/978-981-15-2600-8_12

12.1 Introduction

The development of high-performance low-thrust liquid-propellant rocket engines is among the most important trends in the development of space technology because an increase in the efficiency of low-thrust engines gives an increase in the lifetime of the spacecraft on the orbit. During the design process, the developer faces a number of difficulties arising from the lack of information about the processes occurring in the most thermally stressed area—Combustion Chamber (CC). In the limited volume of CC, it is necessary to organize not only the high-quality mixture but also sufficient film cooling. Another feature to complicate the performance is the pulsed operation mode of the engine, where the non-stationary effects of wall heating, ignition, and pulsations of fuel supply become decisive [1–4].

Taking into account the described difficulties, one can see how important it is for the developer to obtain complete information about the working process in the hot part of the engine in a relatively small amount of time and with the least computational costs. The widely used modern numerical methods in fluid dynamics partially solve this problem and help to understand what most affects the quality of the working process [4].

The purpose of the presented numerical study is to determine the magnitude and nature of the influence of the fuel injection parameters on engine performance.

Chapter is organized as follows. Section 12.2 provides a short review of features of the working process in rocket engines combustion chambers. The experimental test case chosen for this analysis is described in Sect. 12.3. Section 12.4 deals with the numerical setup and models applied in the hot gas simulation, whereas the results of the turbulent combustion of kerosene spray in gaseous oxygen with different injection condition are presented in Sect. 12.5. Finally, Sect. 12.6 gives an overall conclusion and summary of the results.

12.2 Features of the Working Process

The physical picture of the flow in CC is described in detail in [4]. This article describes the most important processes for the development studies. The fundamental role in the organization of the workflow assigns to the fuel injection through the injector head. The main task is the complete and uniform mixing of components across CC. The limited number of nozzle elements due to the small size of the engine complicates sufficient mixing and evaporation. These considerations require the development of nozzles for special structures, optimization of their location, and selection of flow parameters.

Almost overwhelming effect on the completeness of combustion has the atomization of the liquid fuel. Both the quality parameters, such as the type of atomization, the configuration of the devices, and the quantitative parameters, such as the distribution and size of the droplets, are of great importance.

Currently, the prevailing view says that there is a direct relationship between the decrease in the diameter of the average droplet and the increase in the completeness of combustion, which is explained by the rapid evaporation and burning of the droplets. However, the influence of size can be ambiguous. For example, in the case of a wide diameter distribution, the relatively high value of Sauter mean diameter may correspond to a higher degree of combustion than at relatively low values of the mean diameter. The main mechanism for increasing the completeness of combustion when increasing the value of Sauter mean diameter is the action of relatively large droplets that appeared in the spray spectrum. Large drops have more inertia compared to small drops. Large drops stay longer in the combustion chamber and, therefore, travel a longer distance. By evaporating, the drops make a trace of their vapors, which are mixed with the vapors of the second fuel component due to turbulence and diffusion. These phenomena increase the area of evaporation and combustion of fuel. With monodisperous (narrow) distribution of droplets by diameter, small droplets quickly evaporate near the injectors. Small drops make short tracks in the combustion chamber, which may not intersect with each other at all. This means that the fuel and oxidizer mix badly. The rapid evaporation of small droplets leads to a stratification of evaporated propellant and combustion products the combustion chamber volume, resulting in poor mixing of fuel and oxidizer, and, as a result, low the completeness of combustion. This means that for each combustion chamber design there is a distribution of droplets by diameter and a drop injection velocity that ensures maximum combustion performance [5].

Considerable effect on the completeness of combustion has the spray angle and fuel injection rate. If the spray angle is too small, the fuel and oxidizer do not mix properly, and the already reacted mixture stratifies into a separate stream and does not mix with the cold stream, i.e. regions with high or low component ratios appear which reduces the combustion performance. If the spray angle is too wide, medium and large droplets fall on both the mixing head and the wall, wherein the presence of the oxidizer reactions occur, which leads to the wall material burnout.

The only widely used type of wall cooling in modern small size rocket engines is film cooling. Either gaseous or liquid, reducing or oxidizing, its presence always leads to a change in the oxidizer to fuel ratio, which reduces the combustion performance. The use of a liquid film is the most difficult case to evaluate. First, at low feed rates, the laminar (or nearly laminar) film experiences the effects originating in the core flow as stretching and turbulent pulsations. This is reflected in the formation of Kelvin-Helmholtz waves on the surface of the laminar film. Ultimately, the growth of these instabilities can lead to the fragmentation of the film into droplets, thereby increasing the heat transfer from the main flow to the wall. In such a case, it is needed to prevent the early destruction of the film. Second, at higher flow rates a turbulent film, due to the internal phenomena of instability, coupled with the aerodynamic influence of the core flow, can break up into fine droplets faster than laminar film. The presence of a large number of uncertainties leads to the need for constant optimization of the film injection parameters.

This work is aimed to estimate the impact of fuel injection parameters on the efficiency of the working process expressed by the incompleteness of combustion. The phenomena described above, which are associated with the ambiguity of parameters, constitute the main difficulties in determining the optimal values of the variable parameters. The proposed methodology, as well as, the recommendations obtained in the modeling basis, is one of the steps to the improved understanding of the working process in CC of small-sized rocket engines.

12.3 The Studied Object

In the working process study, an experimental kerosene-GOx DMTMAI-200 with a nominal thrust of 200 N was selected as the test object (Fig. 12.1). The experimental study of the engine was carried out earlier in [6].

The injector head contains one central and six peripheral open-type centrifugal two-component spray nozzles, where the fuel and oxidizer nozzles are arranged so that the fuel flowing out of the nozzle enters the oxidant flow, mixes, and flows into the combustion chamber.

The wall in CC is protected from overheating by kerosene film cooling. The cooling kerosene is fed through slots between the injector head and CC wall. The liquid film forms a shroud that flows along the wall of the combustion chamber.

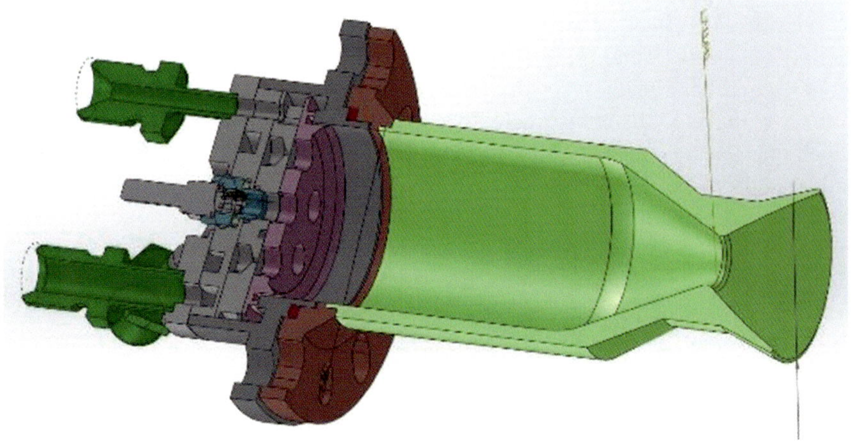

Fig. 12.1 View of the model of experimental rocket engine

12.4 Numerical Modeling of the Processes in Combustion Chamber

Here, the commercial Computational Fluid Dynamics (CFD) code Ansys CFX is used that supports a large number of mathematical models of physical processes. It should be noted that it is widely used to evaluate the working process in engines and power plants including rocket engines.

The numerical study was carried out using a mathematical model built on the basis of Navier-Stokes equations averaged according to Favre [7].

Assumptions of the numerical model. The numerical model is based on the following assumptions:

- The combustion products and fuel components are ideal gases. They have the constant viscosities and heat capacities depending on temperature.
- To simulate the liquid film cooling, Euler-Lagrange method is used, in which the liquid component is represented as a set of liquid drops injected from the film slots.
- The calculation is carried out in a stationary problem formulation.
- The model takes the buoyancy into account.
- The walls in CC, nozzle, and walls of injector head are adiabatic.
- Turbulence model based on Boussinesq hypothesis is used.
- Simulation liquid fuel injection into the flow core is performed using Euler-Lagrange approach with the initial parameters of size and distribution of the liquid spray.
- The radiative heat transfer is not taken into account.

Methods for modeling of the droplets motion, evaporation, break-up, and heat transfer. Modeling the movement of kerosene droplets is done using the classical approach. The approach for drag modeling (Schiller-Naumann (S-N) correlation) uses the enhance to dynamically model the changing shape of the drops (Liu correlation) [8], which is acceptable for the studied processes. As a model for heat transfer, Ranz-Marshall correlation [9, 10] is included. Antoine equation for mass transfer between the phases used reference factors for the *n*-decane $C_{10}H_{22}$ since, in the case of kerosene of JET-A type, it is assumed that under the observed conditions the largest evaporating fraction is the *n*-decane [11].

Combustion simulation. In this study, the flamelet combustion model was used [12]. The model is based on the assumption that the combustion process occurs in thin layers with an internal structure. The flame itself is a complex joint structure of one-dimensional stretched laminar flames called as flamelets. One of the assumptions of this model is the constant Lewis number during the flamelet formation. CFX-RIF library was used to create the flamelets. The reasons for using the flamelet model were its relative robustness, a small number of applied equations, and notable practice of application in CFD codes.

Secondary break-up models. The secondary break-up of liquid can strongly influence the atomization parameters. Currently, there are many approaches to the calculation of such processes. Based on a preliminary study of the models available in Ansys CFX and the recommendations in the open literature, it was decided to use one of the most sophisticated models for the secondary breakup in CFX, CAB model, to simulate the secondary fragmentation of liquid droplets [13–16].

General considerations. The main goal of the numerical study was to determine the influence of various parameters on the engine performance, which is defined as the pressure in CC and the characteristic velocity β defined by Eq. 12.1, where p_{cc} is the pressure in CC, F_{th} is the throat area, $\dot{m}_{\Sigma}$ is the total mass flow rate [2].

$$\beta = \frac{p_{CC} \cdot F_{th}}{\dot{m}_{\Sigma}} \tag{12.1}$$

Numerical domain and grid. The calculation domain is the 60° sector of the internal volume CC of the experimental engine. Gas-liquid centrifugal two-component kerosene-oxygen nozzles are represented by circles on the surface of the injector head. The diameter of the holes is 8 mm. The simulation domain and boundary conditions are shown in Fig. 12.2. The properties of the fuel components are shown in Table 12.1.

The numerical grid is shown in Fig. 12.3. A preliminary analysis of grid convergence based on pressure values in CC and characteristic velocity revealed that the grid convergence for these two parameters is achieved at the number of grid elements of about 300,000.

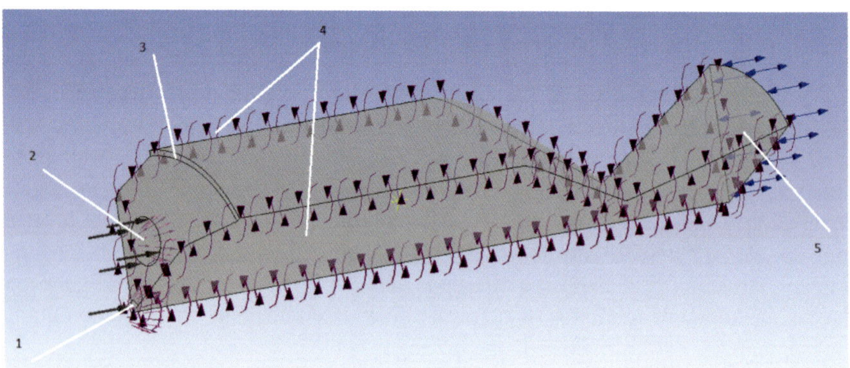

Fig. 12.2 Simulation domain with boundary conditions. 1—the mass flow of the central nozzle, 2—the mass flow of the peripheral nozzle, 3—the mass flow rate of the liquid film, 4—cyclic boundary conditions, 5—the boundary condition of the *opening* type with the setting of the ambient pressure

Table 12.1 Properties of the propellant components [11]

Property	Kerosene (liquid)	Kerosene (gaseous)	Oxygen
Density (kg/m^3)	727	Ideal gas	Ideal gas
Specific heat capacity (J/kg K)	2192.4	Dependence from the NIST database [17]	Dependency from the NIST database [17]
Dynamic viscosity (Pa s)	0.00212	5.28767e−06	2.06594e−05
Surface tension (N/m)	0.027	–	–
Thermal conductivity (W/m K)	0.1218	0.00907355	0.0254141

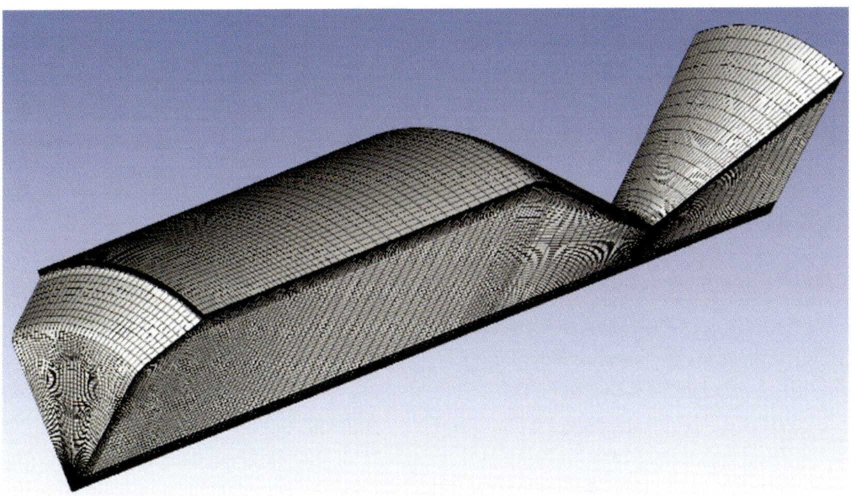

Fig. 12.3 Numerical grid

12.5 Study of the Fuel Injection

Study of the fuel injection parameters influence on the core flow. At the first stage of the study, only core-flow fuel injection simulations took place without accounting for film cooling.

The effect of the injection parameters was determined by varying the components of the velocity of the injected fuel droplets and flow of gaseous oxygen. The velocity components are given in a cylindrical frame with a reference point located in the center of the injector circle and represent the radial, axial, and tangential component of the velocity vector.

The ratio of the axial and radial component determines the angle of the atomization cone. In the case when the ratio of axial and radial components is much less than 1, this will correspond to a wide atomization angle [>45° (angle between the side and the

Table 12.2 Fuel injection parameters

Parameter	Value (kg/s)
Total kerosene mass flow	0.01827
Total oxygen mass flow	0.04972
Total mass flow (O + F)	0.068

axis)]. In the opposite situation, the flow will be stretched along the axis, which will affect mixing and efficiency. The tangential component reflects the degree of swirling, which due to high turbulent intensity promotes a mixing of the fuel components. In addition, the evaluation of the effect of fuel injection at various ratios of the velocity components is important in studying the effect of flow rate pulsations in the nozzle. Such non-stationary phenomena may occur due to some instabilities, both in the injector itself and in the feed system.

The ranges of fuel injection parameter variations are the following:

- The Rosin-Rammler size parameter [18] for the droplet diameter: 3–127 μm.
- The Rosin-Rammler spread parameter: 1.7–18.
- The droplet average Sauter diameter: 2.23–109 μm.
- The axial velocity component: 1.12–4.9 m/s.
- The radial velocity component: 1.12–4.3 m/s.
- The tangential velocity component: 1.12–4.3 m/s.

The fuel injection parameters are shown in Table 12.2.

The calculations were performed at a constant total mass flow rate of 0.068 kg/s with an oxidizer excess ratio of 0.8.

In total, 125 simulations with five variable parameters were performed. An important assumption used in the calculations was that the sprayed fuel droplets and oxygen move with the same injection velocity. This assumption is introduced to simplify the physical interpretation of the computational model and is based on a relatively high flow rate of up to 21 m/s and a small diameter of the droplets. Earlier in [19], it was shown that the diameter of droplets in nozzles of this type mainly depend on the parameters of the carrier phase.

The results of the calculations are shown in Figs. 12.4, 12.5, and 12.6. The plots show the increase of the characteristic velocity with the increase of the radial and tangential components of the velocity can be observed. Moreover, in the range of velocity components of 1.12–3 the characteristic velocity grows rapidly. The increase of the radial component with a simultaneous decrease of the axial component leads to the increase of the spraying angle. The increase of the tangential velocity projection with the decrease of the axial velocity depicts the flow swirling degree. The described effects have a significant impact on the processes occurring in the working volume. This indicates a correlation with the physical picture of the workflow described in previously published papers [20].

The dynamics of changes in the characteristic velocity with the increase of the axial component also looks predictable—a small decrease. This is due to the deterioration of mixing flows with increasing axial components.

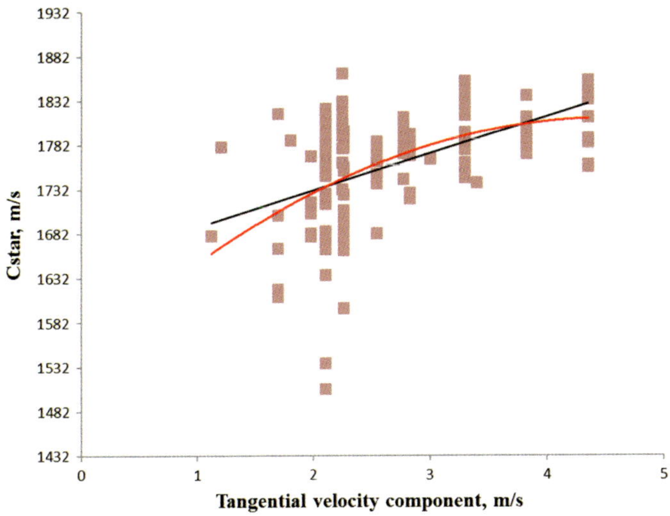

Fig. 12.4 Impact of the tangential velocity on the characteristic velocity (the lines are the lines of trend)

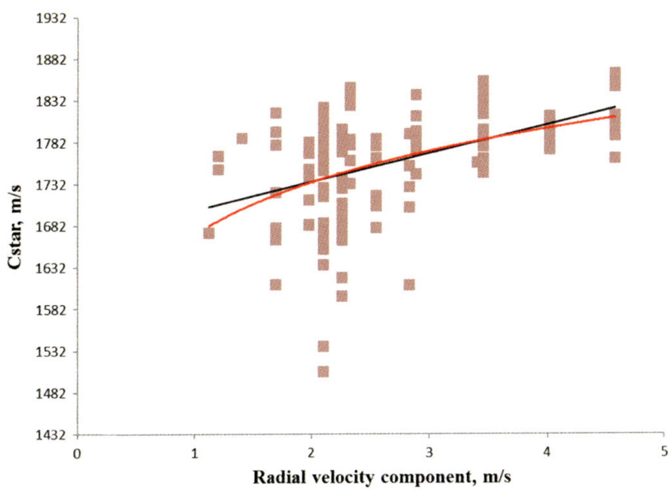

Fig. 12.5 Impact of the radial velocity on the characteristic velocity (the lines are the lines of trend)

The dependence of the characteristic velocity on the axial component is great, but in this case, with an increase of the axial velocity, the values of the characteristic velocity decrease. This is explained by the following: an increase of the axial velocity reduces the residence time of the droplets in CC, which leads to a decrease of the evaporated fuel fraction. On the other hand, the increase of the radial and tangential velocity components leads to better mixing of the fuel components due to

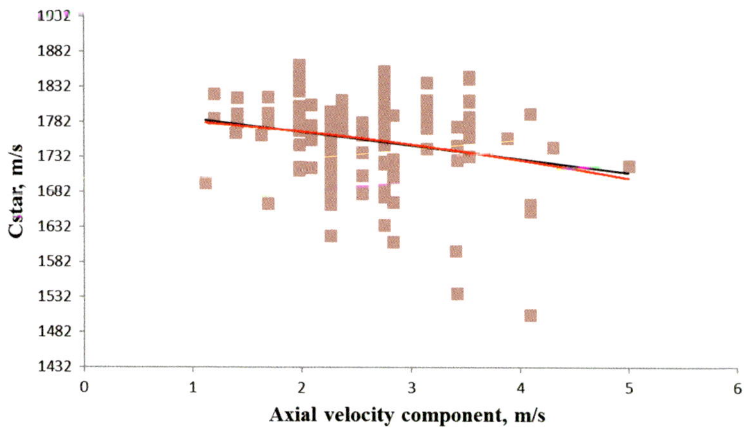

Fig. 12.6 Impact of the axial velocity component on the characteristic velocity (the lines are the lines of trend)

the interaction between adjacent nozzles and the increase of the residence time of the drops in CC due to the curvature and stretching of their traces. The variation of other parameters, i.e. Rosin-Rammler size parameter and the spread parameter have little effect.

The described phenomena will help in the formulation of recommendations for the design of the elements of mixture formation and the geometric and modal appearance of CC with regenerative cooling. For example, we can conclude that the growth of the tangential and radial components with the decrease in the axial component allows to increase greatly the combustion performance.

Study of the injection parameters influence with film cooling. The second part of the research was aimed to model the working process with film cooling by varying a large number of parameters. The liquid component of the fuel was fed through a gap in the peripheral region of the injector head.

To simulate the film cooling, a large number of parcels was injected through the film cooling slots. Thus, the approach is an assumption made to consider computational resources reduction. Moreover, such an approach still takes into account the dynamic, thermal, and mass transfer phenomena. This greatly simplifies the calculation and improves the robustness and time. The choice of this method was made both because of the computational requirements and complexity of other methods of two-phase flow modeling. The disadvantage of this approach is the uncertainty in the diameter of the droplets: the too small drops will evaporate quickly and the too large ones may not evaporate at all. Both in the first and second cases, the working process would be incorrectly interpreted.

The chapter discusses two parameters of fuel component injection into the film: the diameter of kerosene droplets injected as a film cooler and their velocity. Fuel injection parameters are shown in Table 12.3.

The ranges of fuel injection parameter variations:

Table 12.3 Fuel injection parameters

Parameter	Value (kg/s)
Total massflow of kerosene (per core flow)	0.01184
Total oxygen massflow (to the core flow)	0.03221
Kerosene massflow for the film cooling	0.02394
Total massflow (O + F)	0.068

- The Rosin-Rammler size: 12–109 μm.
- The Rosin-Rammler spread: 1–18.
- The axial velocity component: 0.9–3.9.
- The radial velocity component: 0.3–1.7.
- The tangential velocity component: 0.9–3.47.
- The diameter of the liquid droplets injected as the film: 53–250 μm.
- The velocity of the liquid droplets injected as the film: 0.3–2.19 m/s.

As an experimental design plan, a central composite plan extended by additional points was used. The total number of points is 160. The size of the slot used for film cooling component was 1 mm. The results of the study are shown in Figs. 12.7, 12.8, 12.9, 12.10 and 12.11.

According to the simulation results, there are significant differences in the dependence of the completeness of combustion on the magnitude of the velocity components of fuel injection compared to the results obtained in the simulations without film cooling. Increasing the axial velocity component decreases the characteristic velocity to a small extent. The growth of the tangential component also relatively weakly increases the value of the characteristic velocity. The increase in the radial component leads to a significant reduction in combustion performance.

In this case, the following phenomenon occurs. At some values of the radial component, the oxidizer stream and liquid spray begin to fall directly into the film

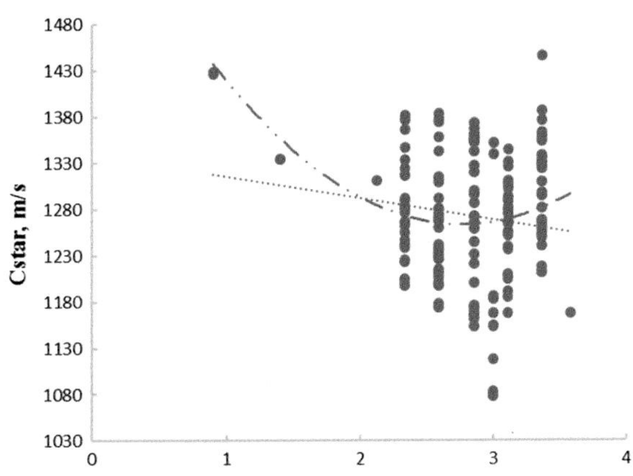

Fig. 12.7 Impact of the axial velocity component on the characteristic velocity (the lines are the lines of trend)

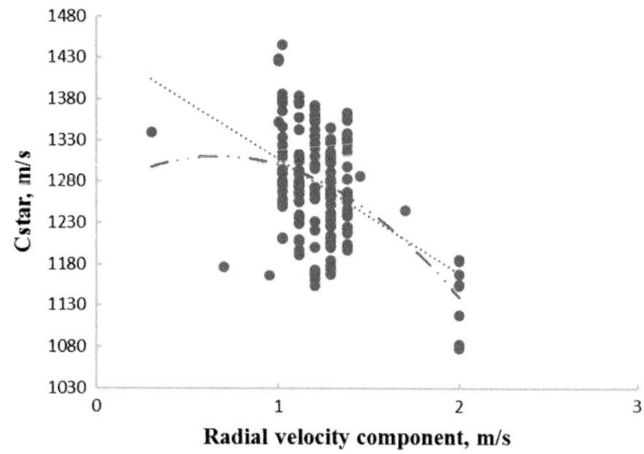

Fig. 12.8 Impact of the radial component on the characteristic velocity (the lines are the lines of trend)

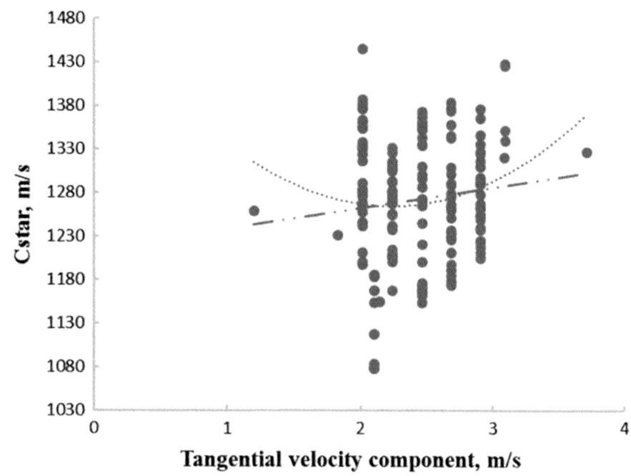

Fig. 12.9 Impact of the tangential component on the characteristic velocity (the lines are the lines of trend)

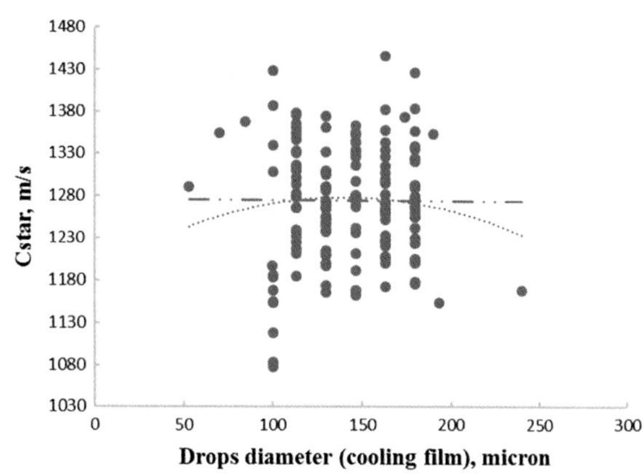

Fig. 12.10 Impact of the diameter of the film droplets on the characteristic velocity (the lines are the lines of trend)

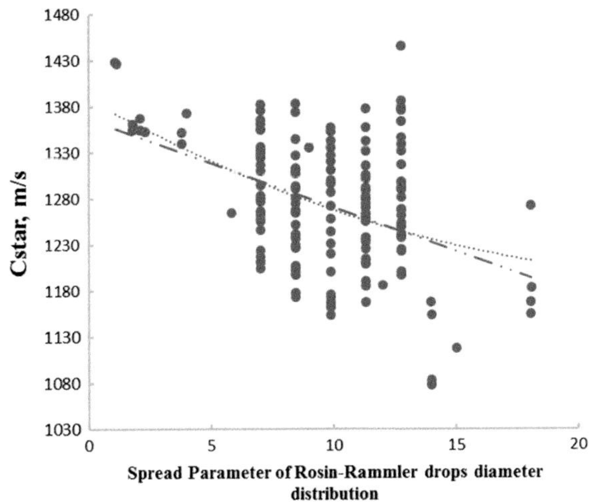

Fig. 12.11 Impact of the spread parameter of Rosin-Rammler distribution in the core flow on the characteristic velocity (the lines are the lines of trend)

area preventing the formation of a homogeneous mixture of fuel and oxidant in the core flow, which reduces the efficiency of mixing, evaporation, and combustion.

In the given variation range, the size parameter of Rosin-Rammler diameter distribution is insignificant. An increase of the spread parameter reduces the characteristic velocity, which is consistent with the results of [5] and can be explained by the conclusions made in this work. An increase in the speed and diameter of the film drops does not almost affect on the efficiency of the workflow in CC.

The results of the simulation of the workflow in CC with film cooling show that the effect of the velocity and diameter of the drops of the film cooling (in these ranges of variation) is not significant at the given flow rate in the film.

For the cases taking the film cooling into account, the characteristic velocity reaches the highest values at low values of the spread parameter, which corresponds to wide distribution. For cases without film cooling, the width of the distribution has little effect on the efficiency of the workflow. This is most likely that the wide distribution in the spectrum provides large and small droplets in the same proportion, which allows to increase the evaporation/combustion areas and the average temperature over the volume of CC. This effect shows itself more significantly in the presence of film cooling, which creates thermal stratification over the cross-section of the combustion chamber and significantly reduces the efficiency of the working process.

To summarize, the numerical study illustrated the effect of various parameters and indicated that mostly the core flow injection has a significant effect. However, the coupled interaction of the effects of the parameters can strengthen or weaken the resulting influence of a single parameter.

12.6 Conclusions

Optimization of the fuel injection parameters can significantly improve engine efficiency and increase combustion performance. Obviously, however, for each design type of CC and the injector head, a certain most efficient spray droplet distribution and combination of velocity components are presented.

An increase of the axial velocity component leads to a decrease in the characteristic velocity. An increase in the radial component, with respect to the axial component, in the case of the non-film cooling leads to the increase of the completeness of combustion, and in the case of the film, to the decrease. The increase of the tangential component allows to achieve a more complete mixing and efficiency of the workflow regardless of the presence of the film.

An increase of the droplet diameter of the film and velocity of the liquid film has little effect on the performance in the studied range.

An increase of the spread parameter reduces the characteristic velocity, which is consistent with the results previously obtained by the authors. The large spread parameter of the droplets diameter distribution means monodispersity of the droplets. With the combustion chamber film cooling, there is a decrease in the performance of combustion process, while increasing the monodispersity of the droplets, as well as, without it. That is, poor mixing of the evaporated propellant leads to a decrease combustion performance to an even greater extent, since film cooling also reduces the combustion performance by itself.

The results of the velocity components impact can describe, among other phenomena, the non-stationary operation, when the instability of the injection parameters (mass flow distribution in the injectors, velocity pulsations, and spray parameters) can significantly affect the efficiency of the engine.

The results obtained may stand as recommendations for the determination of the requirements for mixing elements, the shape of CC, the mode of operation, etc. This is closely related to the goal of creating a workflow modeling methodology, along with obtaining complete information on the qualitative and quantitative parameters of the process. Designers of rocket engines can use the results of this work for considerable simplification of the design process.

Acknowledgements This work was supported by the Russian Ministry of Education and Science (Project 13.7418.2017/8.9).

References

1. Kozlov, A.A., Vorobiev, A.G., Borovik, I.N.: Low-Thrust Liquid-Propellant Rocket Engines. MAI Publishing House, Moscow (in Russian) (2013)
2. Alemasov, V.E., Dregalin, A.F., Tishin, A.P.: The Theory of Rocket Engines. Mashinostroenie, Moscow (in Russian) (1980)

3. Vasiliev, A.P., Kudryavtsev, V.M., Kuznetsov, V.A., Kurpatenkov, V.D., Obelnitskii, A.M.: Fundamentals of the Theory and Calculation of Liquid Rocket Engines. Higher School, Moscow (in Russian) (1993)
4. Koroteev, A.S. (ed.): Work Processes in a Liquid Rocket Engine and Their Modeling. Mashinostroenie, Moscow (in Russian) (2008)
5. Borovik, I.N., Strokach, E.A.: Spray diameter distribution influence on liquid rocket combustion chamber performance, Bulletin PNRPU. Aerosp. Eng. **44**, 45–61 (in Russian) (2016)
6. Tashev, V.P.: Hydrocarbon fuel based on kerosene with additives to increase the energy efficiency of LRE: Ph.D. theses (in Russian) (2013)
7. Yun, A.A., Krylov, B.A.: Calculation and Modeling of Turbulent Flows with Heat Exchange, Mixing, Chemical Reactions and Two-Phase Flows in the Fastest-3D Software Package: Study Guide. MAI Publishing House, Moscow (in Russian) (2007)
8. Liu, B., Mather, D., Reitz, R.D.: Modeling the effects of drop drag and breakup on fuel sprays. SAE Technical Paper 930072 (1993)
9. Ranz, W.E., Marshall, W.R.: Evaporation from drops. Part I. Chem. Eng. Prog. **48**(3), 141–146 (1952)
10. Ranz, W.E., Marshall, W.R.: Evaporation from drops. Part I and Part II. Chem. Eng. Prog. **48**(4), 173–180 (1952)
11. Reid, R.C., Prausnitz, J.M., Sherwood, T.K.: The Properties of Gases and Liquids. McGraw-Hill, New York (1977)
12. Peters, N.: Laminar diffusion flamelet models in non-premixed turbulent combustion. Prog. Energy Combust. Sci. **10**, 319–339 (1984)
13. Spalding, D.B.: Mixing and chemical reaction in steady confined turbulent flames. In: Symposium (Int.) on Combustion, vol. 13, no. 1, pp. 649–657 (1971)
14. Frank, Th., Kumzerova, E. Esch, Th.: Validation of Lagrangian spray formation for use in internal combustion engines. In: 5th Joint FZR & ANSYS Workshop “Multiphase Flows: Simulation, Experiment and Application”, Dresden, Germany, pp. 1–27 (2007)
15. Kumzerova, E., Esch, Th., Menter, F.: Spray simulations: application of various droplet breakup models. In: 6th International Conference on Multiphase Flow, Leipzig, pp. 1–10 (2007)
16. Kumzerova, E., Esch, T.: Extension of the cab for the wide range of Weber number range. In: 22nd European Conference Liquid Atomization & Spray Systems, pp. ILASS08-4-5.1–ILASS08-4-5.7 (2008)
17. NIST (National Institute of Standards and Technology, USA) Chemistry Webbook. http://webbook.nist.gov. Accessed 08 Sept 2019
18. Rosin, P., Rammler, E.: The laws governing the fineness of powdered coal. J. Inst. Fuel **7**, 29–36 (1933)
19. Vasil’ev, A.Y., Maiorova, A.I.: Physical features of liquid atomization when using different methods of spraying. High Temp. **52**(2), 250–258 (2014)
20. Kozlov, A.A., Strokach, E.A.: Investigation of a workflow simulation method in the combustion chamber of liquid rocket motors of small thrust based on the Eddy Dissipation Model, Bulletin PNRPU. Aerosp. Eng. **44**, 27–44 (in Russian) (2016)

Part III
Computational Solid Mechanics

Chapter 13
Methods for Calculating the Dynamics of Layered and Block Media with Nonlinear Contact Conditions

Ilia S. Nikitin, Nikolay G. Burago, Vasily I. Golubev and Alexander D. Nikitin

Abstract Continual models of solid media with a discrete set of slip planes (layered, block media) and with nonlinear type slip conditions at the contact boundaries of structural elements are constructed. The constitutive equations of the resulting system of equations contain a small viscosity parameter in the denominator of nonlinear free terms. For a stable numerical solution of a system of differential equations, an explicit–implicit method is proposed with the explicit approximation of the equations of motion and implicit approximation of the constitutive relations containing a small parameter. From implicit nonlinear difference approximations analytically using the perturbation method, various effective formulas for the correction of stress components after an "elastic" time step are obtained. To calculate the "elastic" step, we used a grid-characteristic method on hexahedral grids, which allowed us to increase significantly the speed of calculations and simulate a non-stationary three-dimensional problem of generating a response from an oriented layered or block cracked cluster located in a homogeneous medium.

I. S. Nikitin (✉) · A. D. Nikitin
Institute of Computer Aided Design of the RAS, 19/18, Vtoraya Brestskaya ul., Moscow 123056, Russian Federation
e-mail: i_nikitin@list.ru

A. D. Nikitin
e-mail: nikitin_alex@bk.ru

N. G. Burago
Ishlinsky Institute for Problems in Mechanics of the RAS, 101, b1, pr. Vernadskogo, Moscow 119526, Russian Federation
e-mail: burago@ipmnet.ru

V. I. Golubev
Moscow Institute of Physics and Technology (National Research University), 9, Institutsriy per., Dolgoprudny, Moscow Region 141700, Russian Federation
e-mail: w.golubev@mail.ru

L. C. Jain et al. (eds.), *Advances in Theory and Practice of Computational Mechanics*, Smart Innovation, Systems and Technologies 173,
https://doi.org/10.1007/978-981-15-2600-8_13

13.1 Introduction

Continual models of deformable solid media with a discrete set of slip planes (layered, block media) and with nonlinear (dry friction or viscous-plastic types) slip conditions at contact boundaries can be obtained with the discrete variant of the slippage theory [1] or with the method of the asymptotic homogenization [2–4]. In all of these cases, the constitutive system of equations with the nonlinear free term and the small stress relaxation time are included. For the stable numerical solution of differential equations, the explicit–implicit method with the explicit approximation of motion equations and the implicit approximation of constitutive equations containing small parameter in the denominator of the free term was proposed. A set of effective formulas for the stress tensor correcting at the "elastic" step was obtained.

The developed model can be used for the numerical simulation of the seismic survey process in complex geological fractured media. Nowadays, a lot of different approaches are created: the finite difference method, Galerkin method, finite element method, etc. [5–7]. There are also various numerical methods actively used to construct hybrid calculation algorithms [8]. Also note that, as the defining system of equations of elasticity that describes the propagation of seismic waves is hyperbolic, its numerical solution can be carried out by the grid-characteristic method. Apparently, the characteristic method was proposed firstly in [9]. It was described in detail for a one-dimensional case in [10] and later generalized for a multidimensional case in [11]. Recently, it was successfully applied for the simulation of waves in the acoustic [12], elastic [13], and fractured [14–17] media. It was also used to solve the inverse geophysical problems, like migration algorithms [18, 19]. Recently, some steps were done to construct the compact schemes with narrow spatial stencils [20]. In this work, it was used to obtain a numerical solution of elastic part of the problem considered. Numerical simulations of the dynamic scattering process for subsurface layered and block objects in elastic 2D and 3D media were carried out using modern high-performance computing systems.

The chapter is organized as follows: Section 13.2 discusses a mathematical model. The layered model system of equations and block model system of equations are presented in Sects. 13.3–13.4, respectively. Numerical method is considered in Sect. 13.5. Section 13.6 provides the simulation results. Section 13.7 concludes the chapter.

13.2 Mathematical Model

Nonlinear interaction conditions between contacting structural elements may be formulated. In Cartesian system x_i $(i = 1, 2, 3)$, the unbounded elastic medium with the oriented system of periodic parallel slip planes is considered. The orientation is set with the unit vector of normal $\mathbf{n}$. The distance between slip planes is constant and equal to ε. The density of the material ρ and Lame moduli λ and μ are known constants. The stress state of the medium is described with the stress tensor $\boldsymbol{\sigma}$. The shear stress vector at the slip plane is equal to $\boldsymbol{\tau} = \boldsymbol{\sigma} \cdot \mathbf{n} - (\mathbf{n} \cdot \boldsymbol{\sigma} \cdot \mathbf{n})\mathbf{n}$ and the normal stress is equal to $\sigma_n = \mathbf{n} \cdot \boldsymbol{\sigma} \cdot \mathbf{n}$. We define the slip velocity vector $\boldsymbol{\gamma}$ and the delamination velocity vector $\boldsymbol{\omega} = \omega\mathbf{n}$ based on the discontinuity of tangential $[\mathbf{V}_\tau]$ and normal $[V_n]$ velocity at contact boundaries: $\boldsymbol{\gamma} = [\mathbf{V}_\tau]/\varepsilon$, $\omega = [V_n]/\varepsilon$.

We assume the presence of thick interlayers between elastic layers $d \ll \varepsilon$; however, we will not take it into account explicitly, but using slip conditions at compressed layers' boundaries. At the contact boundaries the special conditions are specified.

(1) Piecewise linear condition of slip and weak delamination.

When $\sigma_n < 0$ (compressed contact),

$$\boldsymbol{\gamma} = \dot{\boldsymbol{\tau}}/k_\gamma, \ \omega = 0. \tag{13.1}$$

When $\sigma_n \geq 0$ (weak delamination),

$$\boldsymbol{\gamma} = \dot{\boldsymbol{\tau}}/k_\gamma, \ \omega = \dot{\sigma}_n/k_\omega, \ k_\gamma/\mu \ll 1, \ k_\omega/\mu \ll 1. \tag{13.2}$$

(2) The slip condition for Coulomb friction with a low viscous additive and weak delamination.

When $\sigma_n < 0$ (compressed contact), $\boldsymbol{\tau} = q|\sigma_n|(\boldsymbol{\gamma}/|\boldsymbol{\gamma}| + \eta\boldsymbol{\gamma})$ or expressing $\boldsymbol{\gamma}$ through tensions $\boldsymbol{\tau}$ provided by Eq. 13.3.

$$\boldsymbol{\gamma} = \frac{1}{\eta}\frac{\boldsymbol{\tau}}{|\boldsymbol{\tau}|}\langle\frac{|\boldsymbol{\tau}|}{q|\sigma_n|} - 1\rangle \ , \omega = 0 \tag{13.3}$$

When $\sigma_n \geq 0$ (weak delamination),

$$\boldsymbol{\gamma} = \dot{\boldsymbol{\tau}}/k_\gamma, \ \omega = \dot{\sigma}_n/k_\omega, \ k_\gamma/\mu \ll 1, \ k_\omega/\mu \ll 1. \tag{13.4}$$

Here, k_γи k_ω are the coefficient of weak elastic tangential and normal bond of layers, η is the viscosity coefficient, q is the Coulomb friction coefficient, $\langle F(y)\rangle = F(y)H(y)$, $H(y)$ is the Heaviside function, $H(y) = 0$ if $y < 0$, $H(y) = 1$ if $y \geq 0$. The contact plane with the interaction condition defined is called the slip-delamination plane.

It should be noticed that previously in [2] for the delamination regime the "full delamination condition" was used $\Omega \geq 0$: $\boldsymbol{\tau} = \sigma_n = 0$ that is the asymptotic case of the "weak delamination" when $k_\omega \to 0$, $k_\gamma \to 0$, $\Omega = [u_n]/\varepsilon$ is the normalized discontinuity of normal displacements at the contact boundary defined by the equation $\dot{\Omega} = \omega$. Also, for the case of the weak delamination $\omega = \dot{\sigma}_n/k_\omega$, inequalities $\sigma_n \geq 0$ and $\Omega \geq 0$ are equivalent.

From the numerical point of view, small parameters k_γ и k_ω are regularizations that allow us to prevent the oscillations occurrence, when sharply changing the compress boundary to the full delamination condition.

To construct the continuum model with a set of these slip-delamination planes, we are going to deal with $\boldsymbol{\gamma}$ and $\boldsymbol{\omega}$ as discontinuous functions of space and time. Also, we will use main relationships from the slippage theory as many other authors. It allows us to take into account contributions from $\boldsymbol{\gamma}$ and $\boldsymbol{\omega}$ into nonelastic strain tensors $\mathbf{e}^{\gamma}$ and $\mathbf{e}^{\omega}$, respectively:

$$\mathbf{e}^{\gamma} = (\mathbf{n} \otimes \boldsymbol{\gamma} + \boldsymbol{\gamma} \otimes \mathbf{n})/2, \boldsymbol{\gamma} \cdot \mathbf{n} = 0, \tag{13.5}$$

$$\mathbf{e}^{\omega} = (\mathbf{n} \otimes \boldsymbol{\omega} + \boldsymbol{\omega} \otimes \mathbf{n})/2 = \omega \mathbf{n} \otimes \mathbf{n}, \boldsymbol{\omega} = \omega \mathbf{n}. \tag{13.6}$$

The full strain tensor $\mathbf{e}$ equals to the sum of elastic and nonelastic parts provided by Eq. 13.7.

$$\mathbf{e} = \mathbf{e}^{e} + \mathbf{e}^{\gamma} + \mathbf{e}^{\omega}, \mathbf{e} = (\nabla \mathbf{v} + \nabla \mathbf{v}^{T})/2 \tag{13.7}$$

Here, $\mathbf{v}$ is « macroscopic » velocity of medium particles, $\mathbf{e}^{e}$ is the elastic strain tensor accordingly to Hooke's law:

$$\dot{\boldsymbol{\sigma}} = \lambda(\mathbf{e}^{e} : \mathbf{I})\mathbf{I} + 2\mu \mathbf{e}^{e}. \tag{13.8}$$

The final equation is the motion equation in the view of Eq. 13.9.

$$\rho \dot{\mathbf{v}} = \nabla \cdot \boldsymbol{\sigma} \tag{13.9}$$

13.3 Layered Model System of Equations

In layered medium containing a set of elastic layers, the only system of slip-delamination planes are possible with the normal $\mathbf{n}$. If the normal to contact boundaries $\mathbf{n}$ is oriented along the x_2, then its components are $n_j = \delta_j^2$.

Through conditions for $\boldsymbol{\gamma}$ and ω corresponding to the local contact conditions (Eq. 13.1), we can write Eq. 13.10.

$$\gamma_j = \dot{\sigma}_{2j}/k_{\gamma} \ , \omega = \dot{\sigma}_{22} H(\sigma_{22})/k_{\omega} \tag{13.10}$$

Through conditions for $\boldsymbol{\gamma}$ and ω corresponding to the local contact conditions (Eq. 13.2), we can write Eqs. 13.11–13.12.

$$\gamma_j = \frac{1}{\eta} \frac{\sigma_{2j}}{|\boldsymbol{\tau}|} \langle \frac{|\boldsymbol{\tau}|}{q|\sigma_{22}|} - 1 \rangle (1 - H(\sigma_{22})) + \dot{\sigma}_{2j} H(\sigma_{22})/k_{\gamma} \tag{13.11}$$

$$\omega = \dot{\sigma}_{22} H(\sigma_{22})/k_\omega \ , |\boldsymbol{\tau}| = \sqrt{\sum_{k \neq 2} \sigma_{2k}\sigma_{2k}} \tag{13.12}$$

Based on the chosen normal direction, the final system of equations for this model is represented by Eqs. 13.13–13.15.

$$\rho \dot{v}_i = \sigma_{ij,j} \ , \dot{\sigma}_{ii} \underset{i \neq 2}{=} \lambda v_{k,k} + 2\mu v_{i,i} - \lambda\omega \tag{13.13}$$

$$\dot{\sigma}_{22} = \lambda v_{k,k} + 2\mu v_{2,2} - (\lambda + 2\mu)\omega, \ \dot{\sigma}_{ij} \underset{i,j \neq 2}{=} \mu(v_{i,j} + v_{j,i}) \tag{13.14}$$

$$\dot{\sigma}_{2j} \underset{j \neq 2}{=} \mu(v_{2,j} + v_{j,2}) - \mu\gamma_j \ , i \neq j \tag{13.15}$$

13.4 Block Model System of Equations

The block medium consists of parallelepiped elastic elements with three possible slip-delamination planes. These planes are defined with normals $\mathbf{n}^{(s)}$, $s = 1,2,3$. In this case, the nonelastic strain tensors are calculated by Eqs. 13.16–13.17.

$$\mathbf{e}^{\gamma} = \sum_{s=1}^{3} (\mathbf{n}^{(s)} \otimes \boldsymbol{\gamma}^{(s)} + \boldsymbol{\gamma}^{(s)} \otimes \mathbf{n}^{(s)})/2 \ , \boldsymbol{\gamma}^{(s)} \cdot \mathbf{n}^{(s)} = 0 \tag{13.16}$$

$$\mathbf{e}^{\omega} = (\mathbf{n}^{(s)} \otimes \boldsymbol{\omega}^{(s)} + \boldsymbol{\omega}^{(s)} \otimes \mathbf{n}^{(s)})/2 = \omega^{(s)}\mathbf{n}^{(s)} \otimes \mathbf{n}^{(s)} \ , \boldsymbol{\omega}^{(s)} = \omega^{(s)}\mathbf{n}^{(s)} \tag{13.17}$$

If three normals to slip-detachment planes are oriented along the coordinate axis of Cartesian system, then $n_j^{(s)} = \delta_j^s$, where δ_j^s is the Kronecker's symbol.

Through conditions for $\boldsymbol{\gamma}^{(i)}$ and $\omega^{(i)}$ corresponding to the local contact conditions (Eq. 13.1), formulae included in Eq. 13.18 are given as follows:

$$\gamma_j^{(i)} = \dot{\sigma}_{ij}/k_\gamma, i \neq j, \omega^{(i)} = \dot{\sigma}_{ii} H(\sigma_{ii})/k_\omega. \tag{13.18}$$

Through conditions for $\boldsymbol{\gamma}^{(i)}$ and $\omega^{(i)}$ corresponding to the local contact conditions (Eq. 13.2), formulae included in Eqs. 13.19–13.20 are given as follows:

$$\gamma_j^{(i)} = \frac{1}{\eta} \frac{\sigma_{ij}}{\left|\boldsymbol{\tau}^{(i)}\right|} \langle \frac{\left|\boldsymbol{\tau}^{(i)}\right|}{q|\sigma_{ii}|} - 1\rangle (1 - H(\sigma_{ii})) + \dot{\tau}_{ij} H(\sigma_{ii})/k_\gamma \ , i \neq j, \tag{13.19}$$

$$\omega^{(i)} = \dot{\sigma}_{ii} H(\sigma_{ii})/k_\omega, \left|\boldsymbol{\tau}^{(i)}\right| = \sqrt{\sum_{k \neq i} \sigma_{ik}\sigma_{ik}}. \tag{13.20}$$

As for the layered medium, we can rewrite the main system in the suitable form of Eqs. 13.21–13.22.

$$\rho \dot{v}_i = \sigma_{ij,j}\ ,\dot{\sigma}_{jj} = \lambda v_{k,k} + 2\mu v_{j,j} - \lambda \sum_{l \neq j} \omega^{(l)} - (\lambda + 2\mu)\omega^{(j)} \tag{13.21}$$

$$\dot{\sigma}_{ij} = \mu(v_{i,j} + v_{j,i}) - \mu\gamma_j^{(i)} - \mu\gamma_i^{(j)}\ , i \neq j \tag{13.22}$$

13.5 Numerical Method

Both formulated systems are semi-linear hyperbolic systems, and the numerical solution can be obtained with different explicit schemes. However, the slippage process switches on the nonlinear free term with small viscosity parameter in the denominator. The system transforms into the form with small parameter and ordinary explicit schemes will not be stable. To overcome this problem, the use of explicit–implicit method is proposed. The implicit approximation is used only for equations that contain small term in the denominator. All other equations are approximated with the explicit scheme.

Let us describe this approach for $\dot{\sigma}_{2j}$ for compressed contact case, $\sigma_{22} < 0$ for the layered medium using Eq. 13.23.

$$\dot{\sigma}_{2j} \underset{j \neq 2}{=} \mu(v_{2,j} + v_{j,2}) - \mu\sigma_{2j}\langle|\boldsymbol{\tau}| / (q|\sigma_{22}|) - 1\rangle / (\eta|\boldsymbol{\tau}|) \tag{13.23}$$

Implicit approximation with first-order time approximation has a view of Eqs. 13.24–13.25.

$$(\sigma_{2j}^{n+1} - \sigma_{2j}^{n}) \Big/ \Delta t = \mu(v_{2,j}^{n+1} + v_{j,2}^{n+1}) - \mu\sigma_{2j}^{n+1}\langle\Sigma_e^{n+1} \big/ (q|\sigma_{22}^{n+1}|) - 1\rangle \Big/ (\eta\Sigma_e^{n+1}) \tag{13.24}$$

$$\Sigma_e^{n+1} = \sqrt{\left(\sigma_{12e}^{n+1}\right)^2 + \left(\sigma_{32e}^{n+1}\right)^2} \tag{13.25}$$

Here, indices $n+1$ and n mean the current and next time layers, respectively, Δt is the time step. We assume that values v_i^{n+1} and σ_{jj}^{n+1} were calculated with the explicit schemes for elastic equations. Solving this equation and also the same equation for normal tensions we can obtain correcting formulas.

When $\sigma_{22e}^{n+1} < 0$ (compressed contact),

$$\sigma_{22}^{n+1} = \sigma_{22e}^{n+1},\ \sigma_{ii}^{n+1} = \sigma_{iie}^{n+1},\ \omega = 0, i = 1, 3. \tag{13.26}$$

When $\Sigma_e^{n+1} \geq q\left|\sigma_{22e}^{n+1}\right|$,

$$\sigma_{i2}^{n+1} = q\left|\sigma_{22e}^{n+1}\right|\left(\sigma_{i2e}^{n+1} / \Sigma_e^{n+1}\right)\left(1+\delta\Sigma_e^{n+1}\right) / \left(1+\delta q\left|\sigma_{22e}^{n+1}\right|\right), \tag{13.27}$$

$$\gamma_i = (\sigma_{i2e}^{n+1} - \sigma_{i2}^{n+1})/(\mu\Delta t), i = 1, 3. \tag{13.28}$$

When $\Sigma_e^{n+1} < q\left|\sigma_{22e}^{n+1}\right|$,

$$\sigma_{i2}^{n+1} = \sigma_{i2e}^{n+1},\ \gamma_i^{n+1} = 0,\ i = 1, 3, \delta = \eta/(\mu\Delta t). \tag{13.29}$$

When $\sigma_{22e}^{n+1} \geq 0$ (delamination process),

$$\sigma_{22}^{n+1} = \frac{1}{(1+\beta)}(\beta\sigma_{22e}^{n+1} + \sigma_{22}^{n}),\ \sigma_{ii}^{n+1} = \sigma_{iie}^{n+1} - \frac{1}{(1+\beta)}\frac{\lambda(\sigma_{22e}^{n+1} - \sigma_{22}^{n})}{(\lambda+2\mu)},$$

$$\omega = \frac{1}{(1+\beta)}\frac{(\sigma_{22e}^{n+1} - \sigma_{22}^{n})}{(\lambda+2\mu)\Delta t}, i = 1, 3, \tag{13.30}$$

$$\sigma_{i2}^{n+1} = \frac{1}{(1+\alpha)}(\alpha\sigma_{i2e}^{n+1} + \sigma_{i2}^{n}),\ \gamma_i = \frac{1}{(1+\alpha)}\frac{(\sigma_{i2e}^{n+1} - \sigma_{i2}^{n})}{\mu\Delta t}, i = 1, 3. \tag{13.31}$$

Coefficients for weak shear $\alpha = k_\gamma/\mu$ and stretching $\beta = k_\omega/(\lambda + 2\mu)$, where $\alpha, \beta << 1$, $\sigma_{ij\,e}^{n+1} = \sigma_{ij}^{n} + \left(\lambda v_{k,k}^{n+1}\delta_{ij} + \mu(v_{i,j}^{n+1} + v_{j,i}^{n+1})\right)\Delta t$ is the stress value after the elastic step. This formula, in fact, is the adjustment of "elastic" stresses for the "friction cone" with viscous corrections. We used here the simplest formula for $\sigma_{2j\,e}^{n+1}$ only to illustrate the derivation process.

Let us describe precisely the elastic step of the calculation algorithm. The linear dynamic elasticity equations are given by Eq. 13.32.

$$\rho\dot{v}_i = \sigma_{ij,j}\ , \dot{\sigma}_{ij} = \lambda v_{k,k}\delta_{ij} + \mu(v_{i,j} + v_{j,i}) \tag{13.32}$$

Here, λ and μ are the Lamé constants and δ_{ij} is the Kronecker delta.

The first line in the system presents three equations of motion, while the second line presents six rheological relations. The vector of variables consists of nine components and has the form of Eq. 13.33.

$$\mathbf{u} = (v_1, v_2, v_3, \sigma_{11}, \sigma_{12}, \sigma_{13}, \sigma_{22}, \sigma_{23}, \sigma_{33})^T \tag{13.33}$$

Note that the solid mechanic system can be written in matrix form as Eq. 13.34, where $\mathbf{A_j}$ are 9×9 matrices and (x_1, x_2, x_3) is the orthonormal system of coordinates.

$$\frac{\partial\mathbf{u}}{\partial t} = \sum_{j=1}^{3}\mathbf{A}_j\frac{\partial\mathbf{u}}{\partial x_j} \tag{13.34}$$

This system is solved using the grid-characteristic method on parallelepiped meshes. It splits up into three one-dimensional systems of equations:

$$\frac{\partial \mathbf{u}}{\partial t} = \mathbf{A_j}\frac{\partial \mathbf{u}}{\partial x_j},\ j = 1, 2, 3. \tag{13.35}$$

Each system is a hyperbolic and possesses a complete set of eigenvectors with real eigenvalues. Each of the systems can be rewritten by Eq. 13.36, where the matrix $\mathbf{\Omega_j}$ is composed of the eigenvectors and $\mathbf{\Lambda_j}$ is a diagonal matrix.

$$\frac{\partial \mathbf{u}}{\partial t} = \mathbf{\Omega}_j^{-1}\mathbf{\Lambda}_j\mathbf{\Omega}_j\frac{\partial \mathbf{u}}{\partial \xi_j} \tag{13.36}$$

At the splitting step in the fixed direction, the matrix $\mathbf{\Lambda_j}$ is given by

$$\mathbf{\Lambda}_j = diag(c_1, -c_1, c_2, -c_2, c_2, -c_2, 0, 0, 0), \tag{13.37}$$

where

$$c_1 = \sqrt{(\lambda + 2\mu)/\rho},\ c_2 = \sqrt{\mu/\rho}. \tag{13.38}$$

After changing to the variables $\mathbf{v} = \mathbf{\Omega}\,\mathbf{u}$, each of the systems splits into nine independent scalar advection equations (in what follows, the index $\mathbf{j}$ is omitted wherever possible):

$$\frac{\partial \mathbf{v}}{\partial t} + \mathbf{\Lambda}\frac{\partial \mathbf{v}}{\partial \xi_j} = 0. \tag{13.39}$$

The one-dimensional advection equations are solved using the method of characteristics. After all the components of $\mathbf{v}$ are advected, the solution is recovered using Eq. 13.40.

$$\mathbf{u}^{n+1} = \mathbf{\Omega}^{-1}\mathbf{v}^{n+1} \tag{13.40}$$

The underlying computer code involves schemes of second to fourth orders of accuracy. In this study, we used the fourth-order accurate scheme ($\zeta = \Delta t/h$, h is spatial coordinate step):

$$v_m^{n+1} = v_m^n - \zeta(\Delta_1 - \zeta(\Delta_2 - \zeta(\Delta_3 - \zeta\,\Delta_4))),$$

$$\Delta_1 = (-2v_{m+2}^n + 16v_{m+1}^n - 16v_{m-1}^n + 2v_{m-2}^n)/24,$$

$$\Delta_2 = (-v_{m+2}^n + 16v_{m+1}^n - 30v_m^n + 16v_{m-1}^n - v_{m-2}^n)/24, \tag{13.41}$$

$$\Delta_3 = (2v^n_{m+2} - 4v^n_{m+1} + 4v^n_m - 2v^n_{m-2})/24,$$

$$\Delta_4 = (v^n_{m+2} - 4v^n_{m+1} + 6v^n_m - 4v^n_{m-1} + v^n_{m-2})/24.$$

Additionally, we used a grid-characteristic monotonicity criterion. For positive components of diagonal matrix $\mathbf{\Lambda}_j$, it has the form of Eq. 13.42.

$$\min\{v^n_m, v^n_{m-1}\} \le v^{n+1}_m \le \max\{v^n_m, v^n_{m-1}\} \quad (13.42)$$

For negative components of diagonal matrix $\mathbf{\Lambda}_j$, it is symmetric. In the simplest case when this criterion is violated, the solution is corrected as follows:

$$v^{n+1}_m = \begin{cases} \max\{v^n_m, v^n_{m-1}\}, v^{n+1}_m > \max\{v^n_m, v^n_{m-1}\}, \\ \min\{v^n_m, v^n_{m-1}\}, v^{n+1}_m < \min\{v^n_m, v^n_{m-1}\}, \\ v^{n+1}_m, \min\{v^n_m, v^n_{m-1}\} \le v^{n+1}_m \le \max\{v^n_m, v^n_{m-1}\}. \end{cases} \quad (13.43)$$

This limiter preserves the fourth order of the scheme in domains, where the solution is fairly smooth (the characteristic criterion is satisfied). In the case of high solution gradients, the order of the scheme is reduced to the third. Parallel algorithms for high performance computing systems were used [21].

13.6 Simulation Results

In Fig. 13.1, two 2D wave fields of scattered seismic waves on the layered structure with different orientations for contact condition (Eq. 13.2) are shown.

The influence of the layer orientation on the scattered wave field is clearly seen. Seismic response is nonsymmetric for non-vertical layers orientation. Based on this fact, it is possible to make some assumptions about the main direction in the subsurface fractured medium.

In Figs. 13.2 and 13.3, 3D wave fields of scattered seismic waves on layered and block structures for contact condition (Eq. 13.1) are shown. The whole medium was a

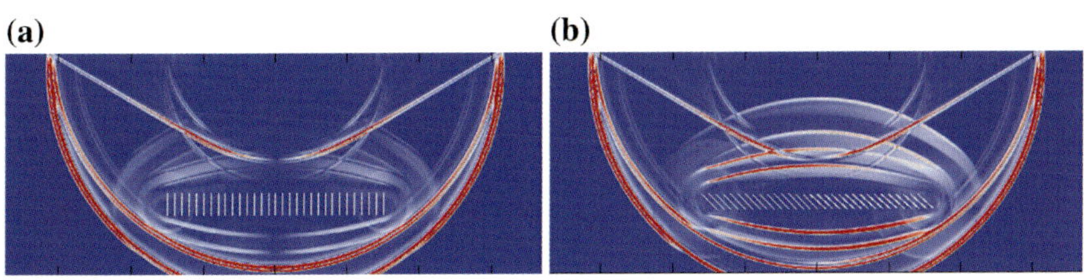

Fig. 13.1 2D-scattered seismic waves on the fractured cluster: **a** vertical cracks orientation, **b** 45° cracks orientation

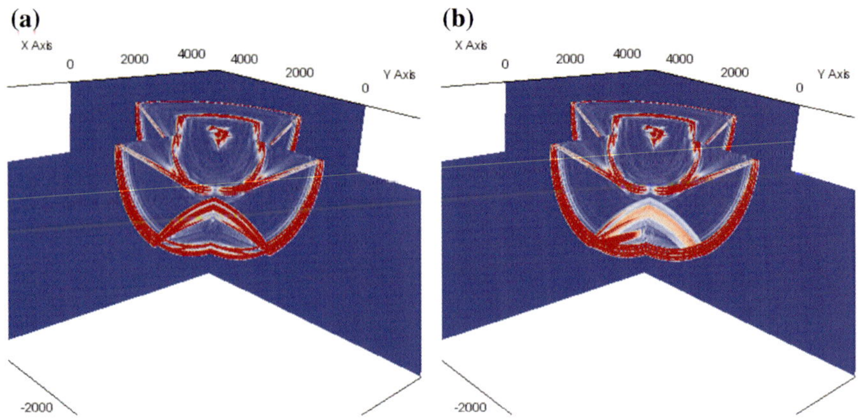

Fig. 13.2 3D-scattered seismic waves for time moment $t = 0.56$ s: **a** on the block, **b** on the layered clusters

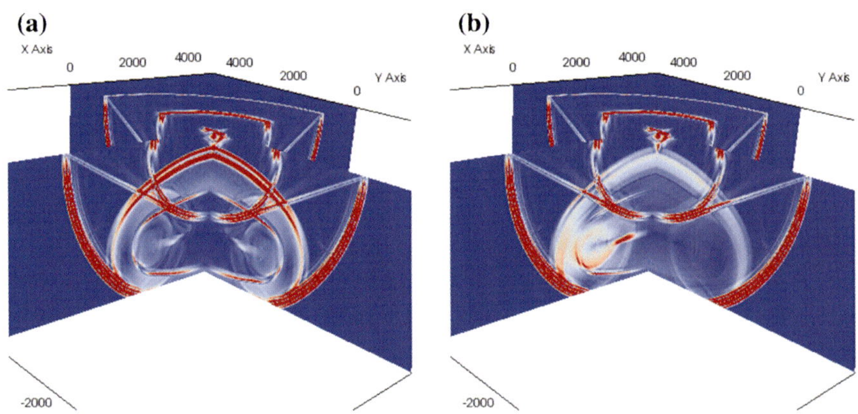

Fig. 13.3 3D-scattered seismic waves for time moment $t = 0.8$ s: **a** on the block, **b** on the layered clusters

parallelepiped with sizes 10 × 10 × 3 km. P-wave velocity was equaled to 4500 m/s, S-wave velocity was equaled to 2250 m/s, density was equaled to 2500 kg/m^3. The day surface condition was used on the upper side. In the center of the medium along axes OX and OY with the depth 50 m the point source with the 30 Hz Ricker time dependence function was applied. At the 2 km depth, the object was set with sizes 3 × 3 × 0.2 km. Two different models were compared: the block and layered with the normal vector along the axis OX. The parallelepiped mesh with 5 m cells containing 2.4 billion of nodes was constructed. The time step was 1 ms, and totally 2,000 steps were done. To visualize results we save each 40 time steps: 3 slices, 2 main vertical slices, and one horizontal equals to the day surface. The total computation time was 5.5 h on 100 computation cores for both mathematical models.

As expected, the amplitude of the seismic response for the block medium is significantly higher due to the existence of horizontal contact planes that reflect the incident wave (see Figs. 13.2a, 13.3a). Also, 3D asymmetry of the response from the layered medium is clearly seen (see Figs. 13.2b, 13.3b). It is explained by the presence of the preferred direction along the normal vector.

We carried out the dynamic analysis of the seismic signal registered at the day surface for both models. Spatial distributions for first, second, third, and fourth responses are depicted in Figs. 13.4, 13.5, and 13.6. The first response is the P-wave reflected from the top boundary of the cluster. It has the smallest arrival time. The second response is the S-wave generated due to the interaction of the initial P-wave with the cluster. The third and fourth responses are the P-wave with two times more depth path length and the S-wave generated while interacting with cluster, respectively. Due to the specific parameters of this model, these two signals are very close, so it is quite difficult to split them. One of the major features of all responses is the symmetry for block model and asymmetry for layered model. It is clearly seen at the wave patterns with OXY plane.

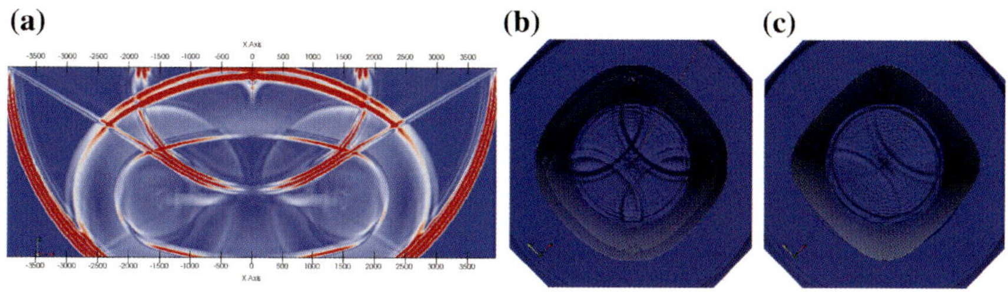

Fig. 13.4 Registration (modulus of velocity) of the first seismic response at the day surface: **a** OXZ slice, **b** OXY slice for block medium, **c** OXY slice for layered medium

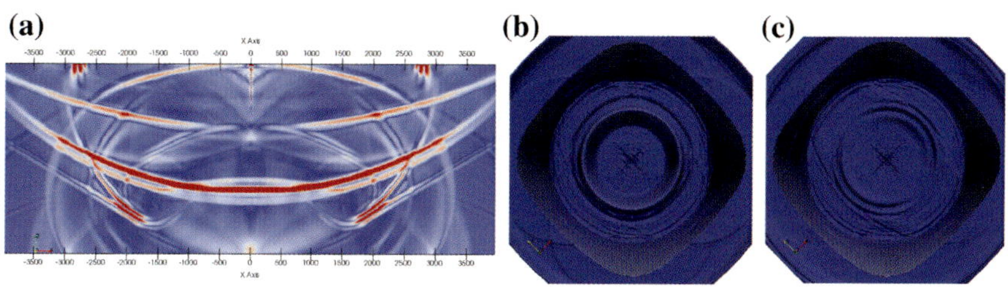

Fig. 13.5 Registration (modulus of velocity) of the second seismic response at the day surface: **a** OXZ slice, **b** OXY slice for block medium, **c** OXY slice for layered medium

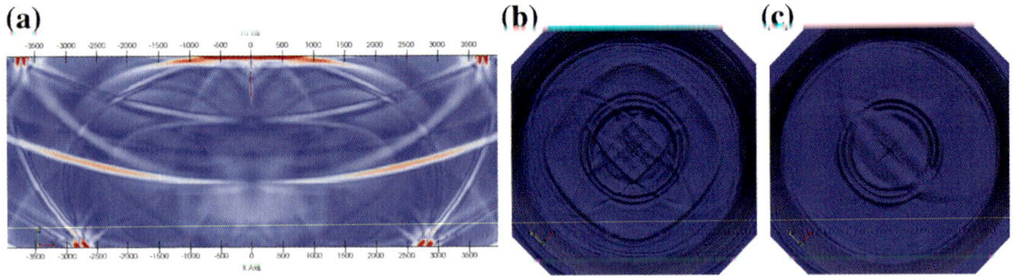

Fig. 13.6 Registration (modulus of velocity) of the third and fourth seismic response at the day surface: **a** OXZ slice, **b** OXY slice for block medium, **c** OXY slice for layered medium

13.7 Conclusions

Continual models of solid media with a discrete set of slip planes (layered, block media) and with nonlinear slip conditions at the contact boundaries of structural elements were constructed. For a stable numerical solution of a system of differential equations, an explicit–implicit method is proposed with an explicit approximation of the equations of motion and an implicit approximation of the constitutive relations containing a small parameter in the denominator. A set of effective formulas for the stress tensor correcting at the "elastic" step was obtained. Numerical simulations of the dynamic scattering process for subsurface layered and block objects in elastic 2D and 3D media were carried out using modern high-performance computing systems.

Acknowledgements This work has been carried out using computing resources of the federal collective usage center Complex for Simulation and Data Processing for Mega-science Facilities at NRC "Kurchatov Institute", http://ckp.nrcki.ru/. This work was carried out with the financial support of the Russian Science Foundation, project no. 19-71-10060.

References

1. Nikitin, I.S.: Dynamic models of layered and block media with slip, friction and separation. Mech. Solids **43**(4), 652–661 (2008)
2. Burago, N.G., Zhuravlev, A.B., Nikitin, I.S.: Continuum model and method of calculating for dynamics of inelastic layered medium. Math. Model. Comput. Simul. **11**(3), 59–74 (2019)
3. Burago, N.G., Nikitin, I.S.: Improved model of a layered medium with slip on the contact boundaries. J. Appl. Math. Mech. **80**(2), 164–172 (2016)
4. Nikitin, I.S., Burago, N.G., Nikitin, A.D.: Continuum model of the layered medium with slippage and nonlinear conditions at the interlayer boundaries. Solid State Phenom. **258**, 137–140 (2017)
5. Virieux, J., Calandra, H., Plessix, R.E: A review of the spectral, pseudo-spectral, finite-difference and finite-element modelling techniques for geophysical imaging. Geophys. Prospect. **59**(5), 794–813 (2011)
6. Carcione, J.M., Herman, C.G., Kroode, A.P.E: Seismic modeling. Geophysics **67**(4), 1304–1325 (2002)

7. Burago, N.G., Nikitin, I.S., Yakushev, V.L.: Hybrid numerical method with adaptive overlapping meshes for solving nonstationary problems in continuum mechanics. Comput. Math. Math. Phys. **56**(6), 1065–1074 (2016)
8. Lisitsa, V., Tcheverda, V., Botter, C.: Combination of the discontinuous Galerkin method with finite differences for simulation of seismic wave propagation. J. Comput. Phys. **311**, 142–157 (2016)
9. Massau, J.: Memoire sur L'integration Graphique des Equations aux Derivess Partielles. F. Meyer-van Loo, Ghent (1899)
10. Zhukov, A.I.: Using the Method of Characteristics for the numerical solution of one-dimensional problems of gas dynamics. Tr. Mat. Inst. Akad. Nauk SSSR **58**, 4–150 (1960)
11. Butler, D.S.: The numerical solution of hyperbolic systems of partial differential equations of three independent variables. Proc. Roy. Soc. London. Ser. A **255**(1281), 232–241 (1960)
12. Beklemysheva, K.A., Vasyukov, A.V., Golubev, V.I., Zhuravlev, Y.I.: On the estimation of seismic resistance of modern composite oil pipeline elements. Dokl. Math. **97**(2), 184–187 (2018)
13. Golubev, V.I., Golubeva, YuA: Full-wave simulation of the earthquake initiation process. CEUR Work. Proc. **2267**, 346–350 (2018)
14. Golubev, V., Khokhlov, N., Grigorievyh, D., Favorskaya, A.: Numerical simulation of destruction processes by the grid-characteristic method. Procedia Comput. Sci. **126**, 1281–1288 (2018)
15. Muratov, M.V., Petrov, I.B.: Application of fractures mathematical models in exploration seismology problems modeling. In: Petrov, I.B., Favorskaya, A.V., Favorskaya, M.N., Simakov, S.S., Jain, L.C. (eds.) Proceedings of the Conference on 50 years of the Development of Grid-characteristic Method, SIST, vol. 133, pp. 120–131. Springer, Berlin (2019)
16. Favorskaya, A.V., Zhdanov, M.S., Khokhlov, N.I., Petrov, I.B.: Modeling the wave phenomena in acoustic and elastic media with sharp variations of physical properties using the grid-characteristic method. Geophys. Prospect. **66**(8), 1485–1502 (2018)
17. Favorskaya, A.V., Kabisov, S.V., Petrov, I.B.: Modeling of ultrasonic waves in fractured rails with an explicit approach. Dokl. Math. **98**(1), 401–404 (2018)
18. Golubev, V.I.: The usage of grid-characteristic method in seismic migration problems. In: Petrov, I.B., Favorskaya, A.V., Favorskaya, M.N., Simakov, S.S., Jain, L.C. (eds.) Proceedings of the Conference on 50 years of the Development of Grid-characteristic Method, SIST, vol. 133, pp. 143–155. Springer, Berlin (2019)
19. Golubev, V.I., Voinov, O.Y., Petrov, I.B.: Migration of seismic data for multi-layered fractured geological media using elastic approach. In: 19th Science and Applied Research Conference Oil and Gas Geological Exploration and Development, pp. 43756.1–43756.5 (in Russian) (2017)
20. Khokhlov, N.I., Golubev, V.I.: On the class of compact grid-characteristic schemes. In: Petrov, I.B., Favorskaya, A.V., Favorskaya, M.N., Simakov, S.S., Jain, L.C. (eds.) Proceedings of the Conference on 50 years of the Development of Grid-characteristic Method, SIST, vol. 133, pp. 64–77. Springer, Berlin (2019)
21. Ivanov, A.M., Khokhlov, N.I.: Efficient inter-process communication in parallel implementation of grid-characteristic method. In: Petrov, I.B., Favorskaya, A.V., Favorskaya, M.N., Simakov, S.S., Jain, L.C. (eds.) Proceedings of the Conference on 50 years of the Development of Grid-characteristic Method, SIST, vol. 133, pp. 91–102. Springer, Berlin (2019)

Chapter 14
Algorithms for Calculating Contact Problems in the Solid Dynamics

Nikolay G. Burago, Ilia S. Nikitin and Alexander D. Nikitin

Abstract The chapter discusses the explicit and implicit non-matrix finite element algorithms for calculating contact interactions between elastic–plastic bodies. We consider Lagrangian contact algorithms that are based on Lagrange multipliers (explicit methods) and penalty functions (implicit methods). Examples of the calculation of contact interactions during high-speed processes of the collision of elastic–plastic bodies and explosion welding of tubular samples are presented.

14.1 Introduction

The development of contact algorithms for elastic–plastic bodies started more than half a century back together with the advent of computers and numerical methods. Reviews of contact algorithms can be found in [1–9].

Here, Lagrangian contact algorithms are considered. Such contact algorithms are the parts of solution method, which are responsible for detection, tracking, and calculation of contact interaction between deformable bodies. In most of the cases, the boundary conditions for problems in solid mechanics belong to the following three types: the boundary conditions with predefined displacements or velocities, boundary conditions with predefined loads from external bodies that are not involved in calculation, and boundary contact conditions that define contact boundaries, their motion, and contact loads during interaction of considered deformable bodies between each

N. G. Burago (✉)
Ishlinsky Institute for Problems in Mechanics of the RAS, pr. Vernadskogo 101, b1, Moscow 119526, Russian Federation
e-mail: burago@ipmnet.ru

I. S. Nikitin · A. D. Nikitin
Institute of Computer Aided Design of the RAS, Vtoraya Brestskaya ul., 19/18, Moscow 123056, Russian Federation
e-mail: i_nikitin@list.ru

A. D. Nikitin
e-mail: nikitin_alex@bk.ru

L. C. Jain et al. (eds.), *Advances in Theory and Practice of Computational Mechanics*, Smart Innovation, Systems and Technologies 173,
https://doi.org/10.1007/978-981-15-2600-8_14

other. In most cases, a contact boundary is not predefined in advance. It can be variable in time and ought to be found and calculated along with and as a part of general problem about deformation of bodies.

Most of proposed contact search algorithms are overviewed in [4]. Search algorithms often use quite a significant part of calculations required to solve the problem. We tested many of such algorithms and the one that was chosen in our practice is considered further.

In most of dynamic incremental (step by step in time) Lagrange algorithms, the contact boundaries are detected by the penetration of boundary nodes through alien boundary cells. That may happen after predictor time step, which is calculated neglecting the contact. The contact zone detection is based on selection of the contacting pairs "boundary cell—alien boundary node" if they are close enough to each other (much less than local spatial cell size) or if a boundary cell is intersected by a track of alien node motion. After contact pairs are found, some so-called "master–slave" algorithm is used during correction stage to prevent penetration with help of normal contact load. In explicit schemes, the correction stage is implemented iteratively by means of circumventing the contact pairs calculating the contact loads from the magnitude of penetration for eliminating it. At each time step for such correction, two run-arounds of boundary are enough. In implicit schemes, the transition from Lagrange multipliers to penalty functions is done by representing Lagrange multiplier (normal contact load) as a product of the distance of penetration by the penalty factor. As a result, the positive definiteness of the discrete operator ("stiffness matrix") is preserved, but it is necessary to take into account that too large values of the penalty factor lead to a worsening of the conditionality of the problem. Therefore, to prevent loss of accuracy and provide convergence of iterative solutions, a preconditioning should be used. For this purpose, the residuals of equations used in iterations are multiplied by an approximate inverse matrix of the system of algebraic equations. In order not to perform time-consuming operations of matrix inversion for preconditioning, it is sufficient to use the matrix of inverse diagonal elements of stiffness matrix, that is, in other words, to use the scaling of unknowns.

Contact pairs algorithms, in our opinion, are the simplest. They are very easily implemented in the case of complex geometry of the solution domain when using a system of Cartesian rectangular coordinates. The implementation within the framework of Galerkin–Petrov variational formulation with reduced requirements for the smoothness of a generalized solution is especially simple (when the equations contain only the first derivatives of the desired functions). Then the simplest piecewise linear approximation on finite elements in the form of rectangular parallelepipeds gives quite good results.

In this chapter, we consider in detail the algorithms of Lagrange multiplier methods for explicit schemes, and algorithms of penalty functions for implicit schemes because in combination with a finite element piecewise linear approximation of the solution. These algorithms are very easy to implement and provide good quality solutions even on personal computers.

This chapter is organized as follows. Statement of general contact problem is considered in Sect. 14.2. Numerical algorithm based on matrix-free finite element method is described in Sect. 14.3. Contact calculation using Lagrange multipliers technique is highlighted in Sect. 14.4. Application of penalty functions to contact computations is presented in Sect. 14.5. The effectiveness of the algorithms is illustrated by two non-trivial examples in Sect. 14.6: the impact of two plates at an angle and axisymmetric welding of two dissimilar tube samples under the action of a detonation wave. Section 14.7 concludes the chapter.

14.2 Mathematical Model

The system of equations describing the behavior of an elastic–plastic medium is used here in the simplified variant [10]. The laws of conservation of mass and momentum as well as the kinematic relations are written as Eq. 14.1.

$$\begin{gathered}\rho = \rho_0 \det(\mathbf{F}^{-1}), \quad \rho \frac{d\mathbf{u}}{dt} = \nabla \cdot \boldsymbol{\sigma} + \rho \mathbf{g} \\ \mathbf{F}^{-1} = \nabla \mathbf{x}^0 \quad \boldsymbol{\varepsilon} = \frac{1}{2}(\mathbf{I} - \mathbf{F}^{-T} \cdot \mathbf{F}^{-1}) \quad \mathbf{e} = \frac{1}{2}(\mathbf{L} + \mathbf{L}^T) \\ \mathbf{e} = \frac{d\boldsymbol{\varepsilon}}{dt} + \boldsymbol{\varepsilon} \cdot \mathbf{L} + \mathbf{L}^T \cdot \boldsymbol{\varepsilon} \quad \mathbf{L} = \nabla \mathbf{u} \quad \frac{d\mathbf{x}}{dt} = \mathbf{u}\end{gathered} \tag{14.1}$$

Here, ρ is the mass density, $\mathbf{u}$ is the velocity, t is the time, $\mathbf{x}$ and $\mathbf{x}^0$ are the radius vectors of material point in actual and initial states, $\mathbf{F}$ is the strain gradient, $\mathbf{L}$ is the velocity gradient, $\boldsymbol{\varepsilon}$ is the Almansi strain tensor, $\mathbf{e}$ is the Eulerian strain rate tensor, $\boldsymbol{\sigma}$ is the Cauchy stress tensor, d/dt is the material time derivative, ∇ is the spatial differentiation operator in actual state, $\mathbf{I}$ is the unity tensor, and $\mathbf{g}$ is the body force density.

Constitutive equations are used according to Prandtl–Reuss plastic flow theory and Mises plasticity condition and expressed by Eq. 14.2.

$$\begin{gathered}\boldsymbol{\sigma} = -p\mathbf{I} + \boldsymbol{\sigma}' \quad \boldsymbol{\sigma}' = 2\mu(\boldsymbol{\varepsilon}' - \boldsymbol{\varepsilon}'_p) \\ p = K\frac{\rho}{\rho_0} \ln \frac{\rho}{\rho_0} \quad \mathbf{e}'_p = H(\Phi_p)\frac{\boldsymbol{\sigma}' : \boldsymbol{e}}{k_s^2}\boldsymbol{\sigma}'\end{gathered} \tag{14.2}$$

Here, p is the pressure, $\boldsymbol{\sigma}'$ is the deviatoric stress, $\boldsymbol{\varepsilon}'$, $\boldsymbol{\varepsilon}'_p$, and $\mathbf{e}'_p$ are the deviatoric strain, plastic strain, and plastic strain rate tensors, respectively, $\Phi_p = \boldsymbol{\sigma}' : \boldsymbol{\sigma}' - k_s^2 \le 0$, H is the Heaviside function, k_s is the yield radius, K and μ are the elastic moduli.

The solution domain in the general case consists of several (spaced apart in space for the problems of dynamics) subdomains corresponding to the deformable bodies under consideration. On a part of the surface, interaction with external bodies is specified in the form of loads:

$$t \ge 0, \quad \mathbf{x} \in S_p \ : \ \boldsymbol{\sigma} \cdot \mathbf{n} = \mathbf{P}^*(\mathbf{x}, t) \tag{14.3}$$

or velocities

$$t \geq 0, \quad \mathbf{x} \in S_u : \mathbf{u} = \mathbf{u}^*(\mathbf{x}, t). \tag{14.4}$$

On the rest of the surface, so-called contact surface $S_{cont} = (S \backslash S_u) \backslash S_p$, velocities and loads are due to the interaction of the bodies in question with each other. Unknown in advance contact surface S_{cont} is the set of all points that have different initial (Lagrangian) coordinates and coincident actual coordinates:

$$\mathbf{x}_-, \mathbf{x}_+ \in S_{cont} : \exists \mathbf{x}_-^0 \neq \mathbf{x}_+^0 | \mathbf{x}_- = \mathbf{x}(\mathbf{x}_-^0, t) = \mathbf{x}_+ = \mathbf{x}(\mathbf{x}_+^0, t).$$

Loads $\mathbf{P}$ and velocities $\mathbf{u}$ on the contact surface S_{cont} are determined by the contact conditions that make up the continuity of the normal velocity, Newton's third law and the law of friction, respectively (Eq. 14.5).

$$\begin{aligned} &(\mathbf{u}^+ - \mathbf{u}^-) \cdot \mathbf{n}^+ = 0 \\ &\mathbf{P}^+ = -\mathbf{P}^- \\ &P_{\tau i} = f(P_n, (\mathbf{u}^+ - \mathbf{u}^-) \cdot \boldsymbol{\tau}_i^+) \end{aligned} \tag{14.5}$$

Here, $\mathbf{n}$, $\boldsymbol{\tau}_i$ are the normal and tangent unity vectors to the surface S, respectively, and

$$\mathbf{n}^+ = -\mathbf{n}^-, \quad P_n = \mathbf{P} \cdot \mathbf{n}, \quad \boldsymbol{\tau}_i^+ = -\boldsymbol{\tau}_i^-, \quad P_{\tau i} = \mathbf{P} \cdot \boldsymbol{\tau}_i, \quad (i = 1, 2).$$

Initial conditions are the following:

$$x \in V, \quad t = 0 : \mathbf{x} = \mathbf{x}^0, \quad \mathbf{u} = \mathbf{u}_0, \quad \boldsymbol{\varepsilon}_p = 0. \tag{14.6}$$

Thus, it is required to solve the initial boundary value problem for the system of equations (Eqs. 14.1, 14.2) with the boundary (Eqs. 14.3–14.5) and initial (Eq. 14.6) conditions.

14.3 Numerical Method

The variation form of the equations of motion (the equation of virtual works) has the form of Eq. 14.7.

$$\int_V \rho(\frac{d\mathbf{u}}{dt} - \mathbf{g}) \cdot \delta \mathbf{u} dV + \int_V \boldsymbol{\sigma} : \nabla \delta \mathbf{u} dV$$

$$= \int\limits_{S_p} \mathbf{p} \cdot \delta \mathbf{u} dS - \delta \int\limits_{S_{cont}} P_n(\mathbf{x}^* - \mathbf{x}) \cdot \mathbf{n} dS - \sum_{i=1}^{2} \int\limits_{S_{cont}} P_{\tau i}(\delta \mathbf{x}^* - \delta \mathbf{x}) \cdot \boldsymbol{\tau}_i dS \quad (14.7)$$

Conditions (Eq. 14.3) are taken into account in the integral over the boundary with a given surface load (the first integral in the right-hand side). Conditions (Eq. 14.4) on a part of the surface with given velocities are taken into account as complementary boundary conditions. The conditions on the contact boundary (Eq. 14.5) are introduced into the variational equation of virtual works using Lagrange multipliers method (the next integrals in the right-hand side). In the integrals over the contact surface in (Eq. 14.7), $\mathbf{x}^* = \mathbf{x}(\mathbf{x}^{0*}, t)$ denotes the intersection point of the continuation of the outer normal $(\mathbf{x} + \alpha \mathbf{n}, \alpha \geq 0)$ with the surface point $\mathbf{x} = \mathbf{x}(\mathbf{x}^0, t)$ $\left(\mathbf{x}^0 \neq \mathbf{x}^{0*}\right)$. If such point $\mathbf{x}^*$ does not exist, i.e., if the line that continues outside normal does not cross S, then assume $\mathbf{x}^* = \mathbf{x} + \mathbf{n}$. For all $\mathbf{x} \in S$, the following inequality must be satisfied:

$$(\mathbf{x}^* - \mathbf{x}) \cdot \mathbf{n} \geq 0. \quad (14.8)$$

This inequality expresses the condition of non-penetration of one part of the surface into another. Equality corresponds to the points of the contact surface $\mathbf{x}^*, \mathbf{x} \in S_{cont}$. The role of Langrage multiplier for inequality (Eq. 14.8) is played by the normal contact load $P_n(x, t)$ to be determined.

In the solution domain, we introduce a grid of elements consisting of tetrahedra, prisms, or parallelepipeds in the three-dimensional case and of triangles and quadrilaterals in the two-dimensional case. Let $\mathbf{x}_i$ $(i = 1, 2, \ldots, N_v)$ be the nodal radius vectors, $C(k, l)$ $(k = 1, 2, \ldots, N_c;\ l = 1, 2, \ldots, M_c)$ be the nodal numbers in elements, and $B(k, l)$ $(k = 1, 2, \ldots, N_b;\ l = 1, 2, \ldots, M_b)$ be the nodal numbers in boundary elements. Let the bypass of the boundaries of the surface elements take place clockwise for the external observer. On the time layer, n introduces the notation: $\mathbf{u}_i^n$, $\mathbf{x}_i^n$ are the velocity and nodal radius-vector, and $[\boldsymbol{\varepsilon}_p]_k^n$ is the plastic strain tensor in the element center. Let ω be the set of nodal numbers, ω_u be the set of boundary nodal numbers with predefined velocities, and Ω be the set of finite element numbers.

Piecewise linear finite element approximation of coordinates, displacements, and velocities is used. The stress, strain, and strain rate tensors are calculated in the centers of the elements ($[\boldsymbol{\sigma}]_k^n$, $[\boldsymbol{\varepsilon}]_k^n$, $[\mathbf{e}_p]_k^n$, and $[\mathbf{e}]_k^n$). Integrals of a variational equation that do not contain time derivatives are replaced by sums of integrals over finite elements by quadrature formulas with Gaussian points at the centers of finite elements. For integrals with time derivatives in finite elements, the quadrature formulas with Gaussian points at the nodes of the grid are used. This makes the mass matrix diagonal for fast processes and provides the use of explicit finite element time integration schemes. Thus, we obtain the discrete form of Eqs. 14.1, 14.2 provided by Eq. 14.9.

$$M_i(\mathbf{u}_i^{n+1} - \mathbf{u}_i^n \gamma_i^n - (1 - \gamma_i^n)\widetilde{\mathbf{u}}_i^n) = \mathbf{f}_i^n \Delta t_n \quad i \in \omega \backslash \omega_u$$

$$\mathbf{u}_i^{n+1} = \mathbf{u}_{*i}^{n+1} \quad i \in \omega_u$$
$$\mathbf{x}_i^{n+1} = \mathbf{x}_i^n + \mathbf{u}_i^{n+1}\Delta t_n \quad i \in \omega$$
$$[\boldsymbol{\varepsilon}'_p]_k^{n+1} = [\boldsymbol{\varepsilon}']_k^{n+1} - [\boldsymbol{\sigma}']_k^{n+1}/2/\mu \quad k \in \Omega \tag{14.9}$$

To calculate new stress–strain state in each finite element $k \in \Omega$, the operations are performed using Eq. 14.10.

$$[\mathbf{F}^{-1}]_k^n = \sum_{j=1}^{M_C} [\nabla]_{kj}^n \mathbf{x}_{C(k,j)}^0 \quad [\boldsymbol{\varepsilon}]_k^n = \frac{1}{2}(\mathbf{I} - [\mathbf{F}^{-T}]_k^{n+1} \cdot [\mathbf{F}^{-1}]_k^{n+1})$$
$$[\mathbf{L}]_k^n = \sum_{j=1}^{M_C} [\nabla]_{kj}^n \mathbf{u}_{C(k,j)}^n \quad [\mathbf{e}]_k^n = \frac{1}{2}([\mathbf{L}]_k^n + [\mathbf{L}^T]_k^n)$$
$$[\mathbf{L}]_k^n = \sum_{j=1}^{M_C} [\nabla]_{kj}^n \mathbf{u}_{C(k,j)}^n \quad [\mathbf{e}]_k^n = \frac{1}{2}([\mathbf{L}]_k^n + [\mathbf{L}^T]_k^n)$$
$$[\boldsymbol{\varepsilon}']_k^n = [\boldsymbol{\varepsilon}]_k^n - \mathbf{I}[\boldsymbol{\varepsilon}]_k^n : \mathbf{I}/30 \quad [\rho]_k^n = \rho_0 \det([\mathbf{F}^{-1}]_k^n)$$
$$[p]_k^n = K\frac{[\rho]_k^n}{\rho_0} \ln \frac{[\rho]_k^n}{\rho_0} 0 \quad [\boldsymbol{\sigma}']_k^n = 2\mu([\boldsymbol{\varepsilon}']_k^n - [\boldsymbol{\varepsilon}'_p]_k^n)$$
$$[\boldsymbol{\sigma}]_k^n = -[p]_k^n \mathbf{I} + [\boldsymbol{\sigma}']_k^n \quad [\tilde{\boldsymbol{\sigma}}']_k^{n+1} = [\boldsymbol{\sigma}']_k^n + 2\mu([\mathbf{e}']_k^n - [\mathbf{e}'_p]_k^n)\Delta t_n$$
$$\beta_k^n = k_s/\max\{k_s, \sqrt{[\tilde{\boldsymbol{\sigma}}']_k^{n+1} : [\tilde{\boldsymbol{\sigma}}']_k^{n+1}}\} \quad [\boldsymbol{\sigma}']_k^{n+1} = \beta_k^n [\tilde{\boldsymbol{\sigma}}']_k^{n+1}$$
$$[\mathbf{L}]_k^{n+1} = \sum_{j=1}^{M_C} [\nabla]_{kj}^{n+1} \mathbf{u}_{C(k,j)}^{n+1} \quad [\mathbf{e}]_k^{n+1} = \frac{1}{2}([\mathbf{L}]_k^n + [\mathbf{L}^T]_k^n)$$
$$[\boldsymbol{\varepsilon}']_k^{n+1} = [\boldsymbol{\varepsilon}]_k^{n+1} - \mathbf{I}[\boldsymbol{\varepsilon}]_k^{n+1} : \mathbf{I}/3 \tag{14.10}$$

Here, the plastic flow is calculated by using method of Wilkins [1]. As an additional viscosity included on shock waves, a variable Lax viscosity is used in the equation of motion. The average speed on the old time layer n is calculated as follows:

$$i \in \omega,\ j \in \omega_i : \widetilde{\mathbf{u}}_i^n = 0.5(\max_{j\in\omega_i} \mathbf{u}_j^n + \min_{jin\omega_i} \mathbf{u}_j^n).$$

Here ω_i are the numbers of neighbors of node i. The hybrid viscosity parameter that regulates Lax viscosity is given by $\gamma_i^n = \min\{1,\ \gamma_0 + \kappa_0|\widetilde{\mathbf{u}}_i^n - \mathbf{u}_i^n|\}$, where $\gamma_0 = u_s/c$, $\kappa_0 = (0.2u_s)^{-1}$, $u_s = \max_{i\in\omega} \mathbf{u}_i^n - \min_{i\in\omega} \mathbf{u}_i^n$, $c^2 = dP/d\rho + 4/3\mu/\rho$.

Parameter γ_k^n includes Lax viscosity in the vicinity of shock waves and increases it to the nominal value with increasing impact velocity. Value M_i is nodal mass calculated by the following expression:

$$M_i = \sum_{k=1}^{N_C} \sum_{l=1}^{M_C} [V]_k^n \rho_k M_C^{-1} \tilde{H}(i - C(k, l)), \quad \mathrm{i} \in \omega.$$

The right sides of the discrete equations of motion are determined by the contributions from the internal forces, external loads, and contact loads provided by Eq. 14.11.

$$\mathbf{f}_i^n = \sum_{k=1}^{N_c} \sum_{l=1}^{M_c} [\mathbf{g_1}]_{kl}^n \tilde{H}(i - C(k, l)) + \sum_{k=1}^{N_b} \sum_{l=1}^{M_b} [\mathbf{g_2}]_{kl}^n \tilde{H}(i - B(k, l)) + \\ + \sum_{k=1}^{N_D} \sum_{l=1}^{M_D} [\mathbf{g}_3]_{kl}^n \tilde{H}(i - D(k, l)) \tag{14.11}$$

Here, function $\tilde{H}$ is equal to one for zero argument and zero otherwise, N_D, M_D are the numbers of boundary contact elements and the number of nodes in the contact elements. The contributions from the internal and specified external loads are as follows:

$$[\mathbf{g}_1]_{kl}^n = -[V]_k^n [\boldsymbol{\sigma}]_k^n \cdot \nabla_{kl}^n, \quad (k = 1, 2, \ldots, N_C, \quad l = 1, 2, \ldots, M_C),$$
$$[\mathbf{g_2}]_{kl}^n = M_b^{-1} [\mathbf{P}]_k^n S_k^n, \quad (k = 1, 2, \ldots, N_b, \quad l = 1, 2, \ldots, M_b).$$

The calculation of contributions from contact loads $[\mathbf{g}_3]_{kl}^n$ is considered further. Explicit method is stable under usual Courant stability condition:

$$\Delta t_n = \min_{k\ in\Omega} \left(\frac{1}{c_k^n (\max_l (|[\nabla_x]_{kl}^n|, |[\nabla_y]_{kl}^n|, |[\nabla_z]_{kl}^n|))^{-1}} \right), \tag{14.12}$$

where $[\nabla_x]_{kl}^n \mathbf{e}_x + [\nabla_y]_{kl}^n \mathbf{e}_y + [\nabla_z]_{kl}^n \mathbf{e}_z$ is the discrete spatial differentiation operator. At high impact velocities (of the order of the speed of sound), to ensure accuracy, an additional time step restriction was introduced expressing the requirement that deformation increments be sufficiently small:

$$\Delta t_n \leq \Delta\varepsilon\mathrm{max} \left(\max_{k\in\Omega} \left([\mathbf{e}]_k^n : [\mathbf{e}]_k^n \right)^{1/2} \right)^{-1}. \tag{14.13}$$

Here, $\Delta\varepsilon\mathrm{max} = 0.1\varepsilon_s$ is the maximum permissible deformation increment at a time step, ε_s is the yield point deformation.

14.4 Use of Lagrange Multipliers

Let the surface of the bodies be represented by boundary cells: segments in two-dimensional case and triangles in three-dimensional case. Let the local numbering of the nodes in the boundary cells in three-dimensional case be taken clockwise if viewed from the outside of the body, and in two-dimensional case the local numbering corresponds to the bypass of the boundary clockwise. This is necessary to uniquely determine the direction of the outer normal to the boundary.

At the beginning of each time step, a preliminary calculation of the new position of the boundary without taking into account the contact is made. Then, among the "boundary cell—boundary node" pairs, contact ones are selected, that is, such that the normal omitted from the boundary node on the boundary cell plane, crosses this boundary cell, and the countable penetration occurs. If there are several candidates for a given boundary node from the boundary elements for the role of partner in a contact pair, the boundary element that is intersected by the trajectory of this boundary node is uniquely selected. The mathematical record of these selection conditions is given below.

To speed up the process of selecting contact pairs, firstly we reject too far from each other situated potential contact partners. Such selection is carried out first by the difference of coordinates (more economical check), then by distance and only then later by projection.

The contact pair forms a triangle in two-dimensional problems and a tetrahedron in three-dimensional case. That is, the number of nodes in it M_C equals to 3 or 4, respectively. The numbers of these nodes are stored in the information array of contact pairs $C(k, l)$, $(k = 1, \ldots, N_C; l = 1, \ldots, M_C))$. For each contact pair k, first $M_C - 1$ numbers of information array $C(k, l)$ correspond to nodes of contact boundary element and the last M_Cth number corresponds to alien contact boundary node.

The algorithm for calculating the contact load is implemented iteratively. In each contact pair, the contact load normal to the boundary cell is determined from the condition that the volume of the contact pair vanishes. In the process of sequential bypass of contact pairs, the contact load and coordinates of the nodes get better until volumes of all the contact pairs become zero. In explicit schemes when implementing Lagrange multipliers method, this is achieved, as a rule, in two run-arounds of the border (the second run-around is done for control). In implicit schemes when implementing the penalty method (it is described in the next section), the velocity field is iterated to the convergence in the entire solution domain.

Below we write out the formulas for calculating the contact pair. Let $\mathbf{x}_i^n$ be the current radius vectors of the contact pair of nodes on the "old" time layer, $\mathbf{u}_i^{n+1}$ and $(\mathbf{x}_i^{n+1})_s = \mathbf{x}_i^n + (\mathbf{u}_i^{n+1})_s \Delta t_n$ are the desired node values of the speeds and radius vectors on the new time layer, respectively, and $s = 0, 1, 2, \ldots$ is the iteration number. The value $(\mathbf{u}_i^{n+1})_0$ corresponds to the preliminary calculation of the new time layer without taking into account the contact.

In three-dimensional case, the outer normal to the boundary cell is determined by the ratio:

$$(\mathbf{n})_s = ((\mathbf{x}_3^{n+1})_s - (\mathbf{x}_1^{n+1})_s) \times (((\mathbf{x}_2^{n+1})_s - ((\mathbf{x}_1^{n+1})_s)/L.$$

Here, $L = |((\mathbf{x}_3^{n+1})_s - (\mathbf{x}_1^{n+1})_s) \times (((\mathbf{x}_2^{n+1})_s - ((\mathbf{x}_1^{n+1})_s)|$. A pair of boundary node—the boundary cell is a contact pair if the normal projection of the node on the plane of the cell belongs to the cell:

$$L_1 + L_2 + L_3 = 1, \tag{14.14}$$

$$d = \mathbf{n} \cdot ((\mathbf{x}_4^{n+1})_s - (\mathbf{x}_1^{n+1})_s) \leq 0. \tag{14.15}$$

Here, values L_i, $(i = 1, 2, 3)$ are L-coordinates of projection $(\mathbf{x}_c^{n+1})_s = (\mathbf{x}_4^{n+1})_s - d(\mathbf{n})$ of boundary node M_C onto the plane of boundary cell 1-2-3:

$$(L_1)_s = ((\mathbf{x}_3^{n+1})_s - (\mathbf{x}_c^{n+1})_s) \times (((\mathbf{x}_2^{n+1})_s - ((\mathbf{x}_c^{n+1})_s)/L,$$
$$(L_2)_s = ((\mathbf{x}_3^{n+1})_s - (\mathbf{x}_1^{n+1})_s) \times (((\mathbf{x}_c^{n+1})_s - ((\mathbf{x}_1^{n+1})_s)/L,$$
$$(L_3)_s = ((\mathbf{x}_c^{n+1})_s - (\mathbf{x}_1^{n+1})_s) \times (((\mathbf{x}_2^{n+1})_s - ((\mathbf{x}_1^{n+1})_s)/L.$$

For contact cell L-coordinates are non-negative. The contact integrals in the right side of the equation of virtual works are represented in the following form:

$$\delta \int\limits_{S_c^n} (P_n)^n (\mathbf{u}^+ - \mathbf{u}^-) \cdot \mathbf{n}^n dS = \sum_{r=1}^{N_c} (P_n)_r^n \sum_{i=1}^{M_c} (L_i)_r^n \delta \mathbf{u}_{K(r,i)} \cdot \mathbf{n}_r^n S_r^n$$
$$+ \sum_{r=1}^{N_c} \delta (P_n)_r \sum_{i=1}^{M_c} (L_i)_r^n \mathbf{u}_{K(r,i)}^{n+1} \cdot \mathbf{n}_r^n S_r^n,$$

$$\int\limits_{S_c^n} \sum_{\alpha=1}^{2} (P_{\tau\alpha})^n (\delta \mathbf{u}^+ - \delta \mathbf{u}^-) \cdot \boldsymbol{\tau}_\alpha^n dS = \sum_{r=1}^{N_c} \sum_{\alpha=1}^{2} (P_{\tau\alpha})_r^n \sum_{i=1}^{M_c} (L_i)_r^n \delta \mathbf{u}_{K(r,i)} \cdot (\boldsymbol{\tau}_\alpha)_r^n S_r^n.$$

Vectors of contact loads in discrete equations of motion are calculated by formulas:

$$(\mathbf{f_c})_i^n = \sum_{r=1}^{N_c} \sum_{l=1}^{M_c} \left((P_n)_r^n \mathbf{n}_r^n + \sum_{\alpha=1}^{2} (P_{\tau\alpha})_r^n (\boldsymbol{\tau}_\alpha)_r^n \right) S_r^n L_{rl}^n \tilde{H}(i - K(r, l)).$$

Here, function $\tilde{H}$ is equal to one for zero argument and zero otherwise. A consequence of the modified equation of virtual work is also the equality:

$$\sum_{i=1}^{M_c} (L_i)_r^n \mathbf{u}_{K(r,i)}^{n+1} \cdot \mathbf{n}_r^n = 0$$

that is providing continuity of normal velocities at the contact boundary. Accounting for contact leads to the emergence of a new group of unknown quantities—Lagrange multipliers $(P_n)_r^n \quad (r = 1, \ldots, N_c)$ that satisfy N_C additional algebraic relations limiting the possible movements of contact nodes and characterized by a non-diagonal matrix, and, hence, introducing an element of implicitness into explicit schemes. Also the mutual dependence of the unknown contact zone and contact loads makes the contact problem nonlinear and leads to the need for their iterative determination (iteration by nonlinearity).

Normal contact loads must be compressive:

$$(P_n)_r^n \le 0, \quad r = 1, \ldots, N_C.$$

Contact normal loads should act against the direction of the outer normal to the boundary. If this condition is violated, the corresponding contact pair is excluded from the set of contact pairs and recalculation is performed. To solve the system of equations on the contact boundary, the simple Gauss–Seidel-type iterative process is applied.

The algorithm described above is used as an addition to the explicit schemes for calculating the deformation. Note that the contact algorithm does not depend on the choice of a particular explicit scheme. At each time step, only two border bypasses are made (the second is for control). We emphasize that all parts of the border are equal (no master-slave borders). Therefore, the contact of some parts of the body boundary with other parts of the same body boundary is calculated in the same way as the contact with other bodies.

To improve the property of the scheme to maintain symmetry and ensure the independence of the result of the order of bypassing contact pairs, the corrected values of contact loads and speeds are accumulated in a separate array, and the main arrays of coordinates and speeds are updated at the end of the calculation of contact pairs.

14.5 Using Penalty Functions

In the case of implicit schemes, the contact interaction is accounted using the penalty function method. A modified equation of virtual work (Eq. 14.7) is used, in which it is assumed:

$$P_n = \tilde{\lambda}(\mathbf{x}^+ - \mathbf{x}^-) \cdot \mathbf{n}^+, \quad \tilde{\lambda} >> 1.$$

Here, $\tilde{\lambda}$ is the penalty factor. In practical calculations, the penalty factor is taken to be equal to the reciprocal of the "machine epsilon" (for four-byte arithmetic, this is equal approximately 10^6). The approximation of contact additional terms in Eq. 14.7 is carried out in the same way as in Lagrange multipliers method. The influence of contact members is taken into account in the iterative process of conjugate gradient method (a matrix-free implementation of implicit schemes is used [11]). Before calculating the new time layer, the contact zone is determined by conditions (Eqs. 14.14, 14.15), but the condition (Eq. 14.15) is weakened and replaced with the condition of sufficient proximity of the boundaries:

$$d = \mathbf{n} \cdot ((\mathbf{x}_4^{n+1})_s - (\mathbf{x}_1^{n+1})_s) \leq 0.1 h_{\text{min}}. \tag{14.16}$$

Here, h_{min} is the minimum length of the boundary segment of the grid. Accounting for contact interactions in the case of an implicit scheme did not complicate the solving process and was implemented simply as an additional subroutine defining the contact area at each step and calculating contact terms of the equations of motion in iterations using the conjugate gradient method.

14.6 Examples

In this section, the impact of two plates at an angle and axisymmetric welding of two dissimilar tube samples under the action of a detonation wave are considered in Sects. 14.6.1 and 14.6.2, respectively.

14.6.1 The Impact of Two Bodies at an Angle

Let the plate-impactor hits the plate-barrier with velocity $u_0/c^{(1)} = 0.2$, where $c^{(1)}$ is a sound velocity in material of impactor that has the following properties:

$$K^{(1)} = 975, \mu^{(1)} = 369, \sigma_s^{(1)} = 1.0, d\sigma_s^{(1)}/da_p = 0.0, c^{(1)} = 1,$$
$$k_\omega^{(1)} = 0.0, k_\theta^{(1)} = 10^3, F = \varepsilon_{\text{max}} - \varepsilon_{lim}^{(1)}, \varepsilon_{lim}^{(1)} = 0.01.$$

Barrier material has the following properties:

$$K^{(2)} = 243, \mu^{(2)} = 92, \sigma_s^{(2)} = 0.25, d\sigma_s^{(2)}/da_p = 0, c^{(2)} = 1,$$
$$k_\omega^{(2)} = 0.0, k_\theta^{(2)} = 10^6, F = \varepsilon_{\text{max}} - \varepsilon_{lim}^{(2)}, \varepsilon_{lim}^{(2)} = 0.01.$$

Here, ε_{max} is the maximal principal strain.

In Fig. 14.1a, the initial grid is visible. Figure 14.1b, c shows the contact interaction. The target has fractured into three parts that are moving independently. Ricochet

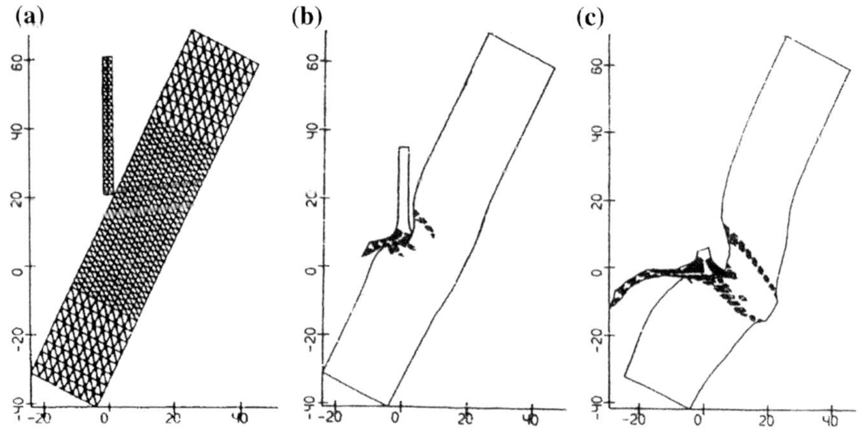

Fig. 14.1 Impact of plates: **a** initial grid, **b** contact evolution, **c** fracture evolution. The target has fractured into three parts. The impactor is destroyed

of impactor that is completely destroyed is seen clearly. The method for calculating destruction is described in [12].

14.6.2 Explosion Welding Problem

Earlier in [13], the problem of explosion welding was numerically solved with allowance for large plastic deformations in order to study the wave formation process at the boundary. Here, axisymmetric welding of two dissimilar metal tube samples under the action of a detonation wave propagating along the outer surface of the tubes is considered. The speed of the traveling detonation wave is comparable to the speed of elastic waves, for example, in titanium and steel, which are used to make bimetallic billets, but it is subsonic, providing a favorable impact mode for welding [14]. The pressure in the detonation wave considerably exceeds the static yield strength of these metals, which makes it possible to use the hydrodynamic approach for approximate solutions [14, 15]. However, it should be noted that the dynamic yield strength at high strain rate can be several times higher than the static one, and the neglect of the strength effects of solids leads to a distorted estimate of the stress–strain state. When modeling the contact interaction, the conditions for Coulomb friction were chosen into which the welding criterion depending on the level of plastic deformations was entered. Such a criterion does not contradict the well-known jet criterion of welding since the formation of a cumulative jet is possible only in the case of developed plastic flows in the vicinity of the collision front.

Figure 14.2 presents the results of the calculation of high-speed collisions of titanium (external) and steel (internal) tubular samples. For better visibility of the process, a regime with a large peak pressure in the detonation wave was chosen,

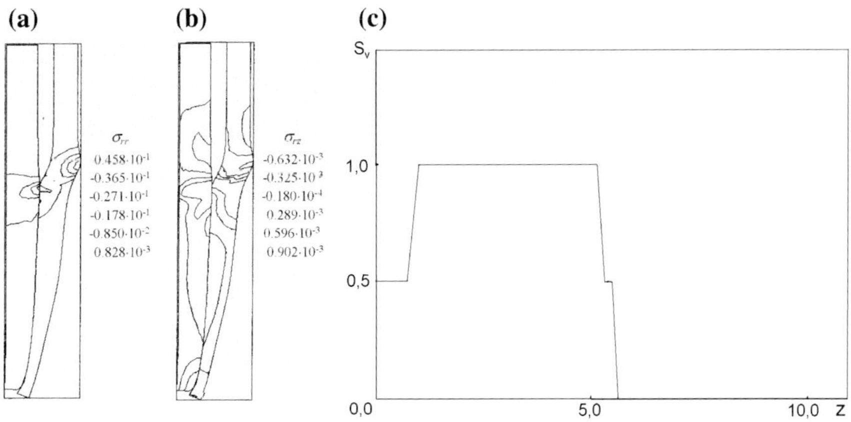

Fig. 14.2 Contour lines: **a** for σ_{rr}, **b** for σ_{rz}, **c** characteristic function S_v along z

which led to severe plastic deformations of the work-pieces, especially in the end zone of the external titanium work-piece.

The following input data were taken. The maximum load in the detonation wave $\sigma_0 = 6.0 \cdot 10^{-2}$, the velocity of the detonation wave $c_d = 0.45$, the duration of the loading pulse $d_{im} = 0.075$, static yield strengths $\tau_{Fe} = 0.0013$, $\tau_{Ti} = 0.0009$, relative thicknesses $d_{Fe}/R = 0.1$, $d_{Ti}/R = 0.075$, the gap between the work-pieces $d_0/R = 0.05$, the critical value of the intensity of plastic deformations $\gamma_{cr} = 0.2$. Figure 14.2a, b shows the contour lines of the stress components σ_{rr} and σ_{rz}.

Quality of welding can be judged by the function S_v (Fig. 14.2c), which is defined at the contact boundary of two bodies and takes values $S_v = 1$ in the contact zone with welding, $S_v = 0.5$ in the contact zone without welding, and $S_v = 0$ outside the contact zone. The function S_v shows that quality of welding of the examined samples is good everywhere except for a small end zone, which is often observed in the experiment (the "melting" of the end, which is usually cut).

14.7 Conclusions

The chapter discusses in detail contact algorithms based on Lagrange multiplier method for explicit finite element schemes and penalty function method for implicit finite element schemes. Examples of the calculation of contact interactions during high-speed processes of the collision of elastic–plastic bodies and explosion welding of tubular samples are presented.

Acknowledgements The study was supported by the Russian Government programs in IPMech RAS and ICAD RAS.

References

1. Wilkins, M.L.: Computer Simulation of Dynamic Phenomena. Springer, New York (1999)
2. Benson, D.J.: Computational methods in Lagrangian and Eulerian hydrocodes. Comput. Meth. Appl. Mech. Engng. **99**, 235–394 (1992)
3. Bourago, N.G.: A survey on contact algorithms. In: International Workshop on Grid Generation and Industrial Applications, Computing Centre of the RAS, pp. 42–59. Moscow (2002)
4. Bourago, N.G., Kukudzhanov, V.N.: A review of contact algorithms. Mech. Solids **40**(1), 35–71 (2005)
5. Fomin, V.M., Gulidov, A.I., Sapozhnikov, G.A., Shabalin, I.I., Babakov, V.A., Kuropatenko, V.F., Kiselev, A.B., Trishin, YuA, Sadyrin, A.I., Kiselev, S.P., Golovlev, I.F.: High velocity interaction of bodies. SB RAS Publ, Novosibirsk (in Russian) (1999)
6. Petrov, I.B., Kholodov, A.S.: Numerical study of some dynamic problems in the solid mechanics by the grid-characteristic method. Comput. Math. Math. Phys. **24**(3), 61–73 (2016)
7. Wriggers, P., Panagiotopoulos, P. (eds.): New developments in contact problems. Springer, Vien GmbH (1999)
8. Barber, J.R., Ciavarella, M.: Contact mechanics. Int. J. Solids Struct. **37**, 29–43 (2000)
9. Golubev, V., Khokhlov, N., Grigorievyh, D., Favorskaya, A.: Numerical simulation of destruction processes by the grid-characteristic method. Procedia Comput. Sci. **126**, 1281–1288 (2018)
10. Burago, N.G., Nikitin, I.S., Nikitin, A.D., Stratula, B.A.: Algorithms for calculation damage processes. Frattura ed Integrità Strutturale **13**(49), 212–224 (2019)
11. Burago, N.G., Nikitin, I.S.: Matrix-free conjugate gradient implementation of implicit schemes. Comput. Math. Math. Phys. **58**(8), 1247–1258 (2018)
12. Burago, N.G.: Modeling of damage in elastic plastic bodies. Comput. Continuum Mechanics **1**(4), 5–20 (2008)
13. Annin, B.D., Sadovskaya, O.V., Sadovskiy, V.M.: Numerical simulation of oblique collision of plates in elastoplastic formulation. Phys. Mesomech. **3**(4), 23–28 (2000)
14. Deribas, A.A.: Physics of Hardening and Explosion Welding. Nauka, Novosibirsk (in Russian) (1980)
15. Godunov, S.K., Zabrodin, A.V., Ivanov, M.Y., Kraiko, A.N., Prokopov, G.P.: Numerical Solution of Multidimensional Problems of Gas Dynamics. Nauka, Moscow (in Russian) (1976)

Chapter 15
Different Approaches for Solving Inverse Seismic Problems in Fractured Media

Vasily I. Golubev, Maxim V. Muratov and Igor B. Petrov

Abstract The inverse seismic problem for oil and gas exploration is investigated. Three different approaches based on the same fundamental grid-characteristic method for solving the direct wave problem were successfully examined. The dynamic behavior of the geological medium is described by the full-wave linear elastic system of equations. Finite difference method on structured meshes was applied. The migration problem in the fractured elastic medium was successfully solved and positions of cracks were recovered. First method calculates the global minimum of the specially constructed functional. Second method uses the modern machine learning approaches for reconstructing the geological model. Third method is based on the traditional migration algorithm with the adjoint operator. It was shown that all examined approaches may be used to solve inverse problems of the seismic survey process.

15.1 Introduction

Information about the properties of geological layers and positions of fractures under the day surface is very important when choosing the places for residential and industrial buildings, exploring oil, and gas deposits [1–3]. Usually methods of acoustics are used to determine these parameters. These methods so-called seismic survey process are cheaper than well drilling.

One of the founders of the seismic migration theory was Claerbout [4, 5]. After the creation of modern high-performance computing systems, significant efforts were

V. I. Golubev (✉) · M. V. Muratov · I. B. Petrov
Moscow Institute of Physics and Technology (National Research University), 9, Institutskii per., Dolgoprudny, Moscow Region 141700, Russian Federation
e-mail: w.golubev@mail.ru

M. V. Muratov
e-mail: max.muratov@gmail.com

I. B. Petrov
e-mail: petrov@mipt.ru

L. C. Jain et al. (eds.), *Advances in Theory and Practice of Computational Mechanics*, Smart Innovation, Systems and Technologies 173,
https://doi.org/10.1007/978-981-15-2600-8_15

made for the development of new precise methods [6, 7]. Initially, all methods were based on the acoustic approach that doesn't take into account the initiation of shear waves and treats the geological medium as the fluid. It was shown, that it leads to the lack of subvertical layer boundaries on migration images. To overcome this drawback, full-wave elastic model was used [8]. Nowadays, the great interest for seismic engineers is the identification of fractured zones. It is associated with the high permeability of the medium and possible availability of hydrocarbons. Mathematical models were developed to take into account the complex structure of media [9–11]. Diffracted waves are under comprehensive study of exploration geophysicists. A lot of research works are devoted to the numerical simulation of seismic responses from fractured media [12–15].

In recent years, machine learning techniques and, in particular, deep neural networks have shown impressive results in many areas, such as computer vision, speech recognition, and machine translation. For example, in the field of computer vision, it was possible to solve many problems previously unsolved, such as the classification problem [16], recognition problem [17], and problem of image generation [18]. One of the significant advantages of deep learning methods is that these methods can be transferred to many other areas related to processing of large amounts of data. One such area is the exploration seismology problems. Several works in this field have already been carried out. In [19], the problem of fault detection in 2D was solved using a deep convolutional neural network. As data for training the neural network, we used synthetic data obtained by solving large direct problems. In [20], a similar problem was solved in 3D. The great advantage that these papers draw attention to the input data for deep learning algorithms, which do not require special processing and, therefore, such methods can be simpler to use than standard exploration seismology methods. Flexibility and relative simplicity make such methods effective for solving practical problems. Thus, in [21] deep neural networks are used to detect CO_2 emissions, and in [22] these methods are used to detect and classify defects in composite materials. The results show that the use of machine learning methods in exploration seismology is important topic for research.

In this research, we investigated three different approaches for solving inverse problems of the seismic survey. Firstly, we constructed the functional of minimization based on synthetic responses from layered and fractured media. All parameters of the model may be estimated by the appropriate minimization procedure. Secondly, we used modern machine learning techniques to reconstruct the fractured structure of the geological medium. Thirdly, we solved the classic migration problem using adjoint operators and the grid-characteristic method on structured meshes.

Chapter is organized as follows. Direct seismic problem solution is discussed in Sect. 15.2. Section 15.3 presents inverse problem solution. Section 15.4 gives the conclusions.

15.2 Direct Seismic Problem Solution

Direct problem solution is an important step in the process of the inverse problem investigation. It allows us to obtain synthetic responses from models with known structures. And it is necessary to have direct solver as a temporary step in the inverse algorithm.

The defining system of equations of a linearly elastic medium can be represented as

$$\rho\frac{\partial V_i}{\partial t} = \frac{\partial \sigma_{ji}}{\partial x_j}, \frac{\partial \sigma_{ij}}{\partial t} = \lambda\left(\sum_k \frac{\partial V_k}{\partial x_k}\right) I_{ij} + \mu\left(\frac{\partial V_i}{\partial x_j} + \frac{\partial V_j}{\partial x_i}\right), \tag{15.1}$$

where V_i is the velocity component, σ_{ij} is the stress tensor, ρ is the density of the medium, λ and μ are the Lame coefficients, I_{ij} is the component of the unit tensor.

Consider the two-dimensional case. Let us introduce the vector of variables $\vec{u} = \{V_x, V_y, \sigma_{xx}, \sigma_{yy}, \sigma_{xy}\}$. Then the system Eq. 15.1 is reduced to the form:

$$\frac{\partial \vec{u}}{\partial t} + \sum_{i=1,2} \boldsymbol{A}_i \frac{\partial \vec{u}}{\partial \xi_i} = 0. \tag{15.2}$$

The numerical solution is found using the grid-characteristic method [23, 24]. Approximation is carried out on a structural rectangular mesh. Values at each point are found using values at mesh reference points $\vec{u}(\vec{r}_{ijkl})$ and weights of these points $p_{ijkl}(\vec{r})$:

$$\vec{u}(\vec{r}) = \sum_{i,j,k,l} p_{ijkl}(\vec{r})\vec{u}(\vec{r}_{ijkl}). \tag{15.3}$$

The boundary condition can be written in the general form as

$$\boldsymbol{D}\vec{u}(\xi_1, \xi_2, t + \tau) = \vec{d}, \tag{15.4}$$

where $\boldsymbol{D}$ is a certain matrix of size 9×3 for the three-dimensional case (5×2 for two-dimensional case), $\vec{d}$ is the vector, $\vec{u}(\xi_1, \xi_2, t + \tau)$ is the value of the required velocity values and components of the stress tensor at the boundary point at the next time step.

At the top boundary of the computational domain, the condition of a free boundary has a view:

$$\boldsymbol{T}\vec{n} = 0. \tag{15.5}$$

To specify the fracture, an infinitely thin fracture model with the condition of fluid-filled fracture was used [25, 26]. Such a fracture is defined as a contact boundary with the condition of free sliding:

$$\vec{v}_a \cdot \vec{n} = \vec{v}_b \cdot \vec{n}, \quad \vec{f}_n^a = -\vec{f}_n^b, \vec{f}_\tau^a = \vec{f}_\tau^b = 0. \tag{15.6}$$

Such a contact boundary completely passes longitudinal oscillations without reflection and fully reflects transverse waves.

The underlying computer code involves schemes of second to fourth orders of accuracy. In this study, we used the fourth-order accurate scheme ($\zeta = \Delta t/h$, h is the spatial coordinate step):

$$\begin{aligned}
v_m^{n+1} &= v_m^n - \zeta(\Delta_1 - \zeta(\Delta_2 - \zeta(\Delta_3 - \zeta\Delta_4))),\\
\Delta_1 &= (-2v_{m+2}^n + 16v_{m+1}^n - 16v_{m-1}^n + 2v_{m-2}^n)/24,\\
\Delta_2 &= (-v_{m+2}^n + 16v_{m+1}^n - 30v_m^n + 16v_{m-1}^n - v_{m-2}^n)/24,\\
\Delta_3 &= (2v_{m+2}^n - 4v_{m+1}^n + 4v_m^n - 2v_{m-2}^n)/24,\\
\Delta_4 &= (v_{m+2}^n - 4v_{m+1}^n + 6v_m^n - 4v_{m-1}^n + v_{m-2}^n)/24.
\end{aligned} \tag{15.7}$$

Additionally, we used the grid-characteristic monotonicity criterion. For positive components of diagonal matrix of eigenvalues, it has the form of Eq. 15.8.

$$\min\{v_m^n, v_{m-1}^n\} \le v_m^{n+1} \le \max\{v_m^n, v_{m-1}^n\} \tag{15.8}$$

For negative components it is symmetric. In the simplest case when this criterion is violated, the solution is corrected as follows:

$$v_m^{n+1} = \begin{cases} \max\{v_m^n, v_{m-1}^n\}, v_m^{n+1} > \max\{v_m^n, v_{m-1}^n\},\\ \min\{v_m^n, v_{m-1}^n\}, v_m^{n+1} < \min\{v_m^n, v_{m-1}^n\},\\ v_m^{n+1}, \min\{v_m^n, v_{m-1}^n\} \le v_m^{n+1} \le \max\{v_m^n, v_{m-1}^n\}. \end{cases} \tag{15.9}$$

This limiter preserves the fourth-order of the scheme in domains, where the solution is fairly smooth (the characteristic criterion is satisfied). In the case of high solution gradients, the order of the scheme is reduced to the third order.

15.3 Inverse Problem Solution

Initially, the inverse problem of determination of number and thicknesses of layers was considered. We assumed that there are a set of layers (from one to three) with known physical properties but unknown thicknesses. Then the vector of parameters includes the number of layers $N = 1, 2, 3$ and the thickness of each layer $H_{\min} \le h_k \le H_{\max}$, $k = 1, 2, 3$. Geological parameters of the first layer are given as follows $C_p = 2000$ m/s, $C_s = 1400$ m/s, $\rho = 2000$ kg/m^3, the parameters of the second layer are given as follows $C_p = 2400$ m/s, $C_s = 1600$ m/s, $\rho = 2500$ kg/m^3, and the

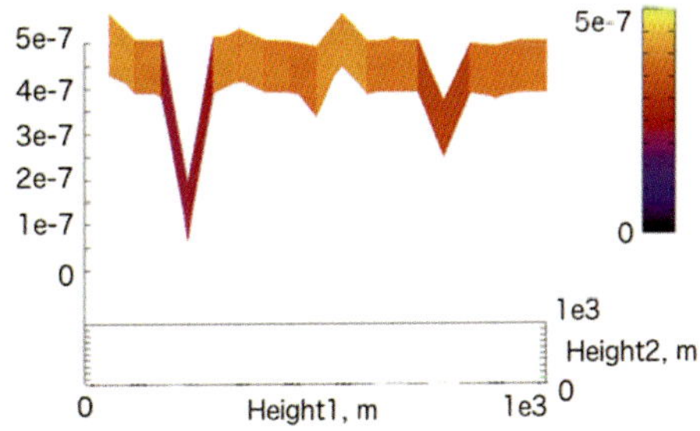

Fig. 15.1 The dependence of $I(z)$ on two-layer heights. One global and many local extremes

parameters of the third layer are given as follows $C_p = 2600$ m/s, $C_s = 1700$ m/s, ρ = 2800 kg/m^3. Berlage source was used with the main frequency of 30 Hz.

Peculiarity of the problem is the fact that information can be obtained only from acoustic measurements. Let us consider 1D case. On the day surface, one geophone is situated which measures vertical velocity $\tilde{V}_y(x_i, t_j)$ of the ground resulting from wave reflections from layers boundaries. And we can try to found the value of $\vec{z}$ when the numerical response $V_y(\vec{z}, x_i, t_j)$ will be as close as possible to $\tilde{V}_y(x_i, t_j)$. Seismic registrations were carried out with spatial step of 10 m.

The mathematical problem can be formulated as the optimization problem of least squares:

$$\min I(\vec{z}), \vec{z} \in \{1, 2, 3\} \times [H_{\min}, H_{\max}]^{\{1,2,3\}}, \tag{15.10}$$

$$I(\vec{z}) = \sum_i \sum_j \left[V_y(\vec{z}, x_i, t_j) - \tilde{V}_y(x_i, t_j) \right]^2. \tag{15.11}$$

Equation 15.11 has no analytical form, and each its value can be obtained only as a result of rather time-consuming numerical calculations. For solving the problem (Eqs. 15.10 and 15.11), we use the direct methods of sorting. The results are presented in Fig. 15.1.

We have found that functional (Eq. 15.11) has one strong global minimum and some small local minimums. This functional is more sensitive to the thickness of the first layer than to the second layer. The main reason for this fact is that the amplitude of the first response exceeds the amplitude of other responses.

The same approach was applied to the seismic survey problem of fractured media. We used the homogeneous half-space model with the free day surface. It contained a single linear fluid-filled crack. Then the vector of parameters includes the crack depth h, $h_1 \leq h \leq h_2$, and the orientation angle α, $\alpha_1 \leq \alpha \leq \alpha_2$. The illustration is represented in Fig. 15.2.

The homogeneous medium had density 2500 kg/m^3, P-wave velocity 3000 m/s, S-wave velocity 1700 m/s. The length of the crack was 100 m. The initial perturbation was P-wave with the length of 100 m. Vertical component of velocity was registered with the spatial distance of 10 m at the day surface.

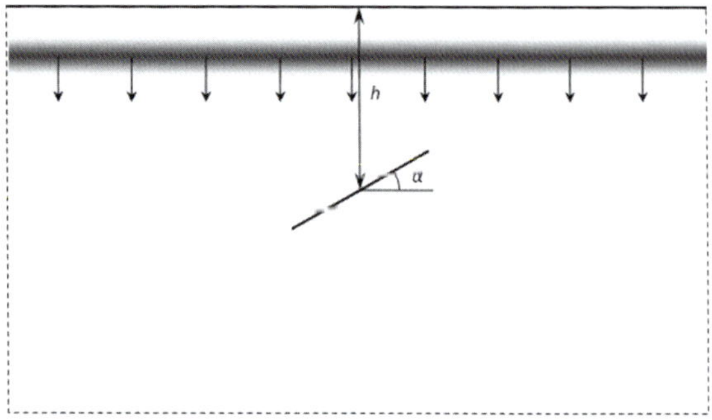

Fig. 15.2 The experiment for the crack identification

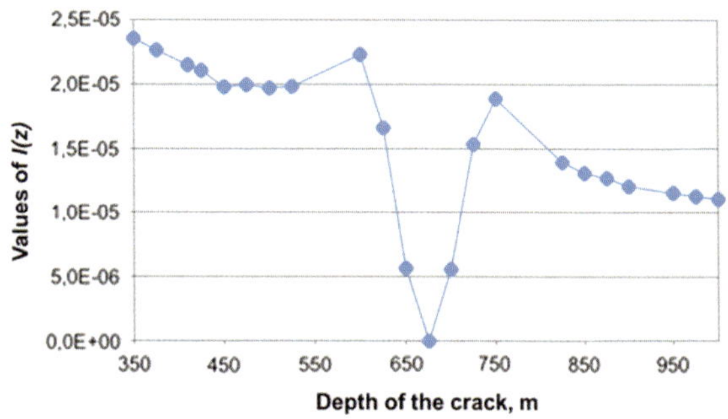

Fig. 15.3 The cross section of minimization functional for fixed crack angle

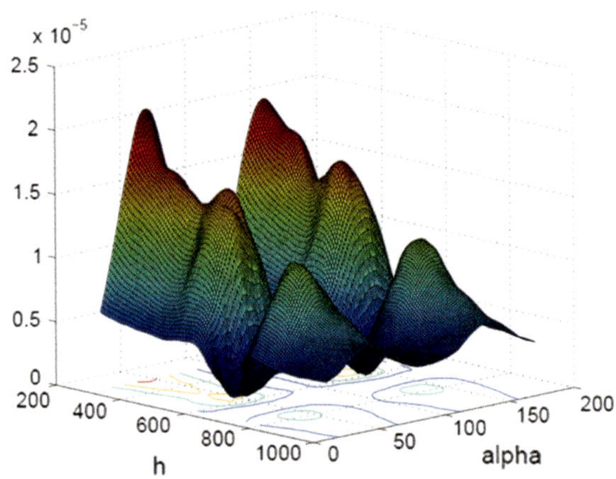

Fig. 15.4 The minimization functional as a function of crack angle (alpha) and crack depth (h)

The mathematical problem can be formulated as the optimization problem of least squares (Eq. 15.11) on the special domain:

$$\min I(\vec{z}), \mathbf{z} \in D = [h_1; h_2] \times [\alpha_1; \alpha_2]. \tag{15.12}$$

In Figs. 15.3 and 15.4, the cross section as 2D plot and 3D plot are represented. As in the previous example, a single global minimum exists.

Another approach based on machine learning methods was applied to the problem of cracks identification. The process of solving the problem of recognizing the spatial position of a fracture in the elastic medium consists of two stages: training of the neural network and recognizing of a control sample of seismic data. To create a training sample, the direct problems are solved with different fracture parameters.

A two-dimensional problem is considered, in which it is proposed to find the spatial position and angle of inclination of a single fracture of a fixed size (100 m) using seismic data. The fracture is in a homogeneous elastic medium with the following elastic characteristics: $C_p = 4500$ m/s, $C_s = 2500$ m/s, $\rho = 2500$ kg/m^3. The size of the computational domain is 2 km × 2 km (Fig. 15.5). The position of the fracture varies in the range of 1000 m vertically and horizontally. The angle of inclination is in the range of ±15° (subvertical fracture). In the middle of the border of the study area, a sinusoidal elastic pulse consisting of 5 periods (a wavelength of 100 m) is excited. The values of the vertical component of the velocity of the reflected waves are recorded on seismic receivers uniformly located on the excitation surface of the wave pulse (65 receivers in total).

Keras deep learning library based on Tensorflow library and CUDA parallel computing architecture was used. Keras library was chosen because of its simplicity of use and possibilities sufficient to solve the problem. In this chapter, a two-dimensional inverse problem was solved, but the proposed method can be extended to the case of 3D. To solve the problem, a neural network consisting of 3 convolutional layers and two fully connected layers was proposed. The training set consisted of pairs (X, y),

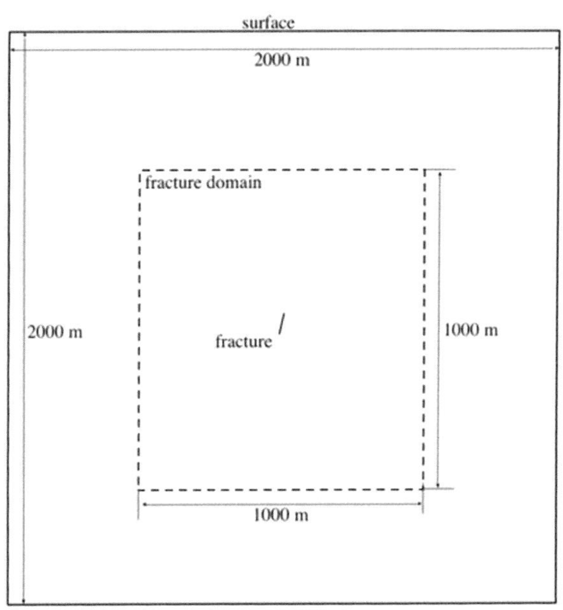

Fig. 15.5 The scheme of fracture placement

where X is a seismogram (a matrix of real numbers 65×65 in size) and y is a set of parameters defining the position of the fracture. In the case under consideration, y was given by 4 real numbers—the coordinates of the ends of the fractures (the height of the fracture is remained by constant). The network has the folloing architecture: the first convolutional layer (63, 63, 64), the second convolutional layer (29, 29, 128), the third convolutional layer (12, 12, 256), the first fully connected layer (9216, 256), and the second fully connected layer (256, 4). The dimensions of all filters in the convolutional layers are 3×3, the activation function is ReLU. The total number of network parameters: 2.7×106. As the optimizer of the neural network, Adam was chosen with a learning rate of 0.001.

The recognition of a control sample is the process of minimizing the functional:

$$J = \sqrt{\sum_i \left\| y_i^{real} - y_i^{pred} \right\|_{L_2}}. \tag{15.13}$$

With each new era in learning, the value of the functional decreases and tends to a certain value (Fig. 15.6). Therefore, the method can be used to solve this class of problems.

The results of recognition of a single fracture are shown in Fig. 15.7, and Fig. 15.8 shows the seismogram of the wave response obtained by seismic receivers on the upper surface of the studied region. Figure 15.8 shows the location and orientation of the real position of the crack and predicted. It is seen quite a good match.

The last problem investigated in this research was the classic 2D migration problem. The goal was to compare two different approaches taken into account the dependency of P-wave and S-wave velocities on the depth. Firstly, a separate grid was constructed for each geological layer. And each grid had its own elastic properties (Lame parameters). The special glue condition was consistently applied to ensure the continuity of the stress tensor and the velocity vector over the contact boundary. In the second approach, we blurred parameters of all layers along the vertical axis and used the numerical schemes with the matrix elements with the spatial dependency.

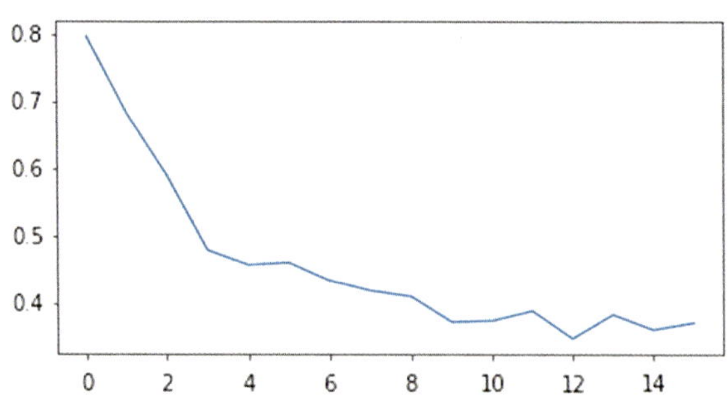

Fig. 15.6 The graph of functional J dependence on period of study

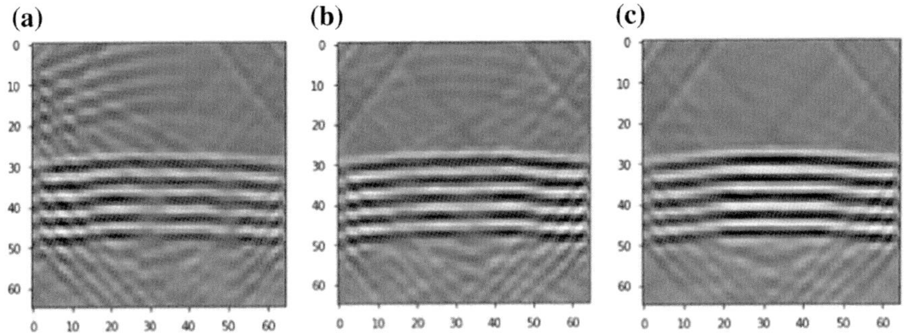

Fig. 15.7 Control seismograms for recognition: **a** vertical cracks, **b** subvertical left rotated crack, **c** subvertical right rotated crack

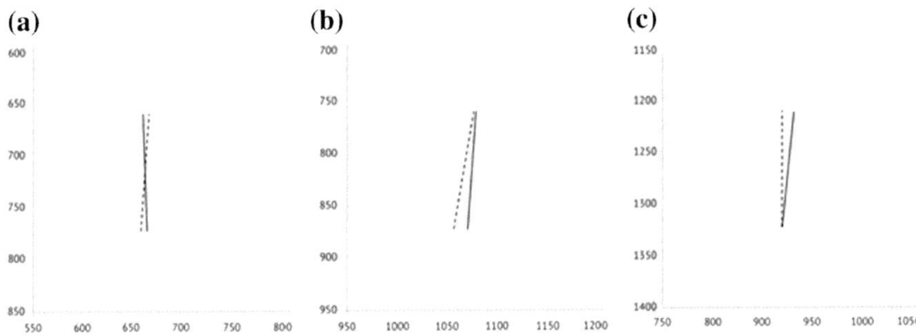

Fig. 15.8 Spatial placement of fractures, real (dotted line), and predicted (solid line): **a** vertical cracks, **b** subvertical left rotated crack, **c** subvertical right rotated crack

With the help of curvilinear structured meshes, we put inside the model subvertical fluid-filled cracks. Eliminating the drastic increase of the problem computational complexity, we decided to use specific boundary conditions rather than solving directly the acoustic equation inside the crack volume. This approach was successfully verified previously on direct seismic problems for different crack orientations.

To obtain the migration image of the layered fractured medium the method proposed in [8] was used. We analyzed the kernel in a form of Eq. 15.14, where $\boldsymbol{v}$ is the velocity vector obtained as a solution of the direct problem and $\boldsymbol{v}^{\dagger}$ is the velocity vector obtained as a solution of the adjoint problem.

$$K_{imp}(\boldsymbol{x}) = -\rho(\boldsymbol{x}) \int \boldsymbol{v}^{\dagger}(\boldsymbol{x}, -t)\boldsymbol{v}(\boldsymbol{x}, t)dt \tag{15.14}$$

Both direct and adjoint problems were solved numerically with the grid-characteristic method.

Table 15.1 Elastic properties of geological layers

Number of layer	Width, m	P-wave velocity, m/s	S-wave velocity, m/s
1	500	3500	1750
2	1000	4500	2250
3	300	5000	2500
4	100	4000	2000
5	300	5500	2750
6	800	5500	2750

In this research, we extended our previous numerical experiments with the migration process of seismic data from layered fractured media presented by [27–29]. We used the same 2D layered geological reference model consisted of six elastic horizontal layers with different geometrical and physical properties. The density was constant along with the depth and was equal to 2500 kg/m^3. We enumerated all layers from 1 (top) to 6 (bottom). It is illustrated at Table 15.1 and Fig. 15.9. Identical elastic parameters for two last layers were used to illustrate the correctness of our special glue contact conditions. At the depth of 1650 m, two fluid-filled subvertical cracks were placed. The length of both cracks was 100 m, and the angle was 10° from vertical. The curvilinear structured grid with the spatial step approximately 5 m and with ~10^6 nodes was used. Ricker point sources with central frequency 30 Hz were placed at each 10 m on the day surface. Three-component receivers were placed at the same places. Totally 3.3 s of physical time were simulated with the time step 0.45 ms.

At the initial step, the direct problem was numerically solved for the reference layered fractured medium. Obtained synthetic seismograms were used in the further processing chain. According to the described above procedure, the migration image was obtained with the layered background model in [27] (Fig. 15.10). Cracks positions are clearly seen on it, and all layer boundaries were filtered naturally.

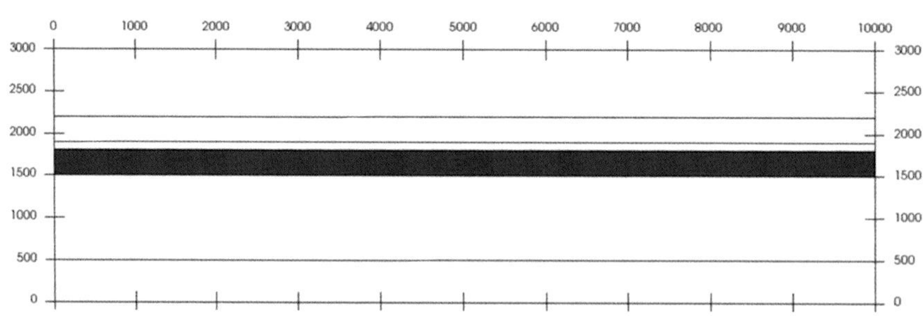

Fig. 15.9 Two-dimensional model of the geological layered medium extended with two subvertical fluid-filled cracks (inside the blue layer). Contact boundaries are depicted with black horizontal lines

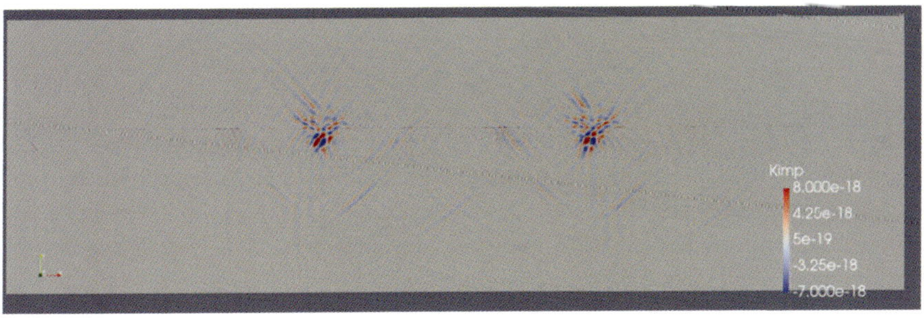

Fig. 15.10 Migration image, obtained with the layered background model [27]. Cracks positions can be identified

Unfortunately, the precise positions of layer boundaries are not known in most cases. After the seismic inversion procedure, the background model is the 3D massive with the dependency of elastic properties on coordinates. To estimate the influence of the background model inaccuracy on the migration image, a new numerical experiment was carried out. We blurred our layered model with Gauss filter along the vertical axis with the 20 m window and used it in the migration pipeline (Fig. 15.11). Due to the discrepancy of background models, the obtained migration image contains not only correct crack positions but layer boundaries too. And their amplitudes are significantly higher than the signal from cracks. The second problem is the presence of two false boundaries with the large enough signal on the final migration image.

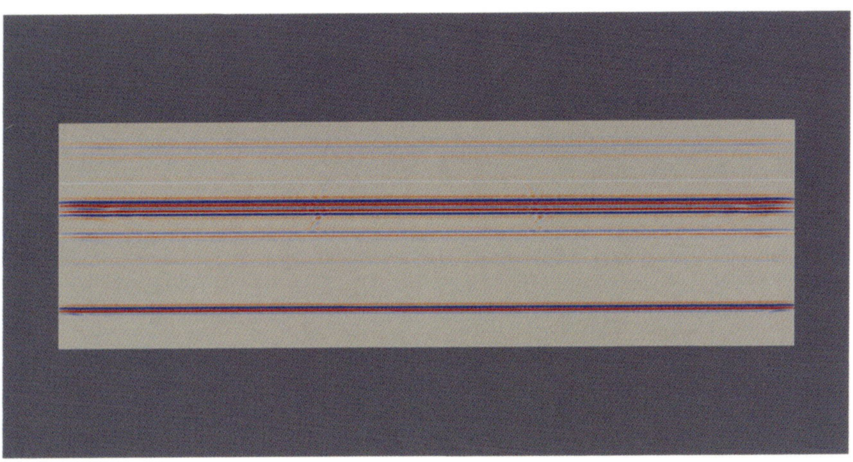

Fig. 15.11 Migration image obtained with the blurred layered background model. Gauss impulse with the 20 m window was applied. True positions of layer boundaries are depicted with white lines

15.4 Conclusions

The migration problem in the fractured elastic medium was successfully solved and positions of cracks were recovered. First method calculated the global minimum of the specially constructed functional. Second method used modern machine learning approaches for reconstructing the geological model. Third method was based on the traditional migration algorithm with the adjoint operator. It was shown that all examined approaches may be used to solve inverse problems of the seismic survey process.

This research is the continuation of the work dedicated for proving the applicability of the grid-characteristic numerical method to migration problems of fractured media. The full-wave simulation with the linear elastic approach was carried out. The blurred background model constructed from the precise layered model was used. The signal from both subvertical fluid-filled cracks was successfully registered. Unfortunately, its amplitude on the migration image is significantly lower than the response from true and false obtained boundaries. The further research may be concentrated on the development of the filtration procedure for eliminating the spatially correlated signal from true layer boundaries and false boundaries too.

Acknowledgements This research has been carried out using computing resources of the federal collective usage center Complex for Simulation and Data Processing for Mega-science Facilities at NRC "Kurchatov Institute", http://ckp.nrcki.ru/. This research was carried out with the financial support of the Russian Science Foundation, project no. 19-11-00023.

References

1. Binnani, N., Khare, R.K., Golubev, V.I., Petrov, I.B.: Probabilistic seismic hazard analysis of punasa dam site in India. In: Petrov, I.B., Favorskaya, A.V., Favorskaya, M.N., Simakov, S.S., Jain, L.C. (eds.) Proceedings of the Conference on 50 Years of the Development of Grid-Characteristic Method, SIST, vol. 133, pp. 105–119. Springer (2019)
2. Stognii, P.V., Khokhlov, N.I.: 2D seismic prospecting of gas pockets. In: Petrov, I.B., Favorskaya, A.V., Favorskaya, M.N., Simakov, S.S., Jain, L.C. (eds.) Proceedings of the Conference on 50 Years of the Development of Grid-Characteristic Method, SIST, vol. 133, pp. 156–166. Springer (2019)
3. Golubev, V.I., Golubeva, Y.A.: Full-wave simulation of the earthquake initiation process. In: CEUR Workshop Proceedings, vol. 2267, pp. 346–350 (2018)
4. Claerbout, J.F.: Coarse grid calculations of waves in inhomogeneous media with application to delineation of complicated seismic structure. Geophysics **36**(3), 407–418 (1970)
5. Claerbout, J.F., Doherty, S.M.: Downward continuation of moveout-corrected seismograms. Geophysics **37**(5), 741–768 (1972)
6. Etgen, J., Gray, S., Zhang, Y.: An overview of depth imaging in exploration geophysics. Geophysics **74**, WCA5–WCA17 (2009)
7. Jiao, K., Huang, W., Vigh, D., Kapoor, J., Coates, R., Starr, W.E., Cheng, X.: Elastic migration for improving salt and subsalt imaging and inversion. In: SEG Las Vegas Annual Meeting, pp. 1–5 (2012)

8. Luo, Y., Tromp, J., Denel, B., Calandra, H.: 3D coupled acoustic-elastic migration with topography and bathymetry based on spectral-element and adjoint methods. Geophysics **78**(4), S193–S202 (2013)
9. Burago, N.G., Nikitin, I.S., Yakushev, V.L.: Hybrid numerical method with adaptive overlapping meshes for solving nonstationary Problems in Continuum Mechanics. Comput. Math. Math. Phys. **56**(6), 1065–1074 (2016)
10. Nikitin, I.S., Burago, N.G., Nikitin, A.D.: Explicit-Implicit schemes for solving the problems of the dynamics of isotropic and anisotropic elastoviscoplastic media. In: IOP Conference Series: Journal of Physics: Conference Series, vol. 1158, pp. 032039.1–032039.8 (2019)
11. Burago, N.G., Nikitin, I.S.: Algorithms of through calculation for damage processes. Comput. Res. Model. **10**(5), 645–666 (2018)
12. Fang, X., Zheng, Y., Fehler, M.C.: Fracture clustering effect on amplitude variation with offset and azimuth analyses. Geophysics **82**(1), N13–N25 (2017)
13. Muratov, M.V., Petrov, I.B.: Application of fractures mathematical models in exploration seismology problems modeling. In: Petrov, I.B., Favorskaya, A.V., Favorskaya, M.N., Simakov, S.S., Jain, L.C. (eds.) Proceedings of the Conference on 50 Years of the Development of Grid-Characteristic Method, SIST, vol. 133, pp. 120–131. Springer (2019)
14. Zheng, Y., Fang, X., Fehler, M.C., Burns, D.R.: Seismic characterization of fractured reservoirs using 3D double beams. In: SEG Technical Program Expanded Abstracts, pp. 1–6 (2012)
15. Hu, H., Zhengm, Y.: 3D seismic characterization of fractures in a dipping layer using the double-beam method. Geophysics **83**, 123–134 (2018)
16. Krizhevsky, A., Sustskever, I., Hinton, G.E.: ImageNet classification with deep convolutional neural networks. In: 25th International Conference on Neural Information Processing Systems, vol. 1, pp. 1097–1105 (2012)
17. Szegedy, C., Toshev, A., Erhan, D.: Deep neural networks for object detection. In: 26th International Conference on Neural Information Processing Systems, vol. 2, pp. 2553–2561 (2013)
18. Goodfellow, I., Pouget-Abadie, J., Mirza, M., Xu, B., Warde-Farley, D., Ozair, S., Courville, A., Bengio, Y.: Generative adversarial nets. In: 27th International Conference Neural Information Processing Systems, vol. 2, pp. 2672–2680 (2014)
19. Zhang, C., Frogner, C., Araya-Polo, M., Hohl, D.: Machine-learning based automated fault detection in seismic traces. In: EAGE Conference and Exhibition, pp. 1–5 (2014)
20. Araya-Polo, M., Dahlke, T., Frogner, C., Zhang, C., Poggio, T., Hohl, D.: Automated fault detection without seismic processing. Lead. Eadge **36**(3), 194–280 (2017)
21. Wu, Y., Lin, Y., Zhou, Z., Delorey, A.: Seismic-net: a deep densely connected neural network to detect seismic events, pp. 1–8. arXiv:1802.02241 (2018)
22. Menga, M., Chua, Y.J., Woutersonb, E., Ong, C.P.K.: Ultrasonic signal classification and imaging system for composite materials via deep convolutional neural networks. Neurocomputing **257**, 128–135 (2017)
23. Ivanov, A.M., Khokhlov, N.I.: Efficient inter-process communication in parallel implementation of grid-characteristic method. In: Petrov, I.B., Favorskaya, A.V., Favorskaya, M.N., Simakov, S.S., Jain, L.C. (eds.) Proceedings of the Conference on 50 Years of the Development of Grid-Characteristic Method, SIST, vol. 133, pp. 91–102. Springer (2019)
24. Khokhlov, N.I., Golubev, V.I.: On the class of compact grid-characteristic schemes. In: Petrov, I.B., Favorskaya, A.V., Favorskaya, M.N., Simakov, S.S., Jain, L.C. (eds.) Proceedings of the Conference on 50 Years of the Development of Grid-Characteristic Method, SIST, vol. 133, pp. 64–77. Springer (2019)
25. Golubev, V.I.: The usage of grid-characteristic method in seismic migration problems. In: Petrov, I.B., Favorskaya, A.V., Favorskaya, M.N., Simakov, S.S., Jain, L.C. (eds.) Proceedings of the Conference on 50 Years of the Development of Grid-Characteristic Method, SIST, vol. 133, pp. 143–155. Springer (2019)
26. Golubev, V., Khokhlov, N., Grigorievyh, D., Favorskaya, A.: Numerical simulation of destruction processes by the grid-characteristic method. Procedia Comput. Sci. **126**, 1281–1288 (2018)

27. Golubev, V.I., Voinov, O.Y., Petrov, I.B.: Migration of seismic data for multi-layered fractured geological media using elastic approach. In: 19th Science and Applied Research Conference Oil and Gas Geological Exploration and Development, pp. 43756.1–43756.5 (in Russian) (2017)
28. Golubev, V.I., Voinov, O.Y., Zhuravlev, Y.I.: On seismic imaging of fractured geological media. Dokl. Math. **96**(2), 514–516 (2017)
29. Golubev, V.I., Voinov, O.Y., Petrov, I.B.: Seismic imaging of fractured elastic media on the basis of the grid-characteristic method. Comput. Math. Math. Phys. **58**(8), 1309–1315 (2018)

Chapter 16
Elastic Wave Scattering on a Gas-Filled Fracture Perpendicular to Plane P-Wave Front

Alena V. Favorskaya

Abstract This chapter discusses the features of the scattering of plane P-waves on gas-filled fractures located along with the motion of the incident wave front. This problem has practical significance in the areas of nondestructive testing and seismic exploration, primarily in the area of railway nondestructive testing. This type of fractures falls into the blind zone. In this chapter, such types of reflected waves are considered that can be registered, and, thus, the blind zone of the recording equipment can be avoided. Analytical expressions for reflected waves amplitudes and scattering angles are obtained. To obtain these expressions, the Wave Logica approach was used. This approach combines the advantages of the analytical study of wave fields and study of the computational solution of the elastic wave equation. Comparison of the analytical expressions with visualized wave fields (wave patterns) at the stage of derivation of these analytical expressions greatly facilitates the study that allows to avoid mistakes and also demonstrates the accuracy of the applied computational method. In this chapter, the grid-characteristic numerical method was used for the numerical solution of the elastic wave equation.

16.1 Introduction

Exact analytical solutions of the wave equation, which are commonly called waves of various types, such as longitudinal P-waves, shear S-waves, Rayleigh, Love, Stoneley, Krauklis, exchangeable, etc., only approximately describe the wave processes observed in real substantially heterogeneous media, as they occur interaction of

A. V. Favorskaya (✉)
Moscow Institute of Physics and Technology (National Research University), 9, Institutsky per., Dolgoprudny, Moscow 141700, Russian Federation
e-mail: aleanera@yandex.ru

National Research Centre "Kurchatov Institute", 1, Akademika Kurchatova pl., Moscow 123182, Russian Federation

Scientific Research Institute for System Analysis of the RAS, 36(1), Nahimovskij pr., Moscow 117218, Russian Federation

L. C. Jain et al. (eds.), *Advances in Theory and Practice of Computational Mechanics*, Smart Innovation, Systems and Technologies 173,
https://doi.org/10.1007/978-981-15-2600-8_16

waves (multiple reflections, interference, etc.). Obtaining an appropriate analytical solution, for example, for Krauklis waves, can be found in [1]. Also, a rigorous analytical approach aimed at studying solutions of hyperbolic systems of equations, most fully presented in [2], is known. Another approach is a full-wave numerical modeling of wave effects [3–7]. In [8–11], a combined approach for studying spatial dynamic wave fields called Wave Logica was proposed.

Note that the approach used in the analysis of spatial dynamic wave fields calculated by full-wave computer simulation describes the dynamics of wave propagation and structure of wave fronts in more detail than the classical approach, which implies the explicit selection of particular analytical solutions of the wave equation and more accurately than the geometric approximation and ray-tracing method [12, 13]. Accordingly, the proposed approach allows to identify the patterns and relationships more accurately that can later be used in applications.

For numerical simulation of ultrasonic nondestructive railway testing, the finite element method [7], the mass-spring-lattice model [14], the finite-difference method [15], and the quasi-analytical finite element method (semi-analytical finite element method) [16] are used.

The infinitely thin gas-filled and fluid-filled fractures are particular limiting cases of Schoenberg family of models [17]. Experimental confirmation of Schoenberg theory was given in [18]. Finite-difference methods have gained wide popularity due to ease of implementation [19]. In this work, the so-called homogeneous approach to finite-difference modeling of dynamic processes in an elastic body was used. In this approach, the boundary conditions for inhomogeneities are explicitly specified, which allows one to simulate an open fracture with a discontinuous (as it passes through its plane) displacement and a continuous stress. The effect of fracture size and density on seismic wave propagation was investigated in [3], while analyzing sections of wave fields were not significantly represented, and a number of qualitative but non-detailed results were made. Also, in the work there is no information about the analysis of the wave front in the dynamics, and a comparison of the sections was made only at a fixed point in time. The effect of the ratio of the wavelength to the length of the fracture on the reflection amplitude was numerically demonstrated in [20]. Three-dimensional calculations in a medium containing two-dimensional fractures with explicit formulation of the boundary conditions were performed [21]. In [22], the dependence of the amplitude of reflection and diffraction on the characteristics of fractures was investigated.

The grid-characteristic method was proposed in [23]. Later a family of numerical methods was developed [24] to simulate the process of ultrasound nondestructive testing of railway tracks [25, 26], composite materials [27], modeling the process of seismic exploration of fractured zones [6, 28], etc.

This chapter is structured as follows. Section 16.2 presents the problem statement and mechanical mathematical model of the problem. Section 16.3 shows the elastic wave patterns, which are visualized elastic wave fields. Section 16.4 deals with the amplitudes of velocity and Cauchy stress tensor components for the wave types under consideration. The derivation of the formulae of scattering amplitudes and angles is discussed in Sect. 16.5. Section 16.6 concludes the chapter.

16.2 Problem Statement

In this section, the problem statement is discussed. The calculations were carried out in mechanical mathematical statements, the scheme of which is shown in Fig. 16.1.

The gas-filled fracture is modeled as two free boundaries along OY axis with conditions according to Eqs. 16.1–16.2.

$$\sigma_{xx} = 0 \tag{16.1}$$

$$\sigma_{xy} = 0 \tag{16.2}$$

In Eqs. 16.1–16.2, σ_{xx} and σ_{xy} are the Cauchy stress tensor components.

Equations 16.1–16.2 are similar with the following formula:

$$\mathbf{f} \equiv \boldsymbol{\sigma} \cdot \mathbf{n} = 0. \tag{16.3}$$

In Eq. 16.3, $\mathbf{f}$ is the traction, $\boldsymbol{\sigma}$ is the Cauchy stress tensor, $\mathbf{n}$ is the normal being out to the borders of the fracture.

Grid-characteristic computational method [24] was used to solve elastic wave equation given by Eqs. 16.4–16.5.

$$\rho \frac{\partial}{\partial t} \mathbf{v}(\mathbf{r}, t) = (\nabla \cdot \boldsymbol{\sigma}(\mathbf{r}, t))^{\mathrm{T}} \tag{16.4}$$

$$\frac{\partial}{\partial t} \boldsymbol{\sigma}(\mathbf{r}, t) = \left(\rho c_{\mathrm{P}}^2 - 2\rho c_{\mathrm{S}}^2\right)(\nabla \cdot \mathbf{v}(\mathbf{r}, t))\mathbf{I} + \rho c_{\mathrm{S}}^2\left(\nabla \otimes \mathbf{v}(\mathbf{r}, t) + (\nabla \otimes \mathbf{v}(\mathbf{r}, t))^{\mathrm{T}}\right) \tag{16.5}$$

In Eqs. 16.4–16.5, $\mathbf{v}(\mathbf{r}, t)$ is the velocity vector-function, $\boldsymbol{\sigma}(\mathbf{r}, t)$ is the symmetric Cauchy stress tensor-function, ρ is the material density, c_{P} is the P-wave speed, c_{S} is the S-wave speed, t is the time, $\mathbf{r}$ is the radius vector.

The size of the integration area and the time were chosen in such a way that it corresponded to the infinite space in Fig. 16.1, and the responses from non-reflecting boundary conditions would not have time to reach the area near the fracture, around which wave processes were studied.

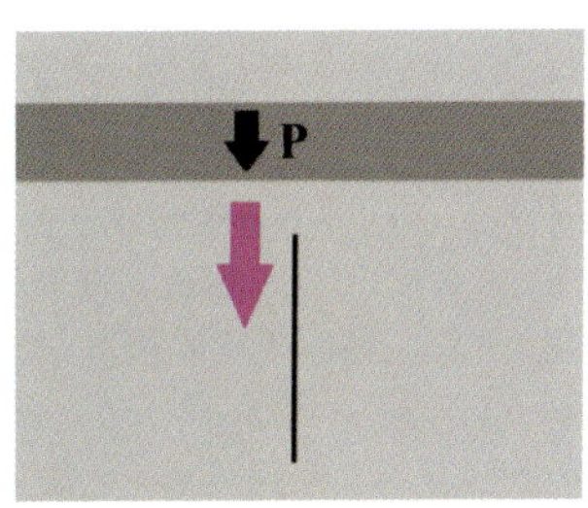

Fig. 16.1 Scheme of mechanical mathematical models, incident P-wave, and gas-filled fracture

Table 16.1 Parameters of elastic media in mechanical mathematical models

Model	P-wave speed (m/s)	S-wave speed (m/s)	Density (kg/m^3)
"cs_norm"	6,250	3,200	7,800
"cs_mini"	6,250	2,000	7,800

Different elastic parameters of the media for two models under consideration are shown in Table 16.1.

The parameters of the computational grid in time and coordinates were chosen in such a way that the visualized wave fields (wave patterns) would coincide with each other upon further refinement of the computational grid step.

16.3 Elastic Wave Patterns

In this section, the elastic wave patterns of various components and different time moments are presented. Figures 16.2 and 16.3 show the results of computational experiments for "cs_mini" and "cs_norm" models, respectively.

In Figs. 16.2, 16.3, the maximum value of the velocity modulus is in red color, middle value is in white color, and zero value is in blue color, while the maximum positive value of velocity and Cauchy stress tensor components is in purple color, maximum negative is in green color, and zero value is in gray color. Note that as maximum values it is not meant the maximum values of the functions, but the maximum values on the scale, in accordance with the principle set forth in the work [29].

Figure 16.4 shows the wave traveling and reflection scheme. This scheme was obtained by analyzing the dynamics of the snapshots of wave fields, the so-called analysis of wave patterns.

In Fig. 16.4 and further in the text, L is the fracture length, D is the scattered S-waves' fronts length, h_P is the incident P-wave wavelet length, h_S is the scattered S-waves wavelet length.

Thus, the scattered S-waves were identified, which are of primary interest. There are also the cylindrical P- and S-waves scattered from the upper and lower edges of the fracture, but they have smaller amplitude.

16.4 Amplitudes of Private Solutions Components

In this section, a derivation of the formulae for the amplitudes of velocity and Cauchy stress tensor components for different private solutions of elastic wave equations figured in the problem are considered. In accordance with Wave Logica approach [8–11], the obtained amplitudes of particular solutions were compared with elastic wave patterns.

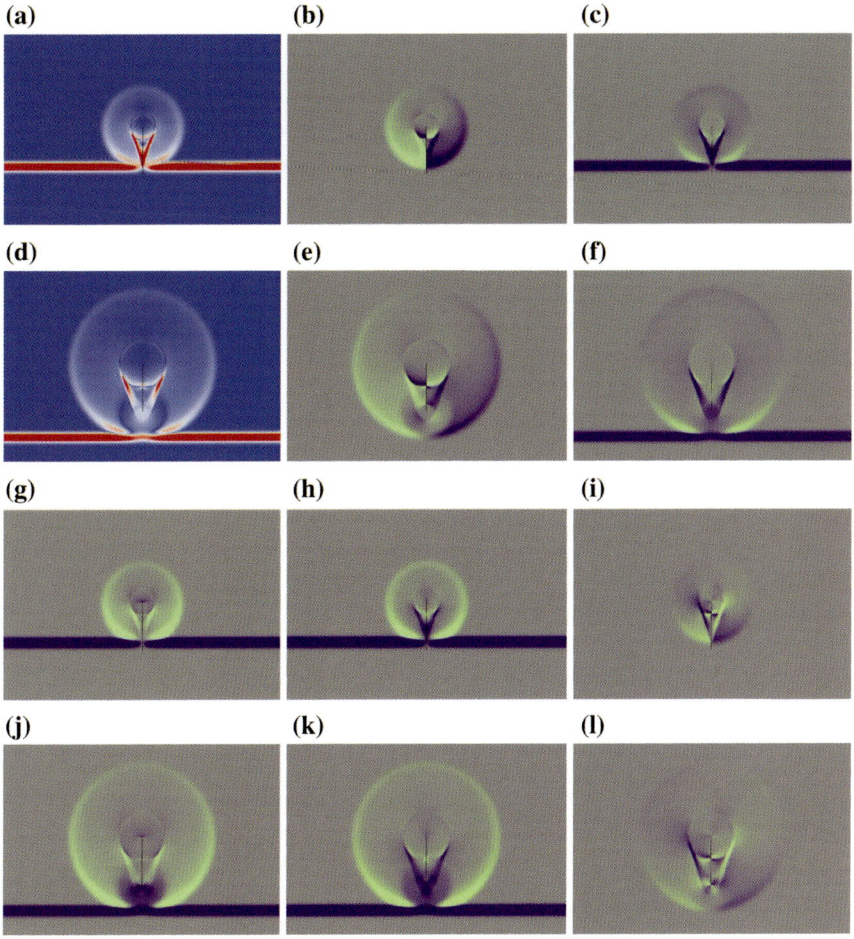

Fig. 16.2 Wave patterns for "cs_mini" model: **a** 21st snapshot, velocity module, **b** 21st snapshot, horizontal component of velocity, **c** 21st snapshot, vertical component of velocity, **d** 35th snapshot, velocity module, **e** 35th snapshot, horizontal component of velocity, **f** 35th snapshot, vertical component of velocity, **g** 21st snapshot, main horizontal component of Cauchy stress tensor XX, **h** 21st snapshot, main vertical component of Cauchy stress tensor YY, **i** 21st snapshot, tangential component of Cauchy stress tensor XY, **j** 35th snapshot, main horizontal component of Cauchy stress tensor XX, **k** 35th snapshot, main vertical component of Cauchy stress tensor YY, **l** 35th snapshot, tangential component of Cauchy stress tensor XY

In accordance with Riemann invariants, for the incident P-wave (Fig. 16.4) traveling oppositely axis OY there are Eqs. 16.6–16.10.

$$v_x^{\mathrm{P}} = 0 \tag{16.6}$$

$$v_y^{\mathrm{P}} = V_{\mathrm{P}} \tag{16.7}$$

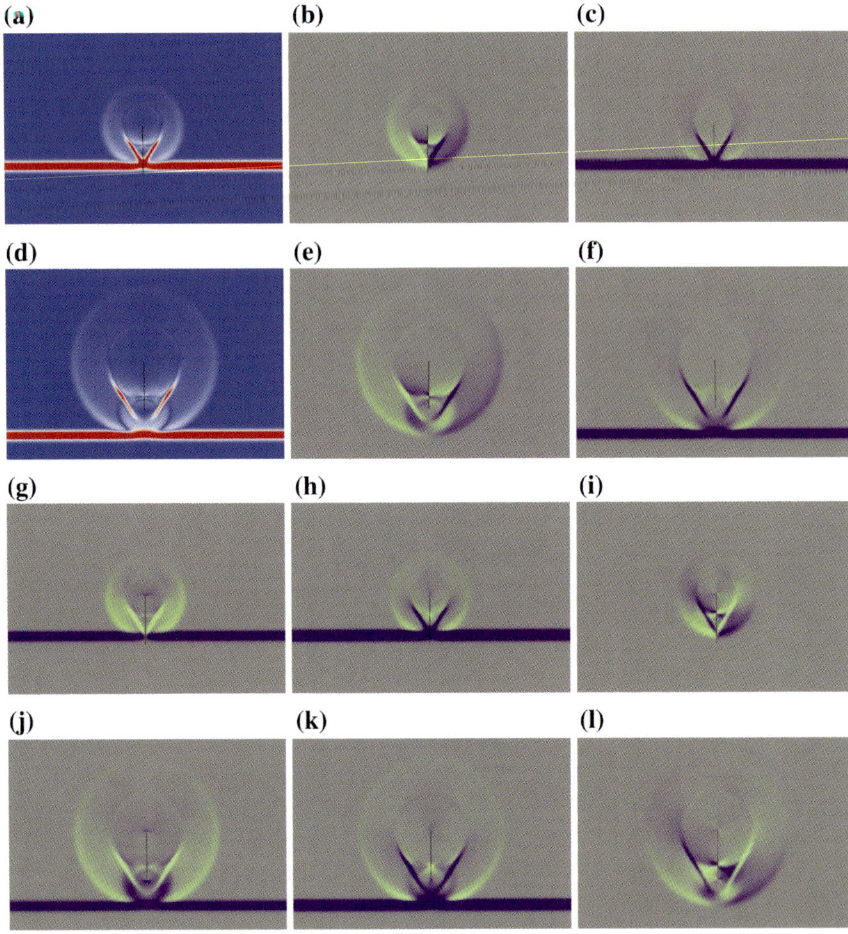

Fig. 16.3 Wave patterns for "cs_norm" model: **a** 21st snapshot, velocity module, **b** 21st snapshot, horizontal component of velocity, **c** 21st snapshot, vertical component of velocity, **d** 35th snapshot, velocity module, **e** 35th snapshot, horizontal component of velocity, **f** 35th snapshot, vertical component of velocity, **g** 21st snapshot, main horizontal component of Cauchy stress tensor XX, **h** 21st snapshot, main vertical component of Cauchy stress tensor YY, **i** 21st snapshot, tangential component of Cauchy stress tensor XY, **j** 35th snapshot, main horizontal component of Cauchy stress tensor XX, **k** 35th snapshot, main vertical component of Cauchy stress tensor YY, **l** 35th snapshot, tangential component of Cauchy stress tensor XY

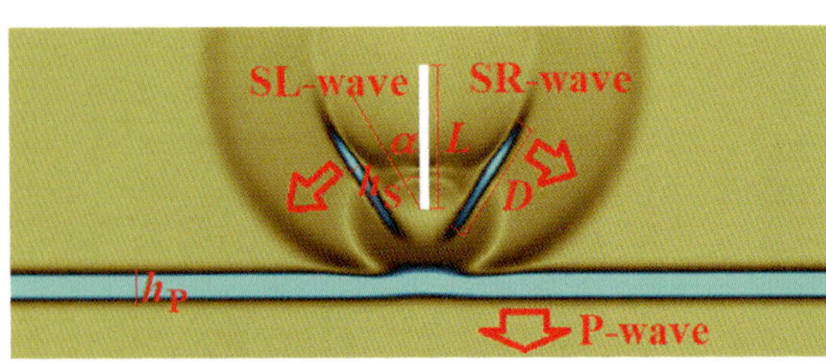

Fig. 16.4 Wave traveling and reflection scheme, maximum value of the velocity modulus is in turquoise, middle value is in black, and zero value is in yellow

$$\sigma_{xx}^{\mathrm{P}} = \left(1 - 2\left(\frac{c_{\mathrm{S}}}{c_{\mathrm{P}}}\right)^2\right)\rho c_{\mathrm{P}} V_{\mathrm{P}} \tag{16.8}$$

$$\sigma_{yy}^{\mathrm{P}} = \rho c_{\mathrm{P}} V_{\mathrm{P}} \tag{16.9}$$

$$\sigma_{xy}^{\mathrm{P}} = 0 \tag{16.10}$$

In Eqs. 16.6–16.10 and further in the text, v_x, v_y are the velocity components, σ_{xx}, σ_{yy}, and σ_{xy} are the Cauchy stress tensor components, ρ is the material density, c_{P} is the P-wave speed, c_{S} is the S-wave speed, V is the proportional to wave amplitude. Note that these expressions are consistent with the signs of wave patterns.

Note that for the models under consideration, a rotation matrix is written as follows for the "right" angle α:

$$\begin{bmatrix} \sin\alpha & \cos\alpha \\ -\cos\alpha & \sin\alpha \end{bmatrix}. \tag{16.11}$$

Then the velocity vector and Cauchy stress tensor will be transformed by Eqs. 16.12–16.16.

$$v_x = \sin\alpha \cdot v_x^{\mathrm{W}} + \cos\alpha \cdot v_y^{\mathrm{W}} \tag{16.12}$$

$$v_y = -\cos\alpha \cdot v_x^{\mathrm{W}} + \sin\alpha \cdot v_y^{\mathrm{W}} \tag{16.13}$$

$$\sigma_{xx} = \sin^2\alpha \cdot \sigma_{xx}^{\mathrm{W}} + \cos^2\alpha \cdot \sigma_{yy}^{\mathrm{W}} + 2\sin\alpha\cos\alpha \cdot \sigma_{xy}^{\mathrm{W}} \tag{16.14}$$

$$\sigma_{yy} = \cos^2\alpha \cdot \sigma_{xx}^{\mathrm{W}} + \sin^2\alpha \cdot \sigma_{yy}^{\mathrm{W}} - 2\sin\alpha\cos\alpha \cdot \sigma_{xy}^{\mathrm{W}} \tag{16.15}$$

$$\sigma_{xy} = -\sin\alpha\cos\alpha \cdot \left(\sigma_{xx}^{\mathrm{W}} - \sigma_{yy}^{\mathrm{W}}\right) - \cos 2\alpha \cdot \sigma_{xy}^{\mathrm{W}} \tag{16.16}$$

In Eqs. 16.12–16.16, the index W corresponds to the components of the velocity and Cauchy stress tensor in the coordinate system, in which the wave moves along the OX axis.

In accordance with Riemann invariants, for S-wave traveling along axis OX there are Eqs. 16.17–16.19.

$$v_y^{\mathrm{S}} = V_{\mathrm{S}} \tag{16.17}$$

$$\sigma_{xy}^{\mathrm{S}} = -\rho c_{\mathrm{S}} v_y^{\mathrm{S}} = -\rho c_{\mathrm{S}} V_{\mathrm{S}} \tag{16.18}$$

$$v_x^{\mathrm{S}} = \sigma_{xx}^{\mathrm{S}} = \sigma_{yy}^{\mathrm{S}} = 0 \tag{16.19}$$

Substituting Eqs. 16.17–16.19 based on Riemann invariants into Eqs. 16.12–16.16, one can obtain the formulae for the shear SR-wave (Fig. 16.4) scattered to the right side of the fracture expressed by Eqs. 16.20–16.24.

$$v_x^{\mathrm{SR}} = \cos\alpha \cdot V_{\mathrm{SR}} \quad (16.20)$$

$$v_y^{\mathrm{SR}} = \sin\alpha \cdot V_{\mathrm{SR}} \quad (16.21)$$

$$\sigma_{xx}^{\mathrm{SR}} = -2\sin\alpha\cos\alpha \cdot \rho c_{\mathrm{S}} V_{\mathrm{SR}} \quad (16.22)$$

$$\sigma_{yy}^{\mathrm{SR}} = 2\sin\alpha\cos\alpha \cdot \rho c_{\mathrm{S}} V_{\mathrm{SR}} \quad (16.23)$$

$$\sigma_{xy}^{\mathrm{SR}} = \cos 2\alpha \cdot \rho c_{\mathrm{S}} V_{\mathrm{SR}} \quad (16.24)$$

After substitution, a comparison is made with the wave patterns in accordance with the signs of the components of the velocity and Cauchy stress tensor.

Substituting Eqs. 16.17–16.19 based on Riemann invariants into Eqs. 16.12–16.16, one can obtain the formulae for the shear SL-wave (Fig. 16.4) scattered to the left side of the fracture expressed by Eqs. 16.24–16.29.

$$v_x^{\mathrm{SL}} = \cos\alpha \cdot V_{\mathrm{SL}} \quad (16.25)$$

$$v_y^{\mathrm{SL}} = -\sin\alpha \cdot V_{\mathrm{SL}} \quad (16.26)$$

$$\sigma_{xx}^{\mathrm{SL}} = 2\sin\alpha\cos\alpha \cdot \rho c_{\mathrm{S}} V_{\mathrm{SL}} \quad (16.27)$$

$$\sigma_{yy}^{\mathrm{SL}} = -2\sin\alpha\cos\alpha \cdot \rho c_{\mathrm{S}} V_{\mathrm{SL}} \quad (16.28)$$

$$\sigma_{xy}^{\mathrm{SL}} = \cos 2\alpha \cdot \rho c_{\mathrm{S}} V_{\mathrm{SL}} \quad (16.29)$$

After substitution, a comparison is made with the wave patterns in accordance with the signs of the components of the velocity and Cauchy stress tensor.

16.5 Scattering Amplitudes and Angles

In this section, a derivation of formulae for scattering amplitudes and angles is discussed. When P-wave moves along the fracture border, the P- and S-waves from the source moving along the fracture appear. This imaginary source moves with the incident P-wave speed and compensate the XX-component of Cauchy stress tensor provided by Eqs. 16.30–16.31.

$$\mathbf{f}_{\mathrm{L}}(y,t) = -\left(1 - 2\left(\frac{c_{\mathrm{S}}}{c_{\mathrm{P}}}\right)^2\right)\rho c_{\mathrm{P}} V_{\mathrm{P}} G\left(t - \frac{y}{c_{\mathrm{P}}}\right)\mathbf{n} \tag{16.30}$$

$$\mathbf{f}_{\mathrm{R}}(y,t) = \left(1 - 2\left(\frac{c_{\mathrm{S}}}{c_{\mathrm{P}}}\right)^2\right)\rho c_{\mathrm{P}} V_{\mathrm{P}} G\left(t - \frac{y}{c_{\mathrm{P}}}\right)\mathbf{n} \tag{16.31}$$

In Eqs. 16.30–16.31, $\mathbf{n}$ is the normal outer to the left border of the fracture, $\mathbf{f}_{\mathrm{L}}(y,t)$ and $\mathbf{f}_{\mathrm{R}}(y,t)$ are the volume density of force for these two sources moving alone left and right borders of the fracture, respectively, $G(t)$ is the shape in the incident P-wave, y is Y coordinate.

That is, the same as in the case of half-space, P-wave reflected from the boundary, and the head H-wave are formed. However, since here the free boundary is localized in the problem, one can obtain the expressions, relating the size of the fracture to the size of the resulting superposition of cylindrical shear waves propagating from a point source, provided by Eqs. 16.32–16.34.

$$\alpha = \arcsin\left(\frac{c_{\mathrm{S}}}{c_{\mathrm{P}}}\right) \tag{16.32}$$

$$h_{\mathrm{S}} = \frac{c_{\mathrm{S}}}{c_{\mathrm{P}}} h_{\mathrm{P}} \tag{16.33}$$

$$D = L \cdot \sin\alpha \tag{16.34}$$

In Eqs. 16.32–16.34, α the angle between S-waves fronts and fracture, L is the fracture length, D is the S-waves fronts extension, h_{P} and h_{S} are the P-wave and S-wave fronts' width, respectively. These values are marked in Fig. 16.4.

Let us single out plane S-waves from this solution. For the right S-wave, one can substitute Eqs. 16.8 and 16.22 into Eq. 16.1 and obtain the following formula:

$$-2\sin\alpha\cos\alpha \cdot \rho c_{\mathrm{S}} V_{\mathrm{SR}} + \left(1 - 2\left(\frac{c_{\mathrm{S}}}{c_{\mathrm{P}}}\right)^2\right)\rho c_{\mathrm{P}} V_{\mathrm{P}} = 0. \tag{16.35}$$

From Eq. 16.35, one can find the amplitude using Eq. 16.32 for the angle α:

$$V_{\mathrm{SR}} = \frac{\left(1 - 2\left(\frac{c_{\mathrm{S}}}{c_{\mathrm{P}}}\right)^2\right)\frac{c_{\mathrm{P}}}{c_{\mathrm{S}}}}{2\sin\alpha\cos\alpha} V_{\mathrm{P}} = -\frac{\cos 2\alpha}{2\sin\alpha} V_{\mathrm{P}}. \tag{16.36}$$

One can also find a nonzero component of shear stresses along the border, which appeared due to the fact that this wave is moving along the free boundary:

$$\sigma_{xy}^{\mathrm{SR}} = -\frac{\cos^2 2\alpha}{2\sin\alpha} \cdot \rho c_{\mathrm{S}} V_{\mathrm{P}}. \tag{16.37}$$

This shear stress will already be compensated by a moving point source of shear stresses since for waves from this point source at the contact boundary; both the shear and main components of the stress tensor will be equal to zero.

Now one can find a similar equation for the left S-wave. Substituting Eqs. 16.8 and 16.27 into Eq. 16.1, one can obtain:

$$2\sin\alpha\cos\alpha\cdot\rho c_S V_{SL}+\left(1-2\left(\frac{c_S}{c_P}\right)^2\right)\rho c_P V_P=0. \tag{16.38}$$

From Eq. 16.38, one can find the amplitude using Eq. 16.32 for the angle α:

$$V_{SL}=-\frac{\left(1-2\left(\frac{c_S}{c_P}\right)^2\right)\frac{c_P}{c_S}}{2\sin\alpha\cos\alpha}V_P=\frac{\cos 2\alpha}{2\sin\alpha}V_P. \tag{16.39}$$

One can also find a nonzero component of shear stresses along the border, which appeared due to the fact that this wave is moving along the free boundary:

$$\sigma_{xy}^{SL}=\frac{\cos^2 2\alpha}{2\sin\alpha}\cdot\rho c_S V_P. \tag{16.40}$$

Note that since Lame parameter λ is positive, the angle α lies in the range from 45° to 90°, which means that the amplitude of the right S-wave has the same sign as the amplitude of the incident P-wave, and the amplitude of the left S-wave is opposite. This corresponds to the calculated wave fields discussed in Sect. 16.3.

Note that S-waves scattered on the fracture do not occur due to scattering only the incident P-wave. These S-waves occur as a result of the interference of the incident P-wave, the cylindrical P-wave scattered from the upper point of the fracture, and waves from a point source of force:

$$\mathbf{f}_c=\rho c_P\left(1-2\left(\frac{c_S}{c_P}\right)^2\right)G(t). \tag{16.41}$$

In Eq. 16.41 and further in the text, $G(t)$ is the wavelet shape function in the incident P-wave, which is equal to zero outside the wave front.

Also note that the scattered S-waves interfere with the cylindrical S-wave from the same point source and with the cylindrical S-wave, which occurs when the result of the interfering of the incident P-wave and the cylindrical P-wave travel from the low point of the fracture. These cylindrical P- and S-waves are formed due to the fact that at the time of the incident plane P-wave travel the upper point of the fracture, for this case a gap in vertical main component of Cauchy stress tensor occurs (Eq. 16.42).

$$\sigma_{yy}=\rho c_P\left(1-2\left(\frac{c_S}{c_P}\right)^2\right)G(t) \tag{16.42}$$

A similar situation is observed for the lower point of the fracture. Thus, the wavelet shape in the generated S-waves is not constant along their front and differs somewhat from the original pulse shape, and not only because these scattered S-waves interfere with the wave from a moving source of shear stresses.

16.6 Conclusions

In the chapter, the analytical formulas for the scattering of plane P-wave on a gas-filled fracture located along the motion of the front of this wave are derived. Scattered S-waves have been identified and studied. These S-waves can be useful for nondestructive testing of fractures perpendicular to the incident P-waves. At the stage of expressions derivation, a comparison was made with visualized elastic wave fields (wave patterns). The effectiveness of Wave Logica approach for applied study of wave fields and obtaining the corresponding analytical expressions is shown.

Acknowledgements The reported study was funded by RFBR according to the research Project No. 18-31-20063.

References

1. Krauklis, P., Krauklis, V.: One type of waves in media with loosely bonded interface. J. Math. Sci. **55**(3), 1725–1732 (1991)
2. Hörmander, L.: The Analysis of Linear Partial Differential Operators I: Distribution Theory and Fourier Analysis. Springer-Verlag (2009)
3. Vlastos, S., Liu, E., Main, I.G., Li, X.-Y.: Numerical simulation of wave propagation in media with discrete distributions of fractures: effect of fracture size and spatial distributions. Geophys. J. Int. **152**(3), 649–668 (2003)
4. Burago, N.G., Nikitin, I.S.: Improved model of a layered medium with slip on the contact boundaries. J. Appl. Math. Mech. **80**(2), 164–172 (2016)
5. Glushko, A.I., Nikitin, I.S.: One method of calculating brittle fracture waves. Mech. Solids **21**(3), 130–135 (1986)
6. Muratov, M.V., Petrov, I.B.: Application of fractures mathematical models in exploration seismology problems modeling. In: Petrov, I.B., Favorskaya, A.V., Favorskaya, M.N., Simakov, S.S., Jain, L.C. (eds.) Proceedings of the Conference on 50 Years of the Development of Grid-Characteristic Method, SIST, vol. 133, pp. 120–131. Springer (2019)
7. Leger, A., Deschamps, M. (eds.): Ultrasonic Wave Propagation in Non Homogeneous Media, vol. 128. Springer Science & Business Media (2009)
8. Favorskaya, A., Kabanova, A., Petrov, I.: Applying Wave Logica method for geophysical prospecting. In: 20th Conference Oil and Gas Geological Exploration and Development, pp. 1–4 (in Russian) (2018)
9. Favorskaya, A.V., Zhdanov, M.S., Khokhlov, N.I., Petrov, I.B.: Modeling the wave phenomena in acoustic and elastic media with sharp variations of physical properties using the grid-characteristic method. Geophys. Prospect. **66**(8), 1485–1502 (2018)
10. Favorskaya, A., Golubev, V., Grigorievyh, D.: Explanation the difference in destructed areas simulated using various failure criteria by the wave dynamics analysis. Procedia Comput. Sci. **126**, 1091–1099 (2018)

11. Favorskaya, A.V.: The use of multiple waves to obtain information on an underlying geological structure. Procedia Comput. Sci. **126**, 1110–1119 (2018)
12. Glassner, A.S. (ed.): An Introduction to Ray Tracing. Elsevier (1989)
13. Julian, B.R., Gubbins, D.: Three-dimensional seismic ray-tracing. J. Geophys. **43**, 95–113 (1977)
14. Zak, A., Krawczuk, M., Ostachowicz, W.: Propagation of in-plane waves in an isotropic panel with a crack. Finite Elem. Anal. Des. **42**(11), 929–941 (2006)
15. Ludwig, R., Lord, W.: A finite-element formulation for the study of ultrasonic NDT systems. IEEE Trans. Ultrason. Ferroelectr. Freq. Control **35**(6), 809–820 (1988)
16. Bartoli, I., Marzani, A., Lanza di Scalea, F., Viola, E.: Modeling wave propagation in damped waveguides of arbitrary cross-section. J. Sound Vibr. **295**(3–5), 685–707 (2006)
17. Schoenberg, M.: Elastic wave behaviour across linear slip interfaces. J. Acoust. Soc. Am. **68**, 1516–1521 (1980)
18. Pyrak-Nolte, L.J., Myer, L.R., Cook, N.G.W.: Anisotropy in seismic velocities and amplitudes from multiple parallel fractures. J. Geophys. Res. **95**, 11345–11358 (1990)
19. Kelly, K.R., Ward, R.W., Treitel, S., Alford, R.M.: Synthetic seismograms finite-difference approach. Geophysics **41**(1), 2–27 (1976)
20. Kruger, L., Oliver, S., Saenger, E.H., Oates, S.J., Shapiro, S.A.: A numerical study on reflection coefficients of fractured media. Geophysics **72**(4), D61–D67 (2007)
21. Zhang, J., Gao, H.: Elastic wave modelling in 3-D fractured media: an explicit approach. Geophys. J. Int. **177**, 1233–1241 (2009)
22. Silvestrov, I., Reda, B., Evgeny, L.: Poststack diffraction imaging using reverse-time migration. Geophys. Prospect. **64**, 129–142 (2016)
23. Magomedov, K.M., Kholodov, A.S.: The construction of difference schemes for hyperbolic equations based on characteristic relations. USSR Comput. Math. Math. Phys. **9**(2), 158–176 (1969)
24. Favorskaya, A.V., Petrov, I.B. Grid-characteristic method. In: Favorskaya, A.V., Petrov, I.B. (eds.) Innovations in Wave Processes Modelling and Decision Making, SIST, vol. 90, pp. 117–160. Springer (2018)
25. Favorskaya, A.V., Kabisov, S.V., Petrov, I.B.: Modeling of ultrasonic waves in fractured rails with an explicit approach. Doklady Math. **98**(1), 401–404 (2018)
26. Favorskaya, A., Khokhlov, N.: Modeling the impact of wheelsets with flat spots on a railway track. Procedia Comput. Sci. **126**, 1100–1109 (2018)
27. Beklemysheva, K.A., Vasyukov, A.V., Petrov, I.B.: Numerical modeling of delamination in fiber-metal laminates caused by low-velocity impact. In: Petrov I.B., Favorskaya, A.V., Favorskaya, M.N., Simakov, S.S., Jain, L.C. (eds.) Proceedings of the Conference on 50 Years of the Development of Grid-Characteristic Method, SIST, vol. 133, pp. 132–142. Springer (2019)
28. Favorskaya, A., Petrov, I., Grinevskiy, A.: Numerical simulation of fracturing in geological medium. Procedia Comput. Sci. **112**, 1216–1224 (2017)
29. Favorskaya, A.V., Petrov, I.B.: The use of full-wave numerical simulation for the investigation of fractured zones. Math. Models Comput. Simul. **11**(4), 518–530 (2019)

Chapter 17
Discrete Element Method Adopting Microstructure Information

Andrew A. Zhuravlev, Karine K. Abgaryan and Dmitry L. Reviznikov

Abstract Multiscale discrete element model adapting information on material microstructure is introduced. Modeled structure is represented by a set of tetrahedral elements bound by their faces. Each element has an associated atomic sample, which represents atomic structure of the element. All element properties are determined from molecular dynamics simulation of its associated sample. Therefore, none of the specific properties of the material are needed aside from its atomic composition and microstructure. The chapter focuses on elastic behavior of modeled structures. Comparisons of the results obtained using multiscale discrete element simulation with molecular dynamics data and known macroscopic material properties show fairly good accuracy of the proposed model.

17.1 Introduction

Growth of available computational resources enables modeling of systems with a high-level time and space resolution. One of the most promising ways in this direction is multiscale modeling [1, 2]. Nevertheless, the traditional methods of computational solid mechanics make use of macroscopic material properties without considering the microstructure. This fact limits applicability of these methods to design new materials with required properties.

A. A. Zhuravlev · K. K. Abgaryan · D. L. Reviznikov (✉)
Moscow Aviation Institute (National Research University), 4, Volokolamskoe Shosse, Moscow 125993, Russian Federation
e-mail: reviznikov@inbox.ru

A. A. Zhuravlev
e-mail: zhuravlyow.andrei@yandex.ru

K. K. Abgaryan
e-mail: kristal83@mail.ru

Federal Research Centre "Information and Control" of the RAS, 44, Vavilova ul, Moscow 119333, Russian Federation

L. C. Jain et al. (eds.), *Advances in Theory and Practice of Computational Mechanics*, Smart Innovation, Systems and Technologies 173,
https://doi.org/10.1007/978-981-15-2600-8_17

Numerical methods for structures simulation can be classified into two major groups. First group contains methods based on continuum equations [3]. This group can be divided into two subgroups. The first one contains methods solving the equations on a mesh. The use of adaptive grids allows to achieve a high spatial resolution and get the detailed pictures of the process [4]. Typical representative of this group is the finite element method. Some approaches have been proposed to make this class of methods applicable to processes with breaking objects [5].

Second subgroup contains meshless methods [6]. These methods search the solution of continuum equations as a combination of basic functions. As an example let us refer to the meshless method [7]. One of the popular methods of this class is the smoothed-particle hydrodynamics, which is based on a set of computational particles approximating the solution with a smoothing kernel function [8, 9]. At the beginning, it was mainly used for fluid simulation but later it was adopted for simulation of solids. Methods of this group are generally computationally expensive but more flexible as it comes to complex processes like those with failure of solids.

Second big group contains the discrete element methods. These methods come in many different variants, a comprehensive review can be found in [10]. Basic methods of this type model the objects as a set of closely packed particles interacting with some potential and moving according to laws of classical dynamics. These methods are relatively simple but in the case of wise potential choice and careful parameters tuning they can be used to simulate quite complex processes [11, 12]. Movable cellular automatons use different approaches for linking and unlinking elements that enable fine-tuning of a model to some known material properties [13]. Dissipative particles dynamics method is one of the first methods trying to adopt information about atomic structure of material through additional dissipative and random forces [14]. Extended discrete element method allows to include the additional properties in simulation like thermodynamic state of element [15]. In general, these methods can work with less information about simulated materials properties compared to the methods based on continuum equations.

The purpose of the presented chapter is to introduce a novel multiscale method for materials modeling, which requires information only from the atomic level (atomic structure and potential of atomic interaction). The latest can be obtained from first principles calculations. Thus, the proposed approach creates the ground for design of new materials with the required properties.

This chapter is structured as follows. Section 17.2 presents the mathematical model and deals with choice of computational parameters required by model. In Sect. 17.3, the results obtained from the proposed model are compared with the results of molecular dynamics simulation and macroscale parameters of selected materials. Section 17.4 concludes the chapter.

17.2 Multiscale Discrete Element Model

Modeled structures are virtually divided into tetrahedral elements. Pairs of elements are stiffly bound by their faces. Each element contains a small but representative sample of atomic structure. For example, for a crystalline material it will be a sample of crystalline lattice oriented in certain direction. The whole system evolution is governed by equations of motion for every element vertex.

Element properties are determined by the atomic sample associated with the element (Fig. 17.1):

$$AS = (\mathbf{B}, \mathbf{M}, \mathbf{P}, \mathbf{V}, \mathbf{Q}, E),$$

where $\mathbf{B} \in \mathbb{R}^{3\times 3}$ is the matrix composed of basis vectors of atomic sample, the sample is considered to be a parallelepiped generated by these vectors, $\mathbf{M} \in \mathbb{R}^{N\times N}$ is the diagonal matrix of atomic masses, $\mathbf{P} \in \mathbb{R}^{3\times N}$ is the matrix of atom positions in local (fractional) coordinates, $\mathbf{V} \in \mathbb{R}^{3\times N}$ is the matrix of atom velocities, $\mathbf{Q} \in \mathbb{R}^{3\times 3}$ is the lattice rotation matrix, $\mathrm{E} : \mathbb{R}^{3\times N} \to \mathbb{R}$ is the potential energy function. Positions of atoms in the element can be calculated using the following formula:

$$\mathbf{P}_{\text{Local}} = \mathbf{Q} \cdot \mathbf{B} \cdot \mathbf{P}.$$

The general scheme of computational process is presented in Fig. 17.2. Let us consider the process in detail starting from element deformation. The deformation for tetrahedral elements is always linear and can be restored using the following formula:

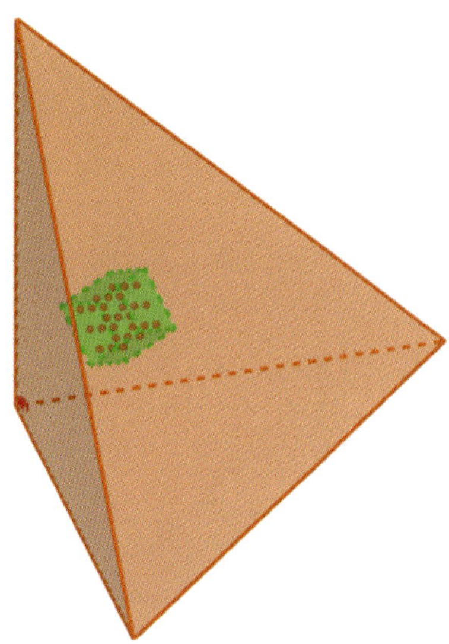

Fig. 17.1 Atomic sample associated with the element

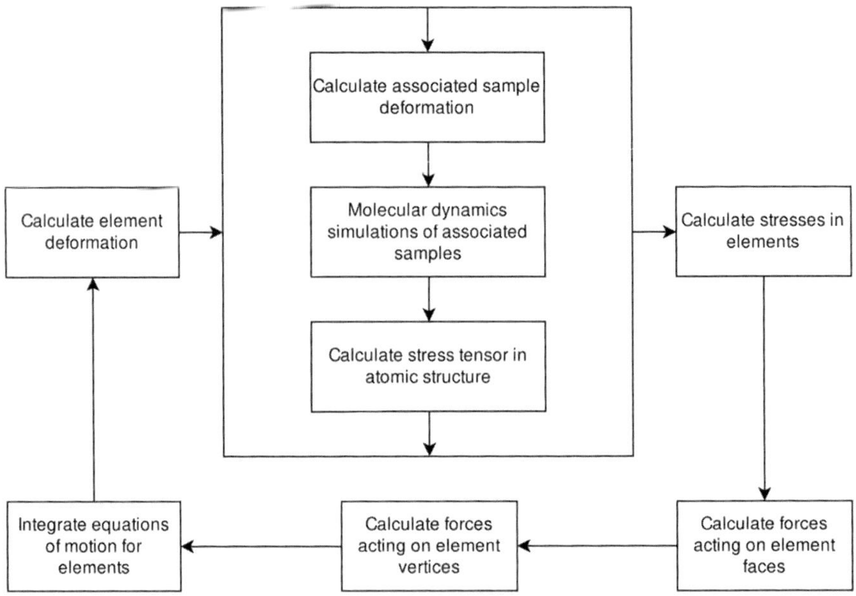

Fig. 17.2 General modeling scheme

$$\mathbf{D}_{\text{Element}} = (\mathbf{v}_1 - \mathbf{v}_0|\mathbf{v}_2 - \mathbf{v}_0|\mathbf{v}_3 - \mathbf{v}_0) \cdot \left(\mathbf{v}_1^* - \mathbf{v}_0^*|\mathbf{v}_2^* - \mathbf{v}_0^*|\mathbf{v}_3^* - \mathbf{v}_0^*\right)^{-1},$$

where $\mathbf{D}_{\text{Element}}$ is the deformation tensor of the element, $\mathbf{v}_i^*$ are the positions of vertices at the beginning of simulation, $\mathbf{v}_i$ are the current positions of vertices.

Then, associated atomic sample for the element is considered to be subjected to the same deformation (Fig. 17.3):

$$\mathbf{D}_{\text{Sample}} = \mathbf{D}_{\text{Element}}.$$

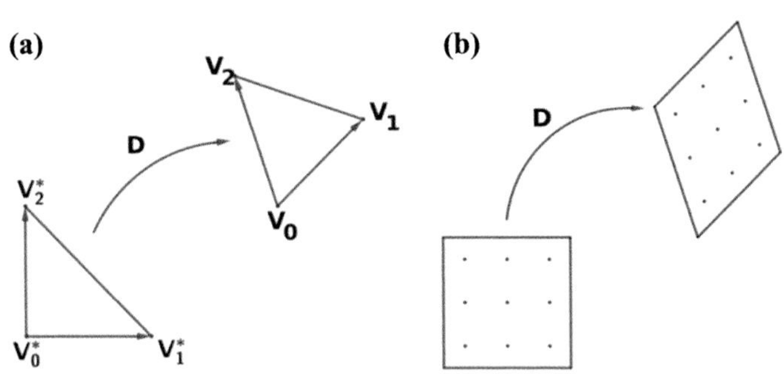

Fig. 17.3 Atomic sample deformation induced by element deformation: **a** element deformation, **b** atomic sample deformation

To get a sample's response to deformation, it is deformed and then a few steps of molecular dynamics are performed in the sample to calculate the resulting lattice structure change due to the deformation. Then real atom positions in global reference frame can be calculated by the following formula:

$$\mathbf{P}_{\text{Global}} = \mathbf{D}_{\text{Sample}} \cdot \mathbf{Q} \cdot \mathbf{B} \cdot \mathbf{P}.$$

Note that atomic samples should not be very large, as they will be involved in computationally expensive molecular dynamics simulations, neither they should be too small, as they must be able to accurately represent properties of material. Samples associated with the element represent its internal structure and can be considered continuously repeated in every direction. To model such behavior, the periodic boundary conditions are imposed on sample, and pairs of atoms are interacting only by minimum distance:

$$\begin{gathered}\boldsymbol{\delta}_{ij} = \underset{\boldsymbol{\delta} \in (-1,0,1)^3}{\arg\min} \left\| \mathbf{r}_i + \mathbf{B}\delta - \mathbf{r}_j \right\|, \\ \mathbf{r}_{ij} = \mathbf{r}_i + \mathbf{B}\delta_{ij} - \mathbf{r}_j.\end{gathered}$$

Motion of atoms in the sample is governed by Newton's law of motion, where forces acting on particles are derived from interatomic potential:

$$\begin{cases} \dot{\mathbf{V}} = \mathbf{F}\mathbf{M}^{-1} = -\nabla E(\mathbf{P}_{\text{Global}})\mathbf{M}^{-1} \\ \dot{\mathbf{P}}_{\text{Global}} = \mathbf{V} \end{cases}.$$

For this chapter, the embedded-atom method is used to compute forces between atoms. It is a many-body potential that gives potential energy in the following form:

$$\begin{gathered} E_i = \mathrm{F}\left(\sum_{j \neq i} \rho(\mathbf{r}_{ij})\right) + \frac{1}{2} \sum_{j \neq i} \varphi(\mathbf{r}_{ij}), \\ E = \sum_i E_i, \end{gathered}$$

where $\varphi(\mathbf{r}_{ij})$ is the pair-wise potential function, $\rho(\mathbf{r}_{ij})$ is the electron density contribution of atom j in position of atom i, F is the embedding function or energy of atom at the point with given electron density. It is a very generic form of potential energy. All functions are tabulated and then interpolated with splines. Forces are calculated according to the potential energy gradient:

$$\begin{gathered} \nabla_{\mathbf{r}_i} E = \sum_{j \neq i} \left(\left(\left. \frac{\partial \mathrm{F}(\rho)}{\partial \rho} \right|_{\rho=\rho_i} + \left. \frac{\partial \mathrm{F}(\rho)}{\partial \rho} \right|_{\rho=\rho_j} \right) \frac{\partial \rho(\mathbf{r}_{ij})}{\partial \mathbf{r}_i} + \frac{\partial \varphi(\mathbf{r}_{ij})}{\partial \mathbf{r}_i} \right) \hat{\mathbf{r}}_{ij}, \\ \rho_i = \sum_{j \neq i} \rho(\mathbf{r}_{ij}). \end{gathered}$$

The forces can be calculated in two steps. At the beginning, the electron density is calculated for every atom in system evaluating density contribution from every neighboring atom. Then, a resulting force can be calculated in a single pass over all neighbors for every particle. It is one of the least computationally expensive potentials aside from pair potentials, and it represents a good trade-off between the computational complexity and physical accuracy of molecular dynamics model.

Equations of motion are integrated by velocity Verlet integration method, which is a common choice for molecular dynamics simulations. For atomic simulations, a general practical rule is to use femtosecond time steps:

$$\begin{cases} \mathbf{P}(t+\mathrm{d}t) = \mathbf{P}(t) + \mathbf{V}(t)\mathrm{d}t - \nabla E(\mathbf{P}(t))\mathbf{M}^{-1}\frac{\mathrm{d}t^2}{2} \\ \mathbf{V}(t+\mathrm{d}t) = \mathbf{V}(t) - (\nabla E(\mathbf{P}(t)) + \nabla E(\mathbf{P}(t+\mathrm{d}t)))\mathbf{M}^{-1}\frac{\mathrm{d}t}{2} \end{cases}.$$

Sample's response on a deformation is given by a virial stress of a molecular system:

$$\mathbf{S}_{\mathrm{Sample}} = \frac{1}{\left|\mathbf{D}_{\mathrm{Sample}}\cdot\mathbf{Q}\cdot\mathbf{B}\right|}\left(-(\mathbf{V}-\bar{\mathbf{V}})\mathbf{M}(\mathbf{V}-\bar{\mathbf{V}})^T + \frac{1}{2}\sum_{i,j\neq i}\mathbf{r}_{ij}\mathbf{f}_{ij}^T\right),$$
$$\bar{\mathbf{V}} = \frac{\mathbf{V}\cdot\mathbf{1}^{N\times 1}\cdot\mathbf{1}^{1\times N}}{N},$$

where $\mathbf{S}_{\mathrm{Sample}}$ is the stress tensor of the sample, $\left|\mathbf{D}_{\mathrm{Sample}}\cdot\mathbf{Q}\cdot\mathbf{B}\right|$ is the sample volume, $\bar{\mathbf{V}}$ is the average sample atom velocity.

Considering that associated sample is a representative fragment of the element and taking into account that inner volume of the element is much greater than boundary volume, we conclude that properties of the element are completely determined by corresponding properties of the associated sample. Therefore, stress tensor of the element is equal to stress tensor of its associated sample:

$$\mathbf{S}_{\mathrm{Element}} = \mathbf{S}_{\mathrm{Sample}}.$$

To determine forces acting on every vertex from the element first of all we ought to determine element response to deformation.

Having a stress tensor for every element, we can calculate their dynamics. Dynamics of every element is fully determined by dynamics of their vertices:

$$\mathbf{f}_{lm} = \mathbf{S}_l\cdot\mathbf{n}_{lm},$$
$$\mathbf{n}_{lm} = \frac{1}{2}\left[\mathbf{v}_{lm_1} - \mathbf{v}_{lm_0}\times\mathbf{v}_{lm_2} - \mathbf{v}_{lm_0}\right],$$
$$\mathbf{f}_{lmk} = \mathbf{f}_{lm}\frac{a_{lmk}}{a_{lm}},$$

where $\mathbf{S}_l$ is the stress tensor of element l, $\mathbf{n}_{lm}$ is the normal to face m of element l, a_{lm} is the face area, $\mathbf{v}_{lm_k}$ is the vertex k of face m of element l, a_{lmk} is the part of face m of element l area closest to vertex $\mathbf{v}_{lm_k}$, $\mathbf{f}_{lmk}$ is the force acting on vertex $\mathbf{v}_{lm_k}$ from face m of element l (Fig. 17.4a).

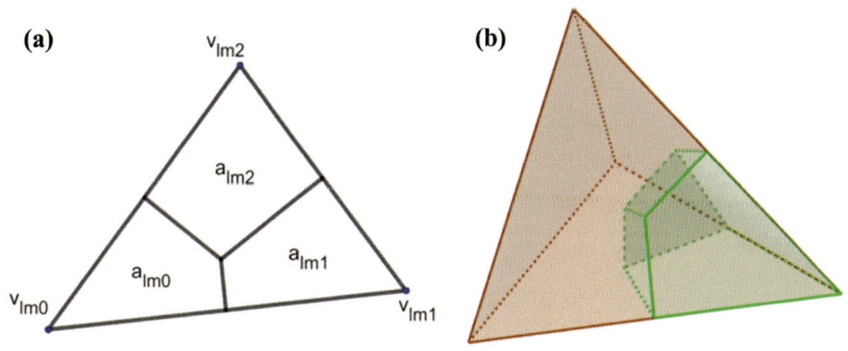

Fig. 17.4 Area and volume partition between vertices: **a** assigned area, **b** assigned volume

Having forces acting on element vertices, we ought to assign masses to vertices to calculate their accelerations. The part of the element closest to the corresponding vertex is associated with the vertex and the element mass is distributed between vertexes according to these volume parts (Fig. 17.4b).

After calculation of accelerations equations of motion are integrated for every vertex using velocity Verlet method:

$$\begin{cases} \mathbf{r}_k(t+\mathrm{d}t) = \mathbf{r}_k(t) + \mathbf{v}_k(t)\mathrm{d}t + \frac{\mathbf{f}_k(t)}{m_k}\frac{\mathrm{d}t^2}{2} \\ \mathbf{v}_k(t+\mathrm{d}t) = \mathbf{v}_k(t) + \frac{\mathbf{f}_k(t)+\mathbf{f}_k(t+\mathrm{d}t)}{2m_k}\mathrm{d}t \end{cases}.$$

To keep the calculations correct an admissible time step is ought to be selected. Time step selection can be guided by the following algorithm:

1. Estimate Young's modulus (E) of material from small molecular dynamics simulation.
2. Estimate speed of sound in material from Young's modulus:

$$c = \sqrt{\frac{E}{\rho}}.$$

3. Select a time step in such a way that no perturbation travels through an entire element in a single step:

$$\begin{aligned} c \cdot \mathrm{d}t &< L, \\ \mathrm{d}t &< \tfrac{L}{c} = L\sqrt{\tfrac{\rho}{E}}, \end{aligned}$$

where L is the characteristic length of elements in model.

There are a few things left to consider before the beginning of simulation. First of all, we have to find out what the minimum possible sample size of the element is. Having periodic boundary conditions imposes the strong restrictions on the sample size. Most interatomic potentials have so called cutoff distance r_{cut}, a maximum distance between interacting atoms; atoms at a larger distance are considered noninteracting. Periodic boundary conditions in three-dimensional space make 27 virtual copies of each atom that can interact with other atoms, but we select only a closest one. Thus, we ought to ensure that for every atom there is no more than one copy of every other atom in cutoff sphere. This is provided by creating a sample large enough to contain cutoff sphere. For cubic lattice it is expressed in the following way:

$$k = \left\lceil \alpha \frac{2r_{\text{cut}}}{a} \right\rceil,$$
$$L = ka,$$

where r_{cut} is the cutoff distance depending on the potential, a is the size of crystal lattice unit cell, α is an expected compression of element, k is the number of crystal unit cells in the sample, and L is the length of the sample side.

In this research for all experiments except the first one, the base rotation of crystal lattice in every element is generated from uniform random distribution over rotation group SO(3). At first, a unit quaternion denoting a chosen rotation is generated in the following way:

$$u_1, u_2, u_3 \sim U(0, 1),$$
$$s = \sqrt{1 - u_1} \sin 2\pi u_2,$$
$$\mathbf{v} = \left(\sqrt{1 - u_1} \cos 2\pi u_2, \sqrt{u_1} \cos 2\pi u_3, \sqrt{u_1} \cos 2\pi u_3\right),$$
$$\mathbf{q} = (s, \mathbf{v}).$$

Then, this quaternion is converted into a rotation matrix for convenience.

Initial velocities are assigned from Maxwell–Boltzmann distribution for selected temperature. Assignment of some initial velocities for atoms in the sample is important due to abnormal behavior of crystals at close to zero temperatures.

Since the proposed method is computationally expensive, it is highly important to use a parallelization. Most expensive parts of computation are handed over to GPU. In this work, all molecular dynamics simulations of samples associated with elements are conducted on GPU using CUDA technology. In CUDA, a computation model the whole GPU task is divided into blocks of threads. Threads in one block can communicate with each other but threads from different blocks have no simple and fast way of communication. Therefore, a computation of molecular dynamics for one element is fully conducted by threads of one block as synchronization between time steps and exchange data on particles movement are needed. Computations in different elements are completely independent of each other, so threads from different blocks do not need any means of communication. This type of computations is well suited for GPU architecture.

17.3 Computational Experiments

Three sets of computational experiments have been performed to validate multiscale discrete element model. The first group of experiments is directed on comparison of the proposed model with direct molecular dynamics simulation. The problem formulation is the following: an ideal crystal of selected material was periodically continued in one direction and stretched in that direction with constant speed. Experimental sample had the size of 120 unit cells in every direction and was stretched to 145% of its initial length in 0.15 ns. Experiments were conducted on the copper and aluminum samples. Both materials have face-centered lattice. Initial sample temperature was set to 300 K. Stress in crystal was studied in this experiment.

Molecular dynamic model contained around 7 millions of atoms. Simulation step was set to 1 femtosecond. Each experiment required around 12 h on Nvidia Tesla P100 GPU, which is modern server GPU. The multiscale discrete element model used 384 elements. The further increase of this value has little effect on the results. All atomic samples contained around 330 thousands of atoms. Note that the amount of atoms in samples is the smallest possible to keep periodic images of every atom from interacting with atom itself. The time step was set to 50 fs. Each experiment required around 2.5 h on Nvidia GeForce GTX 780 GPU, which is a rather old GPU for workstations.

Computational experiments show a good correspondence of the results obtained from proposed model with classical molecular dynamics results, which can be considered as the exact solution (Fig. 17.5). Proposed model reproduces well stresses during elastic deformations and stress after the beginning of sample yielding. However, the model gives slightly different stress at the start of yielding process. This can be explained by the absence of real boundary in the proposed method. All boundary conditions are taken into account only on the top level of model, but atomic samples have no information about sample boundary. This prevents samples from easier dislocations accumulation and breakdown of boundaries.

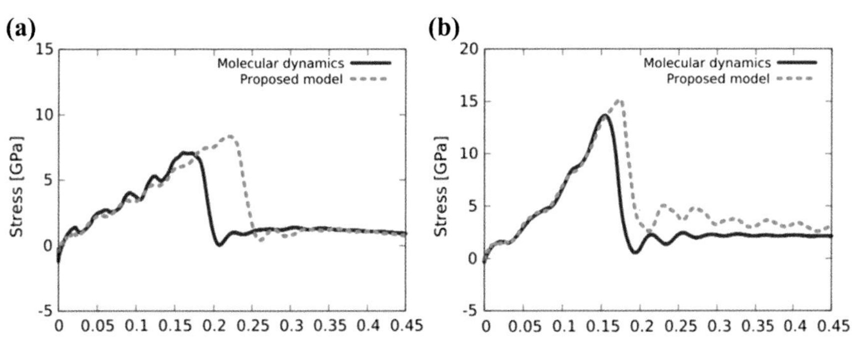

Fig. 17.5 Comparison of stresses obtained from molecular dynamics and proposed models for: **a** aluminum crystals, **b** copper crystals

The second group of experiments focused on comparison of discrete element modeling results with macroscopic material properties. Static material loading is considered. The problem formulation is the following. Polycrystalline sample of selected material was periodically continued in one direction and stretched in that direction with constant speed. Three materials were selected: copper and aluminum having a face-centered crystalline lattice and iron having a body-centered crystalline lattice. Simulation was carried out in quasi-static way; every few steps sample was deformed and then stabilized for a few steps. The object contained 1,536 elements. Each element contained an atomic sample of around 850 atoms for copper and aluminum and around 420 atoms for iron. Stress-deformation dependence was studied to determine Young's modulus. Reference values of Young's modulus for studied materials are: aluminum—70 GPa, copper—130 GPa, and iron—210 GPa.

Experimental results show good correspondence of calculated stress with linear estimation from Young's modulus (Fig. 17.6).

The third group of experiments focused on simulation of dynamic processes in materials. All experimental models contained 3,840 elements. Each element contained an atomic sample of around 850 atoms. In the first computational experiment, a polycrystalline copper plate was affected by instantaneous impulse in a direction normal to its free surface. Pressure wave propagation is presented in Fig. 17.7. The

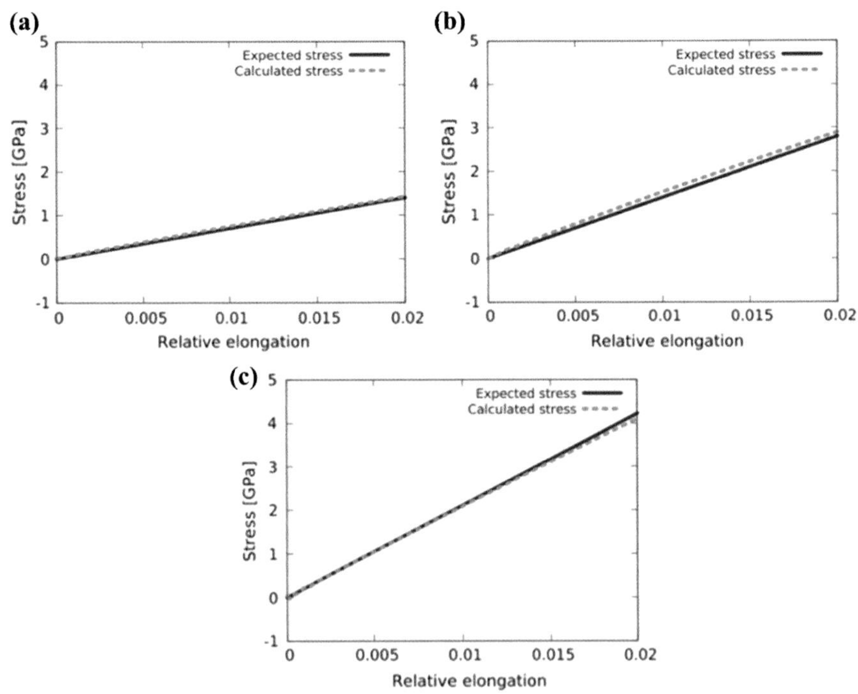

Fig. 17.6 Comparison of stress obtained from the model with stress obtained from linear elasticity theory for examined materials: **a** aluminum, **b** copper, **c** iron

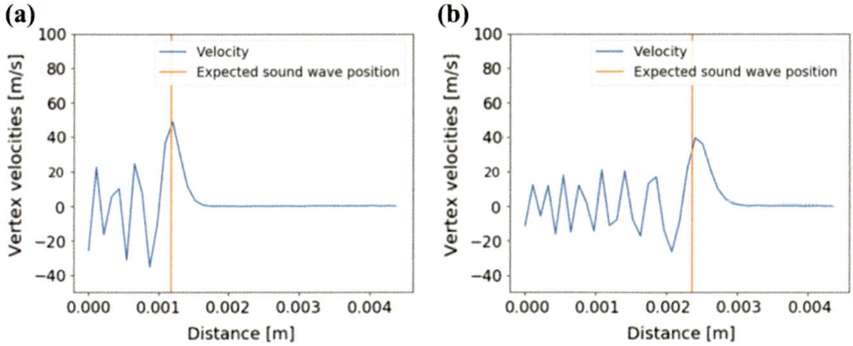

Fig. 17.7 Pressure wave propagation in copper plate at: **a** 0.25 ms, **b** 0.5 ms

reference sound speed here is 4,725 m/s. In the second experiment, a polycrystalline aluminum rod was affected by instantaneous impulse in a direction parallel to it (Fig. 17.8, the reference value is 5,091 m/s). In the third experiment, a polycrystalline aluminum rod was affected by instantaneous impulse in a direction perpendicular to it. Shear wave propagation was studied in this experiment (Fig. 17.9, the reference value is 3,103 m/s).

The expected position of the sound wave calculated taking the reference sound speed in material is given in Figs. 17.7, 17.8, 17.9 by the vertical line. In all cases, the results of computational experiment are in good correspondence with reference data.

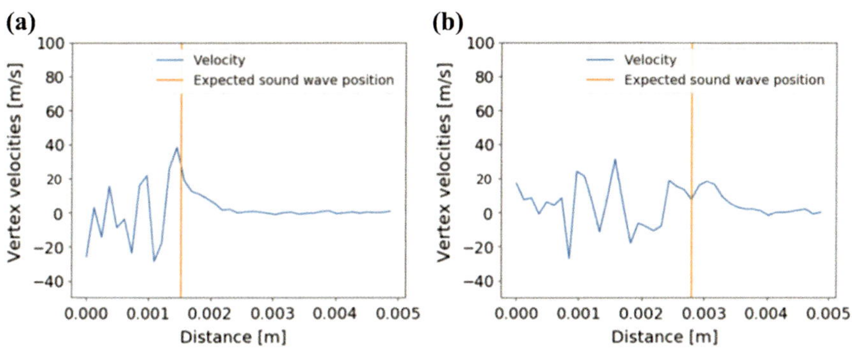

Fig. 17.8 Pressure wave propagation in aluminum rod at: **a** 0.3 ms, **b** 0.55 ms

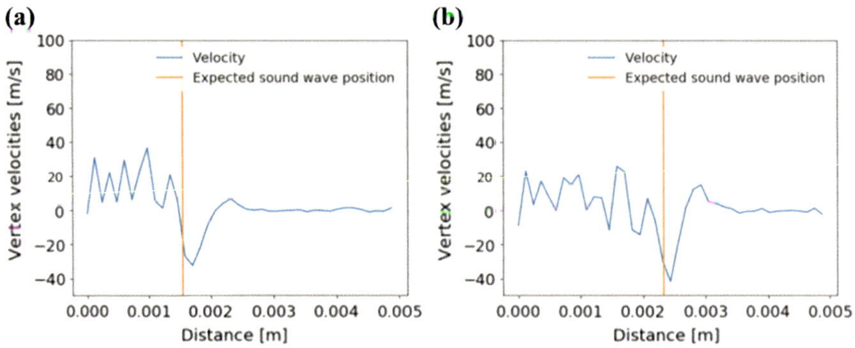

Fig. 17.9 Shear wave propagation in aluminum rod at: **a** 0.5 ms, **b** 0.75 ms

17.4 Conclusions

A novel discrete element method for solids simulation has been presented. The main feature of the proposed method is a combination of macroelements and atomic samples associated with discrete elements. It enables adopting material microstructure into the mathematical model. The proposed method does not need knowledge of any macroscopic properties of material. The required information comes from the atomic level. The results of computational experiments showed a good agreement with molecular dynamics results and the reference data for considered materials. The multiscale discrete element modeling can be applied for design of new materials with required properties.

Acknowledgements The reported study was funded by RFBR, project number 18-08-00703.

References

1. Tadmor, E.B., Miller, R.E.: Modeling Materials. Continuum, Atomistic and Multiscale Techniques. Cambridge University Press, Cambridge (2014)
2. Abgaryan, K.K.: Multiscale Modeling in Material Science Problems. MAKS Press, Moscow (in Russian) (2017)
3. Shabana, A.A.: Computational Continuum Mechanics, 3rd edn. Wiley, Hoboken (2018)
4. Burago, N.G., Nikitin, I.S., Yakushev, V.L.: Hybrid numerical method with adaptive overlapping meshes for solving nonstationary problems in continuum mechanics. Comput. Math. Math. Phys. **56**(6), 1065–1074 (2016)
5. Moës, N., Dolbow, J., Belytschko, T.: A finite element method for crack growth without remeshing. Int. J. Numer. Meth. Engng. **46**, 131–150 (1999)
6. Youping, C., Lee, J.D., Eskandarian, A.: Meshless Methods in Solid Mechanics, 1st edn. Springer, New York (2009)
7. Vasilyev, A.N., Kolbin, I.S., Reviznikov, D.L.: Meshfree computational algorithms based on normalized radial basis functions. In: Cheng, L., Liu, Q., Ronzhin, A. (eds.) Advances in Neural Networks, pp. 583–591. Springer International Publishing (2016)

8. Stellingwerf, R.F., Wingate, C.A.: Impact modeling with smooth particle hydrodynamics. Int. J. Impact Eng. **14**, 707–718 (1993)
9. Connolly, A., Iannucci, L., Hillier, R., Pope, D.: Second order Godunov SPH for high velocity impact dynamics. In: Papadrakakis, M., Kojic, M., Tuncer, I. (Eds.) Proceedings of the 3rd South-East European Conference Computational Mechanics, pp. 13–35 (2013)
10. Chen, Y., Zimmerman, J., Krivtsov, A., McDowell, D.L.: Assessment of atomistic coarse-graining methods. Int. J. Eng. Sci. **49**, 1337–1349 (2011)
11. Abgaryan, K.K., Zhuravlev, A.A., Zagordan, N.L., Reviznikov, D.L.: Discrete-element simulation of a spherical projectile penetration into a massive obstacle. Comput. Res. Model. **7**(1), 71–79 (2015)
12. Abgaryan, K.K., Eliseev, S.V., Zhuravlev, A.A., Reviznikov, D.L.: High-speed penetration. Discrete-element simulation and experiments. Comput. Res. Model. **9**(6), 937–944 (2017)
13. Psakhie, S., Shilko, E., Smolin, A., Astafurov, S., Ovcharenko, V.: Development of a formalism of movable cellular automaton method for numerical modeling of fracture of heterogeneous elastic-plastic materials. Frattura ed Integrita Strutturale **24**, 26–59 (2013)
14. Groot, R.D., Warren, P.B.: Dissipative particle dynamics: bridging the gap between atomistic and mesoscopic simulation. J. Chem. Phys. **107**(11), 4423–4435 (1997)
15. Peters, B.: Measurements and application of a discrete particle model (DPM) to simulate combustion of a packed bed of individual fuel particles. Combust. Flame **131**(1–2), 132–146 (2002)

Chapter 18
Durability Evaluation of Bonded Repairs for the Damaged Metallic Structures Subjected to Mechanical and Thermal Loads

Alexey A. Fedotov and Anton V. Tsipenko

Abstract The analytical calculation model was developed to determine the repair joint parameters and to study the relation between repair design parameters stage by stage. The first stage of calculation is devoted to defining the stress–strain conditions of the joint without influence of the damage—the method of eigenstrains and eigencurvatures based on Eshelby's theory of ellipsoidal inclusion is used. The computations on the second stage include damage geometric parameters and the calculation of the stress intensity factors at the key points of joint is performed. The third stage is to evaluate the damage growth rate within the repair joint taking into account the material properties change under cyclic loads. To figure out the real degradation behavior of carbon fiber plastic material that can be used for a bonded repair patch preparation, the experimental research was performed to study the variation of the longitudinal and transversal elastic moduli at −60, +23, and +80 °C and the variation of Poisson ratio under the cyclic loads at the same values of temperature. The developed model allows to analyze the effectiveness of the load-transfer abilities of the bonded patch and estimate the damage growth rate in the repairable structure. The model can be used as starting point for the analysis of wide number of bonded repair techniques and design variants.

18.1 Introduction

Airplane maintenance process assumes that the airframe structural element is repairable if it has damage that leads to or may lead to decreasing of the residual strength below the defined limits. Cracks and corrosion damage are the widespread types of damage on metallic aircraft structures. The availability of the robust,

A. A. Fedotov (✉) · A. V. Tsipenko
Moscow Aviation Institute (National Research University), 4, Volokolamskoe shosse, Moscow 125993, Russian Federation
e-mail: alexey.a.fedotov@inbox.ru

A. V. Tsipenko
e-mail: tsipenko_av@mail.ru

L. C. Jain et al. (eds.), *Advances in Theory and Practice of Computational Mechanics*, Smart Innovation, Systems and Technologies 173,
https://doi.org/10.1007/978-981-15-2600-8_18

reliable, and justified, technically and economically, repair techniques influences on the effectiveness of serviceability of technical systems, where the aircraft plays the primary role.

Usually, a repair of detected damage is represented by reinforcing metallic patch installed at damage location to restore the mechanical properties of the initial structure: the residual strength, stiffness, fatigue resistance, and level of allowable damage. For commercial airplanes, the basic repair methods are described within Structural Repair Manuals (SRM) delivered with airplane. These methods mainly include the application of bolts and rivets as fastening parts.

At the time of damage detection on the airframe structure based on collected information, the selection of the desired repair alternative is performed:

- The repair procedure is not necessary (damage is inside the outer finish coatings).
- The cosmetic repair procedure (without any patch) and sealing process are required (the size of damage is within the allowable scale defined in SRM).
- The patch repair is required (the damage size will have significant influence on the structural residual strength during the continuous service life).
- The repair is not efficient (the damage component should be replaced).

In general, the repair schema for the restoration of the structural properties should be simple for realization and should have minimal effect on surrounding components. The installed repair parts should not block the movable structural and system parts and should not spoil the aerodynamics and aeroelastics (if it is critical for damaged parts) beyond the established limits.

The typical requirements for repair design are mentioned below [1, 2]:

- Strength restoration of the damaged structure.
- Restriction of the damage growth rate (if the entire damage elimination is not possible or ineffective).
- Minimal change of the initial local stiffness and stress distribution.
- High durability of the installed repair joint subject to design loads and environmental factors.
- Tolerance to additional mechanical damage of the initial structure during the service life.
- Reliable inspection process for installation quality and in-service conditions of the installed repair joint.
- Account of the functional requirements for the structure (aerodynamics, fire protection, electromagnetic effects, etc.).

In addition to listed above, the acceptable repair procedure should meet some extra requirements, such as minimal time of repair installation (i.e., minimal time of the airplane extraction from flight schedule), utilization of the cheap and widespread material and tools, application of simple process steps, and minimal damage to adjacent structural elements.

The mentioned requirements can be met by repair procedures using not only bolted patches but bonded patches as well. Composite bonded repair procedures have some advantages in comparison with bolted repair techniques [1]:

- High stiffness along the defined direction—this property allows us to apply thin patches and to orient them along the load path only.
- Superior fatigue life of composite materials—exceeding the fatigue life of the parent structure.
- Low weight of patch—minimal influence on balancing characteristics for repairs on control surfaces and high lift elements.
- Manufacturing performance to easily produce the patches of different shape, size, and curvature.

Besides the advantages, the composite patch application for bonded repairs has its own problems that need to be addressed at the maintenance planning of aircraft service. One of the primary strength problems is the difference in thermal expansion values between the composite patch material and damaged initial metallic structure. If the adhesives of high curing temperature are used, the local thermal stresses must be evaluated during stress calculation of bonded joints. In many cases of cyclic thermal loading, the resulting stresses may lead to contact failure in the adhesive layer and induce the growth of the existing damage. To minimize the negative effect of the thermal expansion difference between composite patch and metallic parent structure the hybrid metal-polymeric materials (like SIAL or GLARE) may be applied. These materials consist of alternate plies of fiberglass and aluminum and demonstrate the advanced fracture toughness and durability properties [3–5]. The right selection of the adhesive may compensate the difference in thermal strains of the joined dissimilar materials [6–10].

Based on described features, the scope of the bonded repair techniques may be marked out as follows [11, 12]:

1. To decrease the stress intensity:
    - In the area of fatigue crack appearance and propagation.
    - In the area of corrosion cracking and structural materials.
    - In the area, where the allowable damage limits need to be increased (reinforcing patch installation).
2. To restore the strength and stiffness:
    - After the corrosion damage removal beyond SRM limits.
    - After the removal of defects in the material bulk.
    - After machining the surfaces for the reduction of stress intensity.
    - After the thermal damage of structure.
3. To reinforce the weak structure:
    - Deformation decreasing at the locations of concentrated stresses.
    - Decreasing the level of secondary bending moments.
    - Decreasing the vibration level and acoustic damage prevention.

To narrow the search diapason of optimal variants of composite bonded repair patches, it is necessary to have the flexible calculation module to compute the stress–strain condition of the bonded joint at every case of damage and assess the probability of damage growth and damage growth rate taking into account the degradation of material properties used for repair. The computation module described below is built using the inclusion theory [13, 14]. This applied computational module constructs the relatively simple but accurate analytical model suitable enough for preliminary design of the bonded repair joints.

The presented calculations are described in separate sections. Section 18.2 deals with the core analytical model of the bonded joint. This model utilizes the inclusion methodology for heterogeneous materials in the joint and Hart-Smith model of adhesive layer. Section 18.3 contains data achieved for carbon fiber material after the series of tests, where the specimens were loaded by cyclic mechanical force at three values of temperature. Section 18.4 demonstrates the work results for the bonded joint model from Sect. 18.2 augmented by test data from Sect. 18.3. The conclusive graphics in Sect. 18.4 illustrate the variability of the actual stresses at different values of temperature at the edge of repair patch due to the change of the patch material properties. Section 18.5 gives the conclusions.

18.2 Analytical Model of the Bonded Repair

To analyze the bonded repair, the repair schema is split into two areas (Fig. 18.1): the metallic skin with a cutout of the shape same as a patch and inclusion—hybrid plate "patch-adhesive-skin".

For the algorithmization purposes, it is assumed that the first stage of calculation is to assess the patch effect on stress–strain condition of the skin and on the distribution of the internal forces between the skin and patch, while the damage is absent. The second stage defines the damage growth characteristics as a reaction for new stress condition due to the bonded patch existence: Stress Intensity Factors (SIFs) are evaluated by means of fracture mechanics methods.

In this section, two stages are considered. Section 18.2.1 describes stress computation for bonded joint without damage (Stage 1), and computation of SIFs in the skin with damage (Stage 2) is discussed in Sect. 18.2.2.

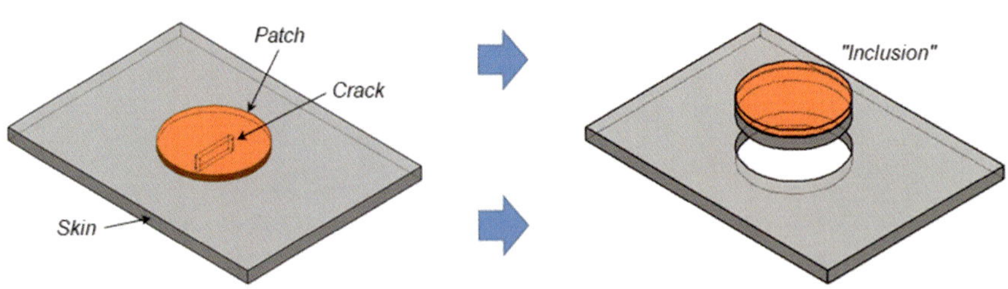

Fig. 18.1 Inclusion schema

18.2.1 Stage 1. Stress Computation for Bonded Joint Without Damage

Stress computation is performed separately for skin with a cutout and for inclusion with a subsequent junction of solutions per strain compatibility conditions. Stresses and strains are defined by external mechanical and thermal loads. One-sided installation of the bonded patch leads to the appearance of the additional bending moment in the bonded joint. Bending moment governs the variable mechanical stresses through the material thickness, this variability is considered as superposition of bending and membrane components.

The computation process is divided into three steps in order to find the stresses in the one-sided bonded system under thermomechanical loads [15].

The first step calculates stresses in the skin at the patch location and stresses in the patch due to temperature differences ΔT after cooling the system "patch-adhesive-skin" down to minimal service temperature. The method of eigenvalues of strains and curvatures by Eshelby [16] and Beom [17] is used to define the fields of strains and curvatures in the skin in the inclusion area and out of it.

The second step allows evaluating the dominance of loading factors within the bonded joint and potentially reducing the amount of calculations. This step finds the virtual value of applied stress σ_{ij}^{f*} to nullify thermal bending stresses in the skin along the patch. Further, the value σ_{ij}^{f*} is compared with initial loads in the skin $\sigma_{\infty ij}$. If the condition $\sigma_{\infty ij} \leq \sigma_{ij}^{f*}$ becomes true, then the case of thermal loads is assumed critical, and SIFs are calculated based on stress found at the first step. Otherwise, the additional third step is performed, and stresses from loads of $\sigma_{\infty ij} - \sigma_{ij}^{f*}$are calculated.

At the third step, previously calculated load of $\sigma_{\infty ij} - \sigma_{ij}^{f*}$ is assumed as initially applied mechanical load, and $\sigma_{ij}^{sstep2} = \sigma_{ij}^{f*}(0)$ represents a thermal load in the center of skin under the patch. Calculation is reduced to two-dimensional case loaded with specific force $P = (\sigma_{\infty 22} - \sigma_{22}^{f*})t_s$. The patch is assumed as free of mechanical stress, and the skin is additionally loaded with internal stress of σ_{ij}^{sstep2} (that is matched to force of $P_0 = \sigma_{22}^{sstep2}(0)t_s$) [18]. The mean and bending stresses in the skin under the patch directly affecting the damage growth rate are calculated at this step.

18.2.2 Stage 2. Computation of Stress Intensity Factors in the Skin with Damage

The stress values acquired at the previous stage are used to define SIFs at the crack tip and evaluate the crack growth parameters. The patch effect is simulated by means of a set of elastic springs bridging the crack. Mean and bending stresses $\overline{\sigma}_{ij}^{s}$ и $\widehat{\sigma}_{ij}^{s}$ are assumed constant along the crack line and equal to values in the center of skin under the patch.

SIF at the crack tip is determined by membrane and bending stress contribution:

$$K_I(z) = K_{\text{mem}} - \frac{2z}{t_s} K_b, \;\; K_{\text{mem}} = \frac{E_s\sqrt{\pi a}}{2}\overline{h}_1(1), \;\; K_b = \frac{3E_s\sqrt{\pi a}}{2}\overline{h}_2(1),$$

where $\overline{h}_{1,2} = \frac{\widetilde{h}_{1,2}}{\sqrt{1-r^2}}$, $r = x/a$, and $\widetilde{h}_{1,2}$ are the solutions of the following normalized integral equations [19, 20]:

$$-\frac{1}{2\pi}\int_{-1}^{1}\frac{\widetilde{h}_1(\eta)}{(r-\eta)^2}d\eta + (k_{tt}a)\widetilde{h}_1(r) + (k_{tb}a)\widetilde{h}_1(r) = \frac{\sigma_m^0}{E_s},$$

$$-\frac{3}{2\pi}\int_{-1}^{1}\frac{\widetilde{h}_2(\eta)}{(r-\eta)^2}d\eta - \frac{15}{2\pi(1+\nu_s)}\left(\frac{a}{t_s}\right)^2\int_{-1}^{1}\widehat{L}\left(\sqrt{10}\frac{a}{t_s}|r-\eta|\right)\widetilde{h}_2(\eta)d\eta$$

$$+ (k_{bt}a)\widetilde{h}_1(r) + (k_{bb}a)\widetilde{h}_2(r) = \frac{\sigma_b^0}{E_s},$$

where k_{ij} are the elastic constants of the springs bridging the crack: $k_{tt} = \frac{d_{tt}}{E_s t_s}$, $k_{tb} = \frac{6d_{tb}}{E_s t_s^2}$, $k_{bt} = \frac{6d_{bt}}{E_s t_s^2}$, $k_{bb} = \frac{36d_{bb}}{E_s t_s^3}$, and d_{ij} are the elements of the inverted compliance matrix c_{ij} described in detail in paper [21], E_s and t_s are the skin elastic modulus and skin thickness, respectively, a is a half of crack length.

The elements of the compliance matrix can be found based on well-known method of bonded joint analysis, where the equations of strains in the adhesive layer are solved (A index is referred to adhesive layer properties) [22] provided by Eq. 18.1.

$$\frac{d^3\gamma_A}{dy^3} - 4\cdot\frac{G_A}{t_A}\cdot\left[\frac{1}{E'_s\cdot t_s} + \frac{1}{E'_p\cdot t_p}\right]\cdot\frac{d\gamma_A}{dy} = 0$$

$$\frac{d^4\varepsilon_A}{dy^4} + \frac{E'_A}{t_A}\cdot\left[\frac{1}{D_s} + \frac{1}{D_p}\right]\cdot\varepsilon_A = 0 \tag{18.1}$$

Here, $D_{s,p}$ is the bending stiffness of the skin and the patch, respectively, $E'_{s,p} = E_{s,p}/(1-\nu_{s,p}^2)$, $E'_A = 2G_A/(1-\nu_A)$.

The rotation of crack surfaces $\widetilde{\theta}_0$ and the opening displacement along the skin centerline $\widetilde{v}_0$ can be found from Eq. 18.1 taking into account the conditions at the boundaries between the adhesive layer and coupling parts.

The link of loading factors and resulting strains in the adhesive layer is described per matrix expression utilizing the compliance parameters of the system of bridging springs:

$$\begin{Bmatrix} \tilde{v}_0 \\ \tilde{\theta}_0 \end{Bmatrix} = \begin{bmatrix} c_{tt} & c_{tb} \\ c_{bt} & c_{bb} \end{bmatrix} \begin{Bmatrix} n_0 \\ m_0 \end{Bmatrix},$$

where $n_0 = N_p\big|_{y=0} = -N_0$, $m_0 = N_s|_{y=0} = -M_0$.

18.3 Elastic Properties Degradation of the Repair Patch Material

The damage propagation in the aluminum skin is defined by applied cyclic load and the material fatigue properties evaluated during the series of standardized tests. In the repair joint, the portion of the applied load is transferred through the repair patch and, thus, the patch loading will be cyclic. In this case, one of the primary factors affecting the repair effectiveness is the elastic properties degradation of the patch material subject to applied cyclic load.

Practical feasible methods to define and control the properties degradation and Polymer Cement Mortars (PCM) fracture parameters are intensively developed [23]. Nevertheless, the researchers are coming to a conclusion that it is necessary to perform the series of experiments with the specimens of required layup and made of required material to get the authentic information about structure behavior under fatigue (cyclic) load taking into account the big scatter of the tracked and logged parameters.

The change of external thermal, moisture, and other climate conditions affects the PCM properties' degradation substantially. This effect noticeably complicates the simulation process of the real PCM service conditions at research facilities. It is noted that even at low temperature in the very beginning of climatic exposure with the case of no mechanical loads, the properties of composites may change due to the relaxation of the material's initial structural nonequilibrium obtained at specimen manufacturing [24–26].

A large number of studies are devoted to alteration of strength limits of composites under the variety of external load factors, while the elastic modulus variation sometimes stands out of scope. For the woven composites, the change of the elastic modulus will be much more visible than for composites made of unidirectional tapes [27]. Patch material elastic modulus is one of the determinant parameters for the bonded repair procedures design. To figure out the degradation behavior of carbon fiber plastic material that is proposed to use for the bonded repair patch preparation, the current research is studied: the variation of the longitudinal and transversal elastic moduli of the carbon fiber plastic specimens at temperatures −60, +23, and +80 °C and variation of Poisson ratio under cyclic load at the same values of temperature. Elastic moduli degradation was evaluated in relation to elastic moduli at the first load cycle at the temperatures mentioned above.

Hereinafter, Sect. 18.3.1 provides a description of the testing procedure and materials. Results of fatigue tests are presented in Sect. 18.3.2. Elastic modulus change evaluation based on fatigue test data is discussed in Sect. 18.3.3.

18.3.1 Testing Procedure and Materials

The study was performed in the laboratory of Kazan National Research Technical University using the servo-hydraulic testing machine ITW BiSS with a climate chamber. Each specimen for fatigue tests was subjected to cyclic tension load with ratio $R = 0.1$. The amplitude of tension load was defined as 67% of mean fracture load for such specimens at current temperature. Bonded strain gauges were used for gathering the strain data due to longitudinal and transversal stretching of the loaded specimens. Tests were performed at three values of temperature: −60, +23, and + 80 °C. Series of 10 specimens were tested at each temperature. General testing procedure matched ASTM D3039/D3479 requirements, specimen geometry for fatigue tests, and location of the strain gauges are shown in Fig. 18.2.

Test specimens were made of plain weave carbon fabric ECC 450 and epoxy resin CHS Epoxy 520 as a matrix. The specimens for fatigue tests were flat plates with 6 plies of carbon fabric with layup $[0/90]_0$, the specimens had the preformed tabs to clamp specimens in the grips of the testing machine. The properties of composite material are shown in Table 18.1.

The specimens were manufactured per Resin Transfer Molding (RTM) process with curing in the oven in accordance with curing cycle recommended by the resin system supplier. The general view of specimens with bonded strain gauges is presented in Fig. 18.3. Since the object of current research was to analyze the elastic properties but not a fracture of specimens, the specimens for fatigue tests were manufactured without stress concentrators in the working zone.

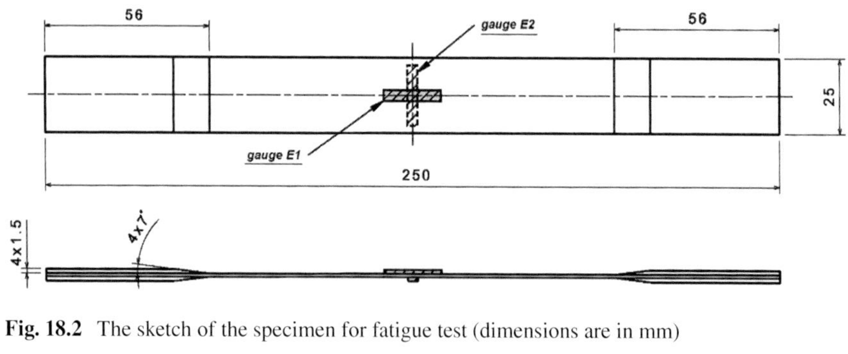

Fig. 18.2 The sketch of the specimen for fatigue test (dimensions are in mm)

Table 18.1 The properties of the ECC 450 material

Property	Value
Number of filaments per yarn	3000
Areal weight, kg/m^3	0.200
Ply thickness, mm	0.327
Fiber type	Torayca T300 J
Fiber ultimate strength, MPa	4210
Fiber elastic modulus, GPa	230
Fiber ultimate strain	1.8%
Fiber density, kg/m^3	1780
CTE, 1/°C	-0.43×10^{-6}

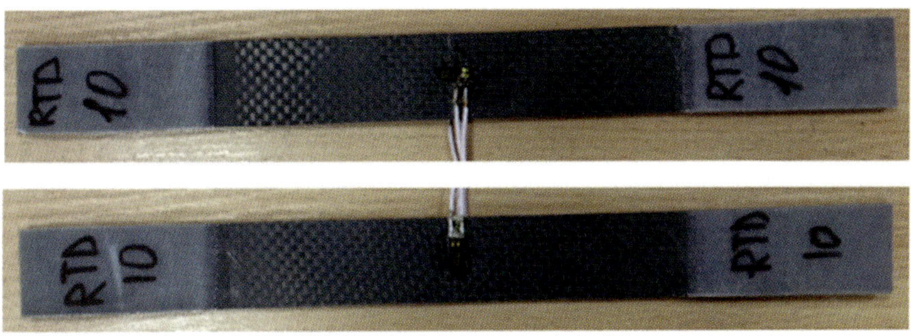

Fig. 18.3 General view of the specimens for fatigue tests

18.3.2 Results of Fatigue Tests

Fatigue tests of carbon specimens were split into two stages. A quasi-static fracture load F_{fracture} at three values of temperature was defined at the first stage. The specimens were elongated with constant load growth rate of 1 mm/min until specimen failure. Fracture stress values for each temperature tested are shown in Fig. 18.4.

The maximum strength (604 ± 30 MPa) was observed for the specimens tested at room temperature. The maximum scatter of measurements (approx. 51 MPa) was found for specimens tested at −60 °C. The increasing influence of random defects in material structure on its strength due to embrittlement during the cooling may account for this effect.

The second stage contained the fatigue tests of specimens under the tension load with 5 Hz frequency and number of cycles 10^5. The frequency value was selected to minimize the time spent for one specimen testing and to reduce the heating level of specimens during the tests [28]. The number of load cycles is stated as a minimal requirement to estimate the fatigue strength by means of fatigue rate. The cyclic load amplitude was set at 0.67 $F_{fracture}$ for each test temperature. This load value matches the patch design ultimate load at the bonded repair procedure creation.

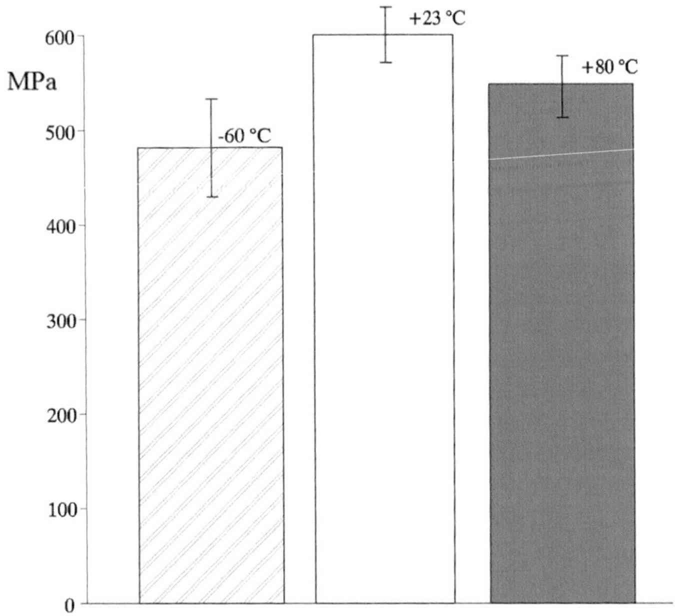

Fig. 18.4 Fracture stresses at static load (vertical lines represent the error bars of measurements)

18.3.3 Elastic Modulus Change Evaluation Based on Fatigue Test Data

The analytical relations satisfying the test data at three values of temperature were constructed after the data statistical processing. Since all series depicted the fast growth of the longitudinal strains on the starting period of tests within 500 load cycles and gently sloping strain growth for the rest time of tests, the results of the experiments are approximated by the following relation with the minimal dispersion:

$$E_1(N) = E_1^{500} + \left(E_1^1 - E_1^{500}\right)k^{-N}, \text{ for } 1 \leq N < 500,$$
$$E_1(N) = E_1^{500} + k_1 N + k_2 N^2, \text{ for } N \geq 500,$$

where N is the number of load cycles, E_1^1 is the mean longitudinal elastic modulus at the first load cycle, E_1^{500} is the mean longitudinal elastic modulus at the 500th load cycle, k, k_i are the relation empirical parameters. Parameter values for the approximating relation are defined in Table 18.2, the approximating relation curves calculated per Table 18.2 data and reduced to elastic modulus at the first load cycle E_1^1 are plotted in Fig. 18.5a, b.

The variation of the transversal elastic modulus was evaluated by change of Poisson ratio on the number of load cycles. Poisson ratio change is well approximated by linear relation:

Table 18.2 Parameters for approximating relation $E_1(N)$

Parameter	−60 °C	+23 °C	+80 °C
$1 \leq N < 500$			
E_1^1, GPa	110.223	106.594	53.402
E_1^{500}, GPa	74.016	70.841	42.508
k	1.056	1.033	1.014
$N \geq 500$			
k_1	6.965×10^{-7}	-4.416×10^{-7}	-1.458×10^{-4}
k_2	-6.965×10^{-10}	3.382×10^{-10}	7.940×10^{-10}

$$\nu_{12}(N) = \nu_{12}^1 + \frac{\nu_{12}^{\text{final}} - \nu_{12}^1}{10^5} N,$$

where ν_{12}^1 is the Poisson ratio at the first load cycle, ν_{12}^{final} is the Poisson ratio at the final 10^5th load cycle. Poisson ratio curves are plotted in Fig. 18.6.

18.4 Bonded Repair Effectiveness Calculation Results

The computation efforts of the repair bonded joint in accordance with the proposed analytical model allow to estimate the effectiveness of the selected repair configuration and narrow the search area to find the optimal design of the system "skin-adhesive-patch". The repair configuration effectiveness estimation assumes the calculation of the following values:

- SIF at the crack tip is directly affecting the rate of the crack growth in the skin under the patch.
- Stress concentration and the skin near the patch edge—additional crack growth condition in the skin along the patch edge will depend on this value.

Figure 18.7 shows the relation $a = f(N)$ between the crack length and number of load cycles for the 2.0 mm thick aluminum skin made of 7075-T6 alloy. The repair patch diameter is 100 mm, and the patch is made of composite materials (carbon fiber, fiberglass, and boron-epoxy fabrics) of quasi-isotropic layup and GLARE 2-3/2-0.2 hybrid metal-polymeric material. Initial crack length is 10 mm. Load diapason is $\sigma_{\max} = 0.4\sigma_{BAl}$, $\sigma_{\min} = 0.04\sigma_{BAl}$. Adhesive material is Cytec FM-73, the mechanical properties are found in the official technical data supplied by manufacturer [29]. Material properties used for calculation are shown in Table 18.3.

The analyzed fiberglass patch showed the negative results as it was not able to block the propagation of crack. The only material demonstrating the crack stopping abilities is boron plastic B(4)/5505.

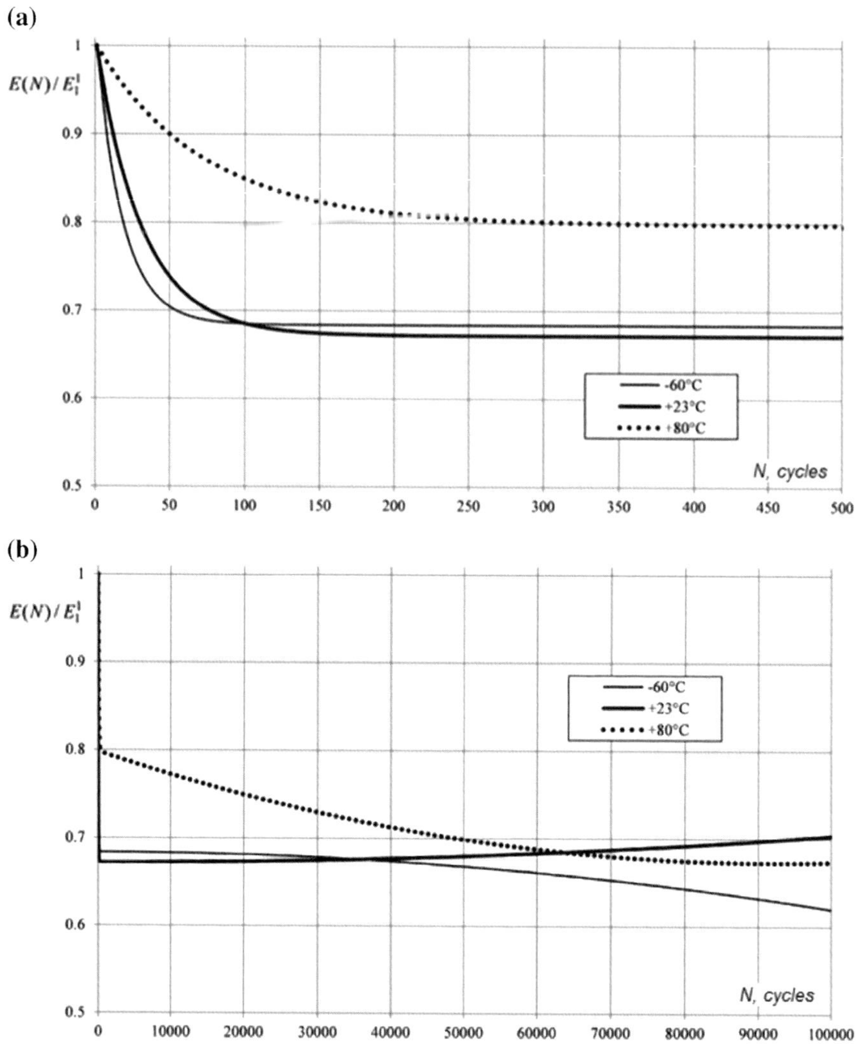

Fig. 18.5 Relationship of the longitudinal elastic modulus on the number of load cycles: **a** the range of cycles shown is 0 to 500, **b** the range of cycles shown is 0 to 10^5

If the patch is made of carbon fiber plastic ECC 450/ CHS Epoxy 520 tested in Sect. 18.2, the damage propagation features will substantially depend on temperature of the bonded joint as shown in Fig. 18.8.

It is evident that the repair patch does not restrict the damage growth in the skin material at +80 °C. Thus, this patch should not be used to repair the structure primarily loaded at the elevated temperature. It is noted that the $a = f(N)$ behavior is changed significantly: the crack growth is intensively increasing at the stating period

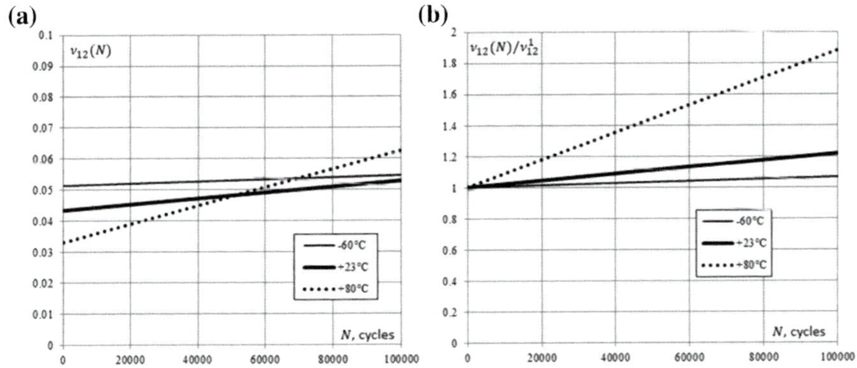

Fig. 18.6 Relationship of Poisson ratio on the number of load cycles: **a** absolute values of Poisson ratios, **b** Poisson ratio change relative to the value at the first loading cycle

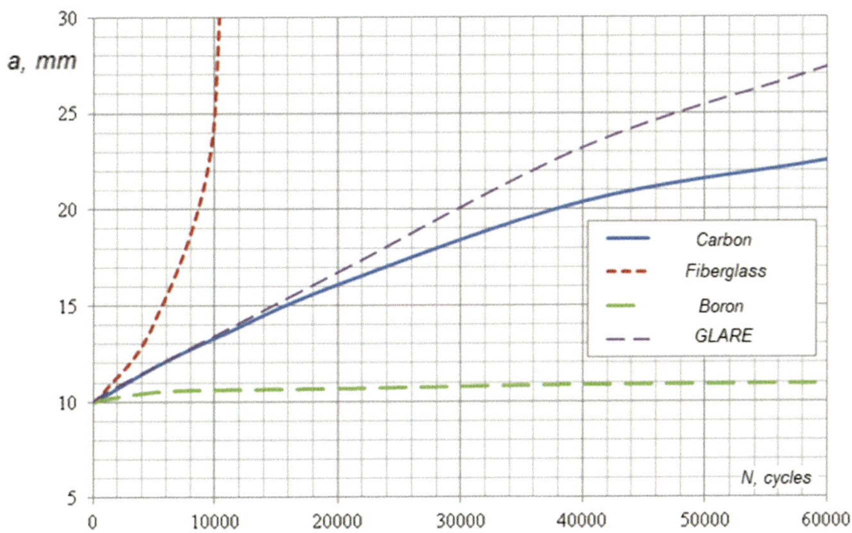

Fig. 18.7 A comparison of the repair effectiveness for the patches made of different materials

of loading unlike the charts shown in Fig. 18.7. Thus, the elastic properties degradation is an essential factor influencing the damage propagation and the inspection interval criteria for the current repair location.

Figures 18.9 and 18.10 show the variation of the stress concentration K_t in the skin along the patch edge depending on stiffness ratio S ($S = \frac{E'_P t_P}{E'_S t_S}$, $E'_{S,P} = \frac{E_{S,P}}{1-\nu_{S,P}}$) at different B/A ratios of the elliptic patches (A and B are the lengths of the ellipse semi-axis).

The higher patch stiffness and the lower patch radius provide the significant increasing of the stress concentration in the skin. This leads to reduction of the repair

Table 18.3 Patch material properties

	E_x, GPa	E_y, GPa	G_{xy}, GPa	α_x, 1/°C	α_y, 1/°C	t, mm
Carbon T300/934	74	74	19.89	1.335×10^{-6}	24.07×10^{-6}	1.58
S-glass/epoxy fiberglass	43	8.9	5.89	11.23×10^{-6}	11.23×10^{-6}	1.95
B(4)/5505 boron	204	18.5	5.59	10.28×10^{-6}	10.28×10^{-6}	1.04
GLARE 2-3/2-0.2	68.9	53.8	15.2	16.38×10^{-6}	24.48×10^{-6}	1.10
7075-T6 sheet (reference)	73.1	73.1	28.0	23.2×10^{-6}	23.2×10^{-6}	2.0

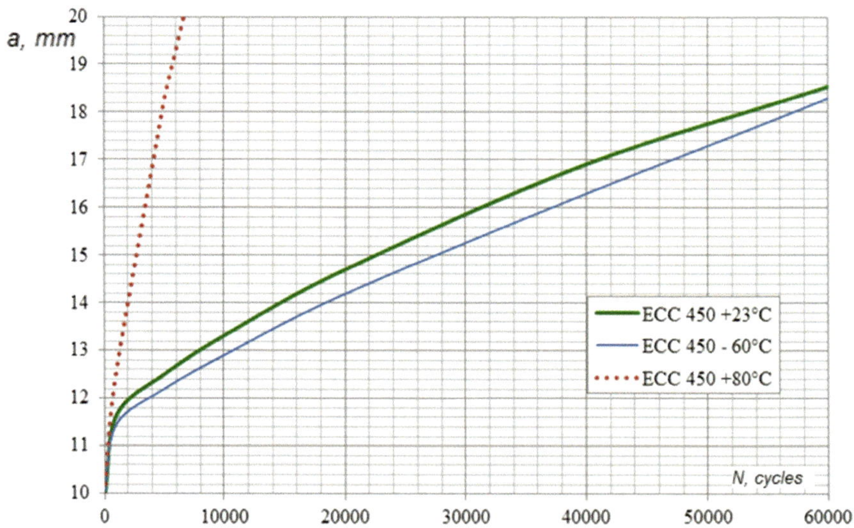

Fig. 18.8 A comparison of the repair effectiveness for the patch made of carbon ECC 450/CHS Epoxy 520 at various temperatures

configuration effectiveness due to additional areas of critical stresses, where the validating stress calculations are required. For the patch geometry within the range $0.5 < B/A < 2$, the utilization of the stiffness value for the degraded patch material gives the underestimation of the stress concentration in the skin near the edge of the patch.

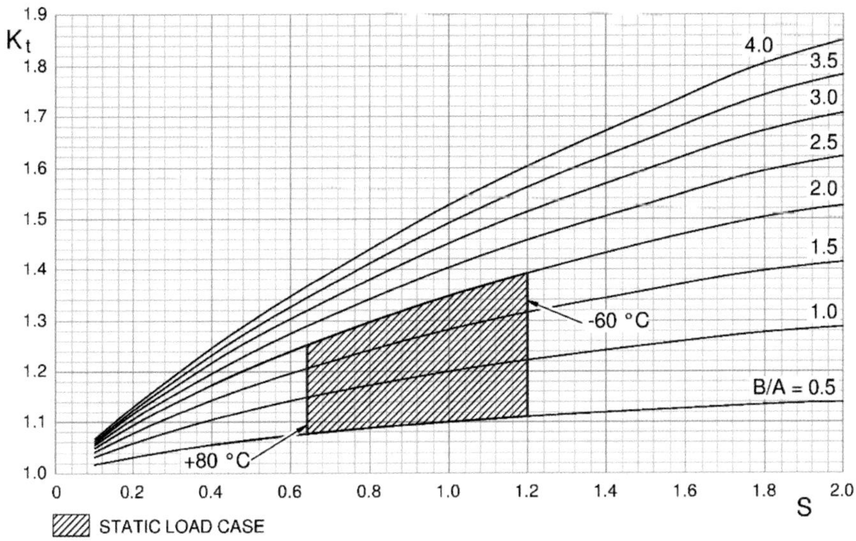

Fig. 18.9 Stress concentration in the skin near the patch edge—static load

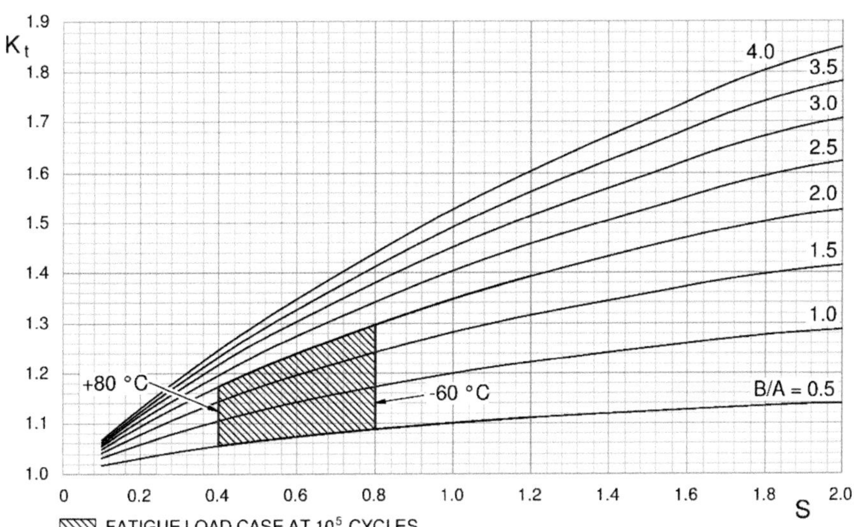

Fig. 18.10 Stress concentration in the skin near the patch edge—cyclic load

18.5 Conclusions

The following statements may conclude the presented chapter:

1. The performed literature survey demonstrates the potential ability to install bonded repairs on the damaged metallic structures of the commercial aircraft. It is possible to develop the repair process that can be realized by airline maintenance personnel in the actual conditions of the design, technological, and environmental restrictions.
2. The experimental study of the fatigue degradation for the composite material mechanical properties was performed. The test specimens were subjected to cyclic loads at various temperatures. The effect of material stiffness degradation on the bonded repair performance was evaluated.
3. The analytical technique for parameter calculation of the bonded repair joint was developed. This technique is suitable for preliminary design and concept review of the repair configurations.
4. The proposed analytical technique can be tuned for specific structural and repair materials and solutions. This technique has a potential for further development and improvement to account several extra features including (but not limited to):

   a. Adhesive behavior under cyclic and long-term loads.
   b. The character of adhesion between the adhesives and coupling surfaces.
   c. Combined influence of the climatic factors on the bonded joint strength and the material mechanical properties.

Also it can be interesting to expand the scope of the proposed analytical technique to the damaged composite structures of modern commercial airplanes like Yak-242, Airbus A350, or Boeing 787 and 777-9X. Such expansion requires revision of the computational core of the technique to simulate the damage growth processes within the complex heterogeneous composite material. The expansion will require performing the intensive experimental and theoretical research.

References

1. Baker, A.A.: Repair of metallic airframe components using fibre-reinforced polymer (FRP) composites. In: Karbhari, V.M. (ed.) Rehabilitation of Metallic Civil Infrastructure Using Fiber Reinforced Polymer (FRP) Composites, pp. 11–59. Elsevier Ltd (2014)
2. Baker, A.A.: A proposed approach for certification of bonded composite repairs to flight-critical airframe structure. Appl. Compos. Mater. **18**, 337–369 (2011)
3. Kablov, E.N., Antipov, V.V., Klochkova, Y.Y.: Aluminium-lithium alloys of new generation and aluminum fiberglass laminates on their basis. Tsvetnye Metally 86–91 (2016)
4. Antipov, V.V., Serebrennikova, NYu., Senatorova, O.G., Morozova, L.V., Lukina, N.F., Nefedova, YuN: Hybrid laminated materials with slow fatigue-crack development. Russian Eng. Res. **37**(3), 195–199 (2017)
5. Shestov, V.V., Antipov, V.V., Ryabov, D.R.: Corrosion resistance and mechanical properties of layered structural material based on aluminum alloy and fiberglass thin sheets. Metallurgist **60**(11–12), 1191–1196 (2017)
6. Petrova, A.P., Lukina, N.F., Dement'eva, L.A., Anikhovskaya, L.I.: Film structural adhesives. Polym. Sci. Ser. D **8**(2), 138–143 (2015)

7. Petrova, A.P., Lukina, N.F., Sharova, I.A.: Assessment of strength of adhesive joints made with epoxy adhesives under the influence of different factors. Polym. Sci. Ser. D **7**(3), 228–232 (2014)
8. Anikhovskaya, L.I., Batizat, D.V., Petrova, A.P.: The VK-50 elastomeric structural film adhesive. Polym. Sci. Ser. D **9**(3), 251–254 (2016)
9. Lukina, N.F., Kotova, E.V., Chursova, L.V., Kirienko, T.A.: Adhesive binders for layered alumopolymer composite materials. Polym. Sci. Ser. D **9**(4), 392–395 (2016)
10. Petrova, A.P., Sharova, I.A., Lukina, N.F., Buznik, V.M.: The applicability of adhesives in Arctic conditions. Polym. Sci. Ser. D. **9**(2), 188–194 (2016)
11. Oakafor, A.C., Bhogpurapu, H.: Design and analysis of adhesively bonded thick composite patch repair of corrosion grind out in 2024T3 clad aging aircraft structures. Compos. Struct. **76**(1–2), 138–150 (2006)
12. Baker, A.A.: Fatigue life recovery in corroded aluminum alloys using bonded composite reinforcements. Appl. Compos. Mater. **13**(3), 127–146 (2006)
13. Rose, L.R.F.: An application of the inclusion analogy. Int. J. Solids Struct. **17**(8), 827–838 (1981)
14. Lee, Y.-G., Zou, W.-N., Ren, H.-H.: Eshelby's problem of inclusion with arbitrary shape in an isotropic elastic half-plane. Int. J. Solids Struct. **81**, 399–410 (2016)
15. Duong, C.N.: An engineering approach to geometrically nonlinear analysis of a one-sided composite repair under thermo-mechanical loading. Compos. Struct. **64**(1), 13–21 (2004)
16. Eshelby, J.D.: The determination of the elastic field of an ellipsoidal inclusion and related problems. Proc. Roy. Soc. (London) **A241**, 376–396 (1957)
17. Beom, H.G.: Analysis of a plate containing an elliptic inclusion with eigencurvatures. Arch. Appl. Mech. **68**(6), 422–432 (1998)
18. Rose, L.R.F.: Theoretical analysis of crack patching. In: Baker, A.A., Jones R. (eds) Bonded Repair of Aircraft Structure, pp. 77–106. Kluwer Academic Publisher (1988)
19. Joseph, P.F., Erdogan, F.: Plates and shells containing a surface crack under general loading conditions. NASA Contractor Report 178323 (1987)
20. Joseph, P.F., Erdogan, F.: Surface crack problems in plates. Int. J. Fract. **41**(2), 105–131 (1989)
21. Wang, C.H., Rose, L.R.F.: A crack bridging model for bonded plates subjected to tension and bending. Int. J. Solids Struct. **36**(13), 1985–2014 (1999)
22. Hart-Smith, L.J.: Adhesive-bonded single-lap joints. NASA report CR-112236 (1973)
23. Eremin, A.V., Byakov, A.V., Lyubutin, P.S., Panin, S.V.: Development of acoustic-optical approach for structural health monitoring of composite materials under cyclic loading. Russ. Phys. J. **59**(7–2), 49–54 (2016)
24. Kablov, E.N., Startsev, O., Krotov, A.S., Kirillov, V.N.: Climatic aging of composite aviation materials: I Aging mechanisms. Russ. Metallurgy (Metally) **2011**(10), 993–1000 (2011)
25. Kablov, E.N., Startsev, O., Krotov, A.S., Kirillov, V.N.: Climatic aging of composite aviation materials: II. Relaxation of the initial structural nonequilibrium and through-thickness gradient of properties. Russian Metall. (Metally) **2011**(10), 1001–1007 (2011)
26. Kablov, E.N., Startsev, O., Krotov, A.S., Kirillov, V.N.: Climatic aging of composite aviation materials: III Significant aging factors. Russ. Metall. (Metally) **2012**(4), 323–329 (2012)
27. Daggumati, S., De Baere, I., Van Paepegem, W., Degrieck, J., Xu, J., Lomov, S.V., Verpoest, I.: Fatigue and post-fatigue stress–strain analysis of a 5-harness satin weave carbon fibre reinforced composite. Compos. Sci. Technol. **74**, 20–27 (2013)
28. De Baere, I., Van Paepegem, W., Degrieck, J.: On the design of end tabs for quasistatic and fatigue testing of fibre-reinforced composites. Polym. Compos. **30**(4), 381–390 (2009)
29. FM-73 Epoxy Film Adhesive. Technical Data Sheet AEAD-00019. Cytec Engineering Materials, https://docplayer.net/37348033-Fm-73-epoxy-film-adhesive.html. Last accessed 10 Aug 2019

Chapter 19
Parametric Identification of Tersoff Potential for Two-Component Materials

Karine K. Abgaryan and Alexander V. Grevtsev

Abstract The chapter is dedicated to the study of the parametric identification of Tersoff potential for one-component and two-component materials. The chapter features a comparison of minimization methods in terms of speed and accuracy, and the results of the implemented software operation for one-component and two-component materials. It was shown that implemented software can do the identification of two-component materials but with slightly less accurate results. However, two-component parametric identification requires more time for computation. All results were compared to the results of the experimental and quantum-mechanical modeling.

19.1 Introduction

Molecular-Dynamic (MD) modeling is applied when a natural experiment is impossible, very complicated, or highly expensive. The molecular-dynamic approach is one of the fields of mathematical modeling that is currently rather frequently used in the problems of material science, which accounts for the relevancy of the work. In MD modeling, the behavior of the interacting atoms of a system is described within the framework of classical dynamics [1]. Their location and speed are determined by means of integration of a system of ordinary differential equations. At the same time, the forces affecting the atoms are determined by interatomic interaction. Usually, its description in a system has a rather complicated form, as it can include interactions of

K. K. Abgaryan (✉) · A. V. Grevtsev
Federal Research Center "Information and Control" of the RAS, 44/2 Vavilova ul, Moscow 119333, Russian Federation
e-mail: kristal83@mail.ru

A. V. Grevtsev
e-mail: alex.grevtsev@gmail.com

K. K. Abgaryan
Moscow Aviation Institute (National Research University), 4, Volokolamskoe shosse, Moscow 125993, Russian Federation

L. C. Jain et al. (eds.), *Advances in Theory and Practice of Computational Mechanics*, Smart Innovation, Systems and Technologies 173,
https://doi.org/10.1007/978-981-15-2600-8_19

various types. The main types of interatomic interactions include the ionic, covalent, metallic, and van-der-Waals interactions. The ion interaction is commonly considered to be caused by a displacement of valence electrons from one atom to another, between which the electrostatic attraction occurs. Covalent interaction occurs when a covalent chemical bond is formed. At the same time, consolidation and concentration of the electron pair occur on the molecular orbital. A metallic bond occurs when shared electron gas interacts with the ion core of a crystalline structure. Van-der-Waals interactions are applied in order to describe intermolecular interaction, in a series of organic compounds, etc.

The chapter considers one- and two-component compounds with the covalent type of chemical bonding described using Tersoff potential, which was proposed in [2] and considered for one-component materials. Parameters of Tersoff potential for silica, which is a one-component material, were calculated in [3].

In this chapter, Tersoff potential was modified for use with two-component materials. Each potential has a certain set of parameters, the values of which are unique for each material. The process of finding such sets represents the parametric identification of the potential. It should be understood that the problem of parametric identification is multiextremal, and, hence, it is necessary to find the global minimum. The work presents a comparison of Monte Carlo and simulated annealing methods for global minimization and Hooke–Jeeves and Radial Granular Search (RGS) methods for local minimization.

Software for parametric identification of certain materials was used to perform calculations shown in the chapter. The software uses parallel calculations in order to reduce the identification time. Various parallelization methods are compared.

Section 19.2 includes the statement of problem of parametric identification of interatomic potential. Section 19.3 presents the comparison of optimization methods used for parametric identification. Section 19.4 describes the computational results for different materials. Section 19.5 includes a description of the implemented software. Section 19.6 presents the conclusion of the chapter.

19.2 Problem Statement

Calculation of the total energy of the modeled material atom system is carried out using Eq. 19.1.

$$U = \frac{1}{2}\sum_{i=0}^{n}\sum_{j=0}^{n} V\left(r_{ij}\right) \tag{19.1}$$

In Eq. 19.1, r_{ij} is the distance between atoms i and j. The interaction energy $V\left(r_{ij}\right)$ is calculated using Eq. 19.2.

$$V\left(r_{ij}\right) = f_c\left(r_{ij}\right)\left[V_R\left(r_{ij}\right) - b_{ij}V_A\left(r_{ij}\right)\right] \tag{19.2}$$

Parameter $V_R(r_{ij})$ is the atom j to atom i repulsion potential, which is calculated by Eq. 19.3.

$$V_R(r_{ij}) = \left[\frac{D_e}{S-1}\right]\exp\left(-\beta\sqrt{2S}\left(r_{ij}-r_e\right)\right) \tag{19.3}$$

Parameter $V_A(r_{ij})$ is the atom j to atom i attraction potential, which is calculated using Eq. 19.4.

$$V_A(r_{ij}) = \left[\frac{SD_e}{S-1}\right]\exp\left(-\beta\sqrt{\frac{2}{S}}\left(r_{ij}-r_e\right)\right) \tag{19.4}$$

Here, f_c is the function of cutoff around the ith atom, which is continuous and differentiable for all r_{ij}. This function allows to reduce a complexity of the calculations and is described by Eq. 19.5.

$$f_c(r) = \begin{cases} 1 & \text{if} \quad r < (R - R_{cut}) \\ \frac{1}{2}\left[1 - \sin\left[\frac{\pi(r-R)}{2R_{cut}}\right]\right] & \text{if } R - R_{cut} < r < R + R_{cut} \\ 0 & \text{if} \quad r > (R + R_{cut}) \end{cases} \tag{19.5}$$

Equation 19.5 has two parameters. Parameter R is the distance parameter, which is measured in angstroms. The cutoff function has the limits of 0 and 1 at the distance of R_{cut} from the value of R in the positive and the negative directions. Parameter b_{ij} is the order of bond between atoms i and j, which depends on the bond angle of atom i and the nuclear environment. The bond order is described according to Eq. 19.6.

$$b_{ij} = \left[1 + \left(\gamma\zeta_{ij}\right)^n\right]^{-\frac{1}{2n}} \tag{19.6}$$

Parameter ζ_{ij} provides weighted calculations for bonds other than i-j. Calculations are carried out for each kth atom using Eq. 19.7.

$$\zeta_{ij} = \sum_{k \neq i,j}^{n} \left(f_c(r_{ik})g\left(\theta_{jik}\right)\omega_{ijk}\right) \tag{19.7}$$

Parameter ω_{ijk} provides a calculation of the efficiency of a coordinate system, which is affected by the distances of various neighboring atoms. Parameter ω_{ijk} can be calculated using Eq. 19.8.

$$\omega_{ijk} = \exp\left[\lambda^3\left(r_{ij} - r_{ik}\right)^3\right] \tag{19.8}$$

Parameter $g(\theta_{jik})$ defines the dependence on the bond angle of atom θ_{jik} with vertex in ith atom, which can be calculated using Eq. 19.9.

$$g(\theta_{jik}) = 1 + \left(\frac{c}{d}\right)^2 - \frac{c^2}{d^2 + (h - \cos(\theta_{jik}))} \tag{19.9}$$

This calculation allows to stabilize the geometry of crystalline grid atoms in shearing operations.

The parametric identification of Tersoff potential requires the determination of the parameter vector $\xi = (\xi_1, \ldots, \xi_k) \in X, X \subseteq R^k$, at which the lowest value of the minimization function (Eq. 19.10) is achieved.

$$F(\xi) = \omega_1 \frac{(f_1(\xi))^2}{\dot{f}_1^2} + \cdots + \omega_n \frac{(f_n(\xi))^2}{\dot{f}_n^2}, \sum_{i=1}^{n} \omega_i = 1 \tag{19.10}$$

In Eq. 19.10, ω_i are the weighting coefficients, the sum of which equals to 1, $\dot{f}_n$ is the reference value of the ith studied material property, $f_i(\xi)$ is the value of the ith studied material property, which is obtained using the specified set of parameters.

The weight factors are usually calculated as $\omega_i = \frac{1}{n}$, where n is the number of studied material properties. Additionally, the squares of the difference between the reference and calculated values of the characteristics are normalized in the minimization function. It allows to obtain more adequate values of the function.

The problem of parametric identification can be formally described as shown in Eq. 19.11.

$$\xi = \arg\min_{\xi \in X} F(\xi) \tag{19.11}$$

It should be noted that there is a vast number of local minima in these types of problems. It is very important that the determined minimum is the global minimum. For this, it is important to ensure the use of minimization method that allows to find the global minimum among the local minima. An increase in the number of potential parameters results in a significant increase in the difficulty of finding a global minimum.

There is a number of minimization methods, which may resolve these types of problems, for example, the simulated annealing or Monte Carlo methods. Each of these methods has a different way of treating the local minima. It is important to understand that such methods have a lower accuracy compared to the local minimization methods.

In a parametric minimization problem, methods allow to find a global minimum with low accuracy should be used first. This provides a possibility to find an area, where the solution is located, faster. Then, high accuracy local minimization methods should be used.

19.3 Comparison of Optimization Methods

Various minimization methods should be compared based in terms of the parameterization time and value of the minimization function. Global minimization methods should be compared first.

Monte Carlo method and simulated annealing method are rather different in terms of their essence. In the first method, numerous sets of parameters are randomly calculated, and the best one of the sets is selected. The accuracy of the result is directly dependent on the number of generated sets. The simulated annealing method has a limited number of iterations, which depends on the temperature reduction function. In the implemented software, the number of iterations of the simulated annealing method is 112. In order to correctly compare the methods in terms of accuracy and time, several runs of each method are required, and the total execution time and the obtained value of the minimization function should be considered.

For comparison, each method was run ten times. Monte Carlo method generates 1,120 sets, and the simulated annealing method includes a total of 1,120 iterations. Table 19.1 demonstrates that Monte Carlo method is approximately 3.5 times faster.

Table 19.2 demonstrates that the behavior of the simulated annealing method is inferior to that of Monte Carlo method. The average value of the minimization function of the simulated annealing method is 0.0839228511, while the average value of the minimization function of Monte Carlo method is 0.0496128700.

The local minimization methods are compared using the resulting sets of parameters obtained under global minimization provided by Monte Carlo method, as this

Table 19.1 Comparison of the time required for global minimization methods

Method	Time, ms
Monte Carlo	66,981
Simulated annealing	238,210

Table 19.2 Comparison by the final value of the minimization function

Set	Monte Carlo method, $F(\xi)$	Simulated annealing method, $F(\xi)$
1	0.0206441	0.185270050549
2	0.026144	0.056386588562
3	0.040383	0.108290095692
4	0.0462078	0.065527332121
5	0.0512995	0.083363670061
6	0.0535324	0.077168913690
7	0.0585524	0.131231663262
8	0.0653591	0.042978009482
9	0.0654594	0.025928630353
10	0.068547	0.063083557049

Table 19.3 Comparison of the time required for local minimization methods

Method	Time, ms
RGS	12,508
Hooke–Jeeves	26,516

Table 19.4 Comparison by the final value of the minimization function

Set	RGS method, $F(\xi)$	Hooke–Jeeves method, $F(\xi)$
1	0.0206441	0.185270050549
2	0.026144	0.056386588562
3	0.040383	0.108290095692
4	0.0462078	0.065527332121
5	0.0512995	0.083363670061
6	0.0535324	0.077168913690
7	0.0585524	0.131231663262
8	0.0653591	0.042978009482
9	0.0654594	0.025928630353
10	0.068547	0.063083557049

method proved itself to be the better of two tested methods. RGS method and Hooke–Jeeves method are compared. Table 19.3 demonstrates that RGS method is almost two times faster than Hooke–Jeeves method.

Table 19.4 demonstrates that the behavior of Hooke–Jeeves method is inferior in terms of the obtained value of the minimization function. The average value of the minimization function using RGS method is 0.0003019762, while the average value of the minimization function using Hooke–Jeeves method is 0.0006873606.

19.4 Results

The computational results for silicon, germanium, aluminum nitride, and boron nitride are represented in Sects. 19.4.1–19.4.4, respectively.

19.4.1 Silicon

Silicon is the most commonly used material in semiconductor devices. Its oxide is rather easily obtained in furnaces with the formation of semiconductor interfaces. Silica has a crystalline structure of a diamond. The valence band in the crystal is completely filled. Silicon retains operation stability at high temperatures. Planetary silica reserves are almost limitless.

Table 19.5 Identified parameters of silicon

Parameter	Value
D_e	5.14592
S	4.69548
β	1.44156
r_e	2.23518
R	2.85
R_{cut}	0.15
c	24,697.3
d	103.132
h	−0.273259
n	1.21149
γ	0.953207
λ	0.00156529

For this material, identification of the potential parameters was carried out using Monte Carlo method. Eight sets with the lowest value of the minimization function are selected from the four thousand randomly generated sets. Then, the selected parameters were refined using RGS method. The calculated parameters are shown in Table 19.5.

The reference values of the properties were calculated on a computer using VASP software package. The experimental data was taken from [1]. The characteristics calculation results with the given potential parameters are provided in Table 19.6.

Table 19.6 Characteristics for silicon

Characteristic	Experiment	VASP	Identification
E_{coh}	−4.63	−4.617291375	−4.59422
a	5.431	5.465408871	5.58357
B	0.9783	0.8936352	0.914167
C'	0.509	0.46909605	0.52863
C_{11}	1.657	1.5190966	1.61829
C_{12}	0.639	0.5809045	0.561208
C_{44}	0.796	0.6277141	0.624638
ζ	0.524	0.522770577	0.517098

19.4.2 Germanium

Germanium is the 32nd element in the periodic table. Under normal conditions, the crystalline grid takes the shape of a diamond and represents a semiconductor material.

Germanium was used in microelectronics for the manufacture of transistors and diodes, until it was replaced by silica. Nevertheless, germanium is still used. According to certain musicians, the use of germanium-based transistors results in a better sound. Currently, germanium is mainly used in UHF devices as an element of the silicon and germanium alloy. This alloy makes it possible to achieve subterahertz frequencies.

With the parameters shown in Table 19.7, the obtained properties deviate by no more than one-tenth from the experimental values. The experimental data was taken from [2]. The characteristics given in Table 19.8 were rather close to those obtained in the VASP software package.

Table 19.7 Identified parameters of germanium

Parameter	Value
D_e	5.60702
S	6.09551
β	1.4484
r_e	2.24235
R	2.95
R_{cut}	0.15
c	33,739
d	125.481
h	−0.524262
n	1.57274
γ	1.6957
λ	17.5114

Table 19.8 Characteristics for germanium

Characteristic	Experiment	VASP	Identification
E_{coh}	−3.85	−3.78168	−3.77807
a	5.658	5.645	5.6753
B	0.7516	0.745452	0.746909
C'	0.403	0.3469163	0.348045
C_{11}	1.2889	1.208007	1.2102
C_{12}	0.4829	0.514174	0.514251
C_{44}	0.671	0.604663	0.602228
ζ	0.521	0.561069	0.562381

19.4.3 Aluminum Nitride

Aluminum nitride is mainly used with a crystalline grid of a wurtzite shape under standard temperature and pressure. Aluminum nitride of a sphalerite shape is also interesting in terms of its optical and physical properties. This type of material is very difficult to manufacture, as it is very reactive and requires a special sterility of raw materials.

This is a two-component material. The number of parameters depends on the number of interactions between various elements. There are three types of interactions for the two elements. Aluminum interacts with aluminum, aluminum interacts with nitrogen, and nitrogen interacts with nitrogen.

Thus, there are three times more parameters. Table 19.9 demonstrates all 3 sets of 12 parameters. R and R_{cut} for each interaction are not identified, but specified at the initial stage.

The obtained characteristics given in Table 19.10 are close to those obtained using VASP software package.

Table 19.9. Identified parameters of aluminum nitride

Parameter	Al-Al value	Al-N value	N-N value
D_e	3.35512	3.91278	2.71983
S	0.835993	1.09552	3.10157
β	1.05919	0.509301	1.5406
r_e	2.13729	2.57726	3.19281
R	2.335	2.335	2.335
R_{cut}	0.8	0.8	0.8
c	85871.1	3201.3	746,383
d	158.496	31.1845	30.3546
h	−0.273204	−0.658417	7.15848
n	20.9305	5.40596	6.0999
γ	0.0416951	0.0623046	0.0649637
λ	1.05287	1.48318	0.585351

Table 19.10 Characteristics for germanium

Characteristic	Experiment	VASP	Identification
E_{coh}	−5.76	−5.75281	−5.73307
a	4.38	4.38	4.22682
B	2.08	2.22859	2.24682
C'	0.72	0.791064	0.794709
C_{11}	3.04	3.283342	3.3049
C_{12}	1.60	1.701214	1.71347
C_{44}	1.93	1.909453	1.8594
ζ	0.55	0.640232	0.627198

The experimental data was taken from [4]. It should be noted that the deviation of VASP characteristics from the experimental values is rather substantial. It should be noted that the number of identifiable parameters increased three-fold. Due to that, a higher deviation of characteristics may be observed.

19.4.4 Boron Nitride

Boron nitride with a sphalerite-type crystalline grid is a stable structure under normal conditions. This material is characterized by exceptional physical properties, excellent strength, and chemical inertness. A wide bandgap, high melting temperature, and low dielectric constant make boron nitride a very good material for microelectronic devices. The use of Tersoff potential for molecular-dynamic modeling of boron nitride is described in [5, 6].

Boron nitride is mainly used as a boundary layer in the growing of GaN on SiC. This material is of great interest for nanotube growing. The optical properties and high transparency make boron nitride a useful material for optical windows and X-ray diaphragms.

At the end of the identification, potential parameters became equal to the values given in Table 19.11. The characteristics for such values are shown in Table 19.12. The experimental data was taken from [4].

Table 19.11 Identified parameters of boron nitride

Parameter	B-B value	B-N value	N-N value
D_e	47.3631	5.99477	2.29434
S	1.9582	6.03235	3.69895
β	0.0337772	1.26559	1.44987
r_e	2.5521	1.6491	2.97387
R	1.95	1.95	1.95
R_{cut}	0.75	0.75	0.75
c	2579.2	34907.1	787946
d	160.381	102.02	26.1181
h	−0.388802	−0.615747	−5.13706
n	53.5382	64.6551	6.47006
γ	0.391766	0.402921	0.0875969
λ	0.217494	0.0118349	0.182635

Table 19.12 Characteristics for boron nitride

Characteristic	Experiment	VASP	Identification
E_{coh}	−6.68	−6.728191125	−6.56826
a	3.6157	3.6157	3.79437
B	4.00	4.206769067	4.10621
C'	3.15	3.3215539	3.3132
C_{11}	8.20	8.6355076	8.63461
C_{12}	1.90	1.9923998	2.00906
C_{44}	4.80	4.7791548	4.87595
ζ	–	0.381397288	0.382457

19.5 Description of Software

The chapter features the results of calculations performed on the developed software, which is designed for the parametric identification of Tersoff potential parameters for various materials. The programming language applied was C++. OpenMP package was used for parallel calculations. The implemented software can be installed on Windows or Linux operating systems.

For clarity of the results, we present the parameters of the processor of the system, on which the calculations were performed: AMD Ryzen 7 1800X Eight-Core Processor 3.60 GHz.

19.6 Conclusions

The chapter presents a study of the molecular-dynamic modeling stage referred to as the parametric identification of potential. A software product for parametric identification was implemented. Tersoff potential was modified, so that it could be used for two-component materials. Due to modification, the number of potential parameters increased three-fold, which in turn led to the decline of the identification accuracy.

Monte Carlo method and simulated annealing method were compared for global minimization. As demonstrated by the results, Monte Carlo method is more accurate and faster than simulated annealing method. Hooke–Jeeves method and RGS method were compared for local minimization. Based on the provided results, RGS method is better.

The work also features the results of implemented software operation. The program is suitable for both one-component and two-component materials. The characteristics calculated with the obtained sets of parameters for one-component and two-component materials are the same or feature a minor deviation from the characteristics calculated in VASP software package. It implies that the parameters were correctly identified.

References

1. Abgaryan, K.K.: Multiscale Modeling in the Problems of Structural Materials Science. MAKS Press (in Russian) (2017)
2. Powell, D.: Lattice dynamics and parameterisation techniques for the Tersoff potential applied to elemental and type III-V semiconductors. Diss. for the degree of Doctor of Philosophy, The University of Sheffield (2006)
3. Abgaryan, K.K., Posypkin, M.A.: Optimization methods as applied to parametric identification of interatomic potentials. Comput. Math. Math. Phys. **54**(12), 1929–1935 (2014)
4. Shimada, K., Sota, T., Suzuki, K.: First-principles study on electronic and elastic properties of BN, AlN, and GaN. J. Appl. Phys. **84**(9), 4951–4958 (1998)
5. Los, J.H., J. Kroes, M.H., Albe, K., Gordillo, R.M., Katsnelson, M.I., Fasolino, A.: Extended Tersoff potential for boron nitride: energetics and elastic properties of pristine and defective h-BN. Phys. Rev. B **96**(18), 184108.1–184108.11 (2017)
6. Albe, K., Möller, W.: Modelling of boron nitride: atomic scale simulations on thin film growth. Comput. Mater. Sci. **10**(1–4), 111–115 (1998)

Part IV
Numerical Study of Dynamic Systems

Chapter 20
Multi-agent Optimization Algorithms for a Single Class of Optimal Deterministic Control Systems

Andrei V. Panteleev and Maria Magdalina S. Karane

Abstract The algorithms and software of three metaheuristic multi-agent methods: fish school search, krill herd, and imperialist competitive algorithm are considered. Recommendations on the parameter selection for each method are given. On the basis of krill herd and imperialist competitive algorithm, a hybrid extremum search algorithm is formulated. An algorithm for finding open-loop control for a single class of dynamic systems based on the use of the described multi-agent algorithms is also formed. Optimal open-loop control has the form of a step function with a given switching set. Software that allows to find the optimal open-loop control, criterion value, and coordinates of switching points of the control law on the basis of the suggested algorithms was formed. A specially selected set of test open-loop optimal control problems was solved. The obtained results confirmed that the numerical solution is close to the optimal one.

20.1 Introduction

The development of multi-agent optimization algorithms has been going on for many years in order to improve the existing classical optimization methods that do not always give a solution that is close to optimal or even replaces them with new algorithms. A feature of multi-agent algorithms is the use of a group of individuals (agents) on a certain set that performs certain actions (operations) in order to reach an extreme point. Using this approach, we can optimize not only multiextremal functions of many variables, but also find a solution for optimal open-loop control problems [1, 2], which is also important due to the practical importance of such problems in aviation and space technology.

A. V. Panteleev (✉) · M. M. S. Karane
Moscow Aviation Institute (National Research University), 4, Volokolamskoe Shosse, Moscow 125993, Russian Federation
e-mail: avpanteleev@inbox.ru

M. M. S. Karane
e-mail: mmarselina@mail.ru

L. C. Jain et al. (eds.), *Advances in Theory and Practice of Computational Mechanics*, Smart Innovation, Systems and Technologies 173,
https://doi.org/10.1007/978-981-15-2600-8_20

The Chapter is organized as follows, Section 20.2 provides a description of multi-agent methods. Application of multi-agent methods for optimal open-loop control problems is discussed in Sect. 20.3. Section 20.4 concludes the chapter.

20.2 Description of Multi-agent Methods

Multi-agent methods for solving the optimization problem are described in this section. Section 20.2.1 provides an optimization problem. Fish school search algorithm, krill herd algorithm, imperialist competitive algorithm, and hybrid multi-agent algorithm are introduced in Sects. 20.2.2–20.2.5, respectively.

20.2.1 Optimization Problem

Given the objective function $f(x) = f(x_1, x_2, \ldots, x_n)$ defined on the set of admissible solutions $D \subseteq R^n$. It is required to find the constrained global maximum of a function $f(x)$ on set D, i.e. ,such a point $x* \in D$, that

$$f(x*) = \max_{x \in D} f(x), \tag{20.1}$$

where $x = (x_1, x_2, \ldots, x_n)^T$, $D = \{x | x_i \in \ [a_i,\ b_i],\ \ i = 1, 2, \ldots, n\}$.

The task of finding the minimum of a function $f(x)$ is reduced to the task of finding the maximum by replacing the sign before the function with the opposite: $f(x*) = \min_{x \in D} f(x) = -\max_{x \in D} [-f(x)]$. Function $f(x)$ can be multiextremal, so the required solution in the general case is not unique.

20.2.2 Fish School Search Algorithm

Solution search strategy. Fish school search algorithm [3–5] uses the results of studying the behavioral features of certain fish species that can exist only within the flock, which reduces the individual freedom of their movements, but increases the intensity of the competition for food. Such a union of fish, as shown by observing them in the oceans and rivers, confirms that the benefits greatly exceed the drawbacks.

The algorithm uses the following main features of the behavior of schools of fish:

- Food (imitation of the natural instinct of fish consisting in the search for food). It is necessary because fish must be fed in order to grow strong and be capable of reproduction. When they get food, the fish gain weight, and, when they swim, they lose weight.

- Swimming (this function is implemented collectively by all fish in flock to search for food).
- Reproduction (imitation of a natural selection mechanism creating new objects to support the search process).

Each fish from the flock has an internal "memory" of the success in the search for food (approaching the extreme point) enclosed in the weight of the fish. A flock evolves by exchanging information between parents as a result of reproduction and also as a result of collective movement. The aquarium is a set of acceptable solutions. The presence of food shows the fish of the aquarium area that defines good regions for finding a solution. In the process of swimming, the idea of a global redirection of all fish to that part of the aquarium, which is considered by all flock fish as the most preferable from the point of view of food search, is realized. Reproduction of fish, in turn, allows to move from a comparative study of the areas of the aquarium to a process that refines the solution within the framework of the found area. For the process of reproduction, fish with the greatest weight are only allowed. In the process of finding solutions, an initial population *NP* of fish is first generated—a school on the set D using a uniform distribution.

Next, fish when searching for food (determining the extremum point) use Food operation, Swimming operation, and Reproduction operation discussed below.

Food operation. Fish are looking for food that is scattered in the aquarium in various concentrations. To search for food, fish perform individual movements. As a result, the fish may gain weight or vice versa depending on the search result. It is assumed that the increment of fish weight is proportional to the normalized increment of the value of the objective function:

$$W^{j,k+1} = W^{j,k} + \frac{f(x^{j,k+1}) - f(x^{j,k})}{\max\limits_{m=1,\ldots,NP}\left\{\left|f(x^{m,k+1}) - f(x^{m,k})\right|\right\}}, \tag{20.2}$$

where $W^{j,k+1}$, $W^{j,k}$ are the fish weights with number j at the $(k+1)$th and kth iterations, respectively, $f(x^{j,k+1}), f(x^{j,k})$ are the objective function values corresponding to the new $x^{j,k+1}$ and current $x^{j,k}$ fish position, respectively. The weight of the fish usually varies from 1 to W_{scale}. For $k = 0$, the weight of all fish is the same and equals $\frac{W_{scale}}{2}$.

Swimming operation.

(a) Each fish moves in a random direction so that the value of the objective function increases. If a new position of the fish is not included in the set D (aquarium), then the movement does not occur. After completing the movement, the fish feeds, the new position of each fish, $j = 1, \ldots, NP$, is calculated according to Eq. 20.3, where $\delta \sim R[-0.5,\ 0.5]$ and $step_{ind}^{k}$ decrease linearly with the increase of iterations number (an individual step is a vector with the same coordinates).

$$x^{j,k+1} = x^{j,k} + step_{ind}^{k} \cdot \delta \tag{20.3}$$

New position is accepted if value of the objective function increases (when the restriction on belonging to the set D is fulfilled). Otherwise, the fish does not move.

(b) After each fish has moved, the weighted average of the individual shifts is calculated by Eq. 20.4, where j is the fish number.

$$\Delta x_{ind}^{j} = x^{j,k+1} - x^{j,k} \tag{20.4}$$

It means that the fish that made a successful movement determines the resulting direction of movement more than others. Each fish in the school changes its position according to Eq. 20.5.

$$x^{j,k+1} = x^{j,k} + \frac{\sum_{m=1}^{NP} \Delta x_{ind}^{m} \cdot \left[f\left(x^{m,k+1}\right) - f\left(x^{m,k}\right)\right]}{\sum_{m=1}^{NP} \left[f\left(x^{m,k+1}\right) - f\left(x^{m,k}\right)\right]} \qquad j = 1, \ldots, NP \tag{20.5}$$

(c) Collective movement is based on the movement of the whole flock. If the flock increased the weight in case of a successful move, the radius of the flock increases. Otherwise, it decreases. The position of each fish varies with respect to the center of gravity of the flock (barycenter) determined at each iteration k:

$$Bari(k) = \frac{\sum_{j=1}^{NP} x^{j,k} \cdot W^{j,k}}{\sum_{j=1}^{NP} W^{j,k}}. \tag{20.6}$$

Besides, the increment vector $step_{vol}$ (vector with the same coordinates linearly decreasing with increasing number of iterations) is used.

If the total weight of the flock has increased, then the fish move toward the barycenter:

$$x^{j,k+1} = x^{j,k} - step_{vol} \cdot rand \cdot \frac{\left[x^{j,k} - Bari(k)\right]}{\left\|x^{j,k} - Bari(k)\right\|}. \tag{20.7}$$

If the total weight of the flock has decreased, then the fish move away from the barycenter:

$$x^{j,k+1} = x^{j,k} + step_{vol} \cdot rand \cdot \frac{\left[x^{j,k} - Bari(k)\right]}{\left\|x^{j,k} - Bari(k)\right\|}, \tag{20.8}$$

where $rand \sim R[0, 1]$.

Reproduction operation. Fishes that have reached the threshold weight ($W^{i,k} > thr$) are selected. For each selected fish, a pair is sought that is, a fish, which corresponds to the maximum ratio of its weight to the distance to the applicant. Each pair of fish with numbers i and j generates a descendant with weight and position:

$$W^{child,k} = \frac{W^{i,k} + W^{j,k}}{2}, \tag{20.9}$$

$$x^{child,k} = \frac{x^{i,k} + x^{j,k}}{2}. \tag{20.10}$$

Next, parents and descendants are ranked by weight. To preserve the size of the population, all fish that have the lowest weight (the population with size NP remains in the flock) are removed. The process ends, when a certain number of iterations $ITER$ is reached. The solution is to choose the position of the fish with the highest weight.

Recommendations on the selection of parameters. The population size NP determines the number of calculations of the objective function at each iteration. For a problem with a large range of admissible solutions, it is recommended to take a larger parameter value NP. Recommended parameter values $NP \in [30,\ 40]$.

The number of iterations $ITER$ determines how long the search for new solutions will continue. As a rule, the more $ITER$, the more accurate the answer. For a standard set of functions, the recommended values depend on the complexity of the function $ITER \in [1000,\ 10000]$.

The maximum weight W_{scale} is that an individual can gain. With the passing of the search time, the fish with the maximum weight W_{scale} is selected, and its position is taken as the answer. The recommended value of this parameter $W_{scale} \in [1000,\ 5000]$. The threshold weight W determines those fish that will be allowed to reproduce. Recommended value $W \in [900,\ 4500]$.

Individual step $step_{ind0}$ is a vector with the same components, which decreases linearly with increasing number of iterations to the value $step_{ind1}$, $step_{vol}$ is the vector with the same coordinates linearly decreasing with increasing number of iterations. Recommended values are the following: $step_{ind0} = 0.1$, $step_{ind1} = 0.01$, and $step_{vol} = 0.001$.

20.2.3 Krill Herd Algorithm

Solution search strategy. Krill herd method [3, 6, 7] refers to bioinspiration because it is based on the results of the analysis of the behavior of krill packs resembling shrimps. Their positions change under the influence of three factors: the presence of other members of the population, need to search for food, and random walks. Usually. the movement of krill population is determined by two goals: the increase in the density of krill and attainment of food.

At the beginning of the process, a population NP is generated from individuals on a set of admissible solutions D using a uniform distribution. It is assumed that the motion of the jth member of the population occurs according to Eq. 20.11, where x^j is the position, V^j is the speed, which consists of three components.

$$\frac{dx^j}{dt} = V^j \tag{20.11}$$

The first component is determined by the influence of neighbors (members of the population that belong to a certain neighborhood of jth element of a certain radius), the best element in the entire population, and information about its former speed. The second component is determined by the movement toward the food source (the "center of mass" of the population is taken for it), information about the former speed in search of food, and memory of its best result for all the iterations. The third component imitates the random walks of the individual, which decreases with the increasing number of iterations. To revive the search process, the cross and mutation operations used in other evolutionary methods and method of differential evolution are applied. The search procedure ends when the specified number of iterations is reached.

Solution search algorithm. Solution search algorithm includes the following steps.

Step 1. Set the method parameters: NP is the number of krill in the population, $S_{\max}$ is the maximum krill speed, μ is the small positive number, $I_{\max}$ is the maximum number of iterations, V_f is the maximum speed of movement to the food source, $D_{\max}$ is the maximum diffusion speed of krill. Let $I = 1$ (iteration count).

Step 2. Generate the initial krill population on a set D using the uniform distribution law: $x^1, \ldots, x^{NP}$. Calculate the values of the objective function $f(x^1), \ldots, f(x^{NP})$. Find the best and worst solutions x^{best}, x^{worst}.

Step 3. For each element of the population, $j = 1, \ldots, NP$, perform the following steps.

Step 3.1. Find the radius of the neighborhood around the solution x^j:

$$d_{\varepsilon_j} = \frac{1}{5 \cdot NP} \sum_{k=1}^{NP} d_{j,k}, \tag{20.12}$$

where $d_{j,k} = \sqrt{\sum_{i=1}^{n} \left(x_i^j - x_i^k\right)^2}$. Determine the number of neighbors S_j of the solution x^j from the condition $d_{j,k} \leq d_{\varepsilon_j}$.

Step 3.2. Find:

$$\Delta \bar{x}^{j,k} = \frac{x^k - x^j}{d_{j,k} + \mu}, k = 1, \ldots, S_j \quad \text{(for all neighbors)}, \tag{20.13}$$

$$\Delta \bar{f}^{j,k} = \frac{f(x^j) - f(x^k)}{f(x^{wrost}) - f(x^{best})}, k = 1, \ldots, N_j, \alpha_j^{local} = \sum_{k=1}^{N_j} \Delta \bar{f}^{j,k} \cdot \Delta \bar{x}^{j,k}. \tag{20.14}$$

Step 3.3. Find:

$$c^{best} = 2 \cdot \left(rand + \frac{I}{I_{\max}} \right), rand \sim U[0; 1], \Delta \bar{f}^{j,best} = \frac{f(x^j) - f(x^{best})}{f(x^{wrost}) - f(x^{best})}, \tag{20.15}$$

$$\Delta \bar{x}^{j,best} = \frac{x^{best} - x^j}{d_{j,best} + \mu}, \alpha_j^{target} = c^{best} \cdot \Delta \bar{f}^{j,best} \cdot \Delta \bar{x}^{j,best}. \tag{20.16}$$

Step 3.4. Find:

$$\alpha^j = \alpha_j^{local} + \alpha_j^{target}. \tag{20.17}$$

Step 3.5. Define:

$$S^{j,New} = S_{\max} \cdot \alpha^j + \omega \cdot S^{j,old}, \tag{20.18}$$

where $\omega \sim U[0; 1]$, $S^{j,old}$ is the old speed generated by the other members of the population (under $I = 0$, the speed is assumed to be equal to the zero vector).

Step 4. For any element of the population, $j = 1, \ldots, NP$, perform the following steps.

Step 4.1. Find the position of the food source:

$$x^{food} = \frac{\sum_{j=1}^{NP} \frac{x^j}{f(x^j)}}{\sum_{j=1}^{NP} \frac{1}{f(x^j)}}. \tag{20.19}$$

Step 4.2. Find:

$$\Delta \bar{f}^{j,food} = \frac{f(x^j) - f(x^{food})}{f(x^{wrost}) - f(x^{food})}, \Delta \bar{x}^{j,food} = \frac{x^{food} - x^j}{d_{j,food} + \mu}, \tag{20.20}$$

$$c^{food} = 2 \cdot \left(1 - \frac{I}{I_{\max}}\right), \beta_j^{food} = c^{food} \cdot \Delta \bar{f}^{j,food} \cdot \Delta \bar{x}^{j,food}. \tag{20.21}$$

Step 4.3. Find:

$$\Delta \bar{f}^{j,jbest} = \frac{f(x^j) - f(x^{jbest})}{f(x^{wrost}) - f(x^{jbest})}, \tag{20.22}$$

where x^{jbest} is the best position of jth population element,

$$\Delta \bar{x}^{j,jbest} = \frac{x^{jbest} - x^j}{d_{j,jbest} + \mu}, \beta_j^{best} = \Delta \bar{f}^{j,jbest} \cdot \Delta \bar{x}^{j,jbest}. \tag{20.23}$$

Step 4.4. Find:

$$\beta^j = \beta_j^{food} + \beta_j^{best}. \tag{20.24}$$

Step 4.5. Define:

$$F^{j,New} = V_f \cdot \beta^j + \omega_f \cdot F^{j,old}, \tag{20.25}$$

where V_f is the maximum speed of movement to the food source, $\omega_f \sim U[0,\ 1]$, $F^{j,old}$ is the old speed generated by the movement to the food source (under $I = 0$, the speed is assumed to be equal to the zero vector).
Step 5. Define:

$$D^j = D_{\max} \cdot \left(1 - \frac{I}{I_{\max}}\right) \cdot \delta, \tag{20.26}$$

where δ is n-dimensional vector with components $\delta_i \sim U[-1,\ 1]$.

Step 6. For any $j = 1, \ldots, NP$ define:

$$V^j = S^{j,New} + F^{j,New} + D^j. \tag{20.27}$$

Step 7. For any $j = 1, \ldots, NP$ define:

$$x^{j,New} = x^{j,old} + V^j \cdot \Delta t, \tag{20.28}$$

where $\Delta t = c_i \cdot \sum_{i=1}^{\infty} (b_i - a_i)$, c_i is the number on the interval $[0,\ 2]$. If $x_i^{j,New} \notin [a_i,\ b_i]$, then let $x_i^{j,New} = a_i + \chi \cdot [b_i - a_i]$, $\chi \sim U[0,\ 1]$.
Step 8. Crossbreeding. For any $j = 1, \ldots, NP$ perform the following steps:

Step 8.1. Find:

$$Cr = 0.2 \cdot \wedge\bar{f}^{j,best}. \quad (20.29)$$

Step 8.2. Define:

$$x_i^{j,Cr} = \begin{cases} x_r^{j,New} & \text{if } rand_i < Cr \\ x_i^{j,New} & \text{overwise} \end{cases} \quad r \in \{1, \ldots, j-1, j+1, \ldots, n\}, i = 1, \ldots, n, \quad (20.30)$$

where r is the random integer from set $\{1, \ldots, j-1, j+1, \ldots, n\}$.
As a result, solutions $x^{1,Cr}, \ldots, x^{NP,Cr}$ are found.

Step 9. Mutation. For any $j = 1, \ldots, NP$ perform.

Step 9.1. Find:

$$Mu = 0.05 \cdot \Delta\bar{f}^{j,best}. \quad (20.31)$$

Step 9.2. Define:

$$x_i^{j,Mu} = \begin{cases} x_i^{best} + \upsilon\left(x_i^p + x_i^q\right) & \text{if } rand_i < Mu \\ x_i^{j,Cr} & \text{else} \end{cases} \quad i = 1, \ldots, n, \quad (20.32)$$

Step 9.3. Let $x^j = x^{j,Mu}, j = 1, \ldots, NP$.
where $\upsilon \sim U[0,\ 1]$. If $x_i^{j,Mu} \notin [a_i,\ b_i]$, then let $x_i^{j,Mu} = a_i + \chi \cdot [b_i - a_i]$.

Step 10. Calculate the values of the objective function. Find the best and worst solutions x^{best}, x^{worst}. For each solution x^j, find the best position x^{jbest} for all past iterations.
Step 11. Check fulfillment of termination condition. If $I < I_{\max}$, then let $I = I + 1$ and go to Step 3. If $I = I_{\max}$, then process is complete. As an approximate solution of the problem, select $x^{best}, f(x^{best})$.

Recommendations on the selection of parameters. The population size NP determines the number of calculations of the objective function at each iteration. For a problem with a large range of admissible solutions, it is recommended to take a larger parameter value NP. Recommended parameter values $NP \in [40,\ 50]$.

The number of iterations $I_{\max}$ determines how long the search for new solutions will continue. As a rule, the more $I_{\max}$, the more accurate the answer. For a standard set of functions, the recommended values depend on the complexity of the function, i.e., $I_{\max} \in [1000,\ 10{,}000]$.

The maximum krill speed $S_{\max}$ is used to determine the speed of each member of the flock (see Step 3.5). The recommended value of this parameter $S_{\max} = 0.01$. Small positive number μ corrects the change in the position of the$j$th member of the population (see Steps 3.2, 3.3, 4.3). The recommended value $\mu = 0.3$.

The maximum speed of movement to the food source V_f is used to determine the speed of movement to the food of each member of the flock (see Step 4.5). The recommended value of this parameter $V_f = 0.02$. The maximum diffusion speed of krill is $D_{\max}$. The recommended value of this parameter $D_{\max} \in [0.002,\ 0.01]$.

20.2.4 Imperialist Competitive Algorithm

Solution search strategy. The strategy of imperialist competitive algorithm [3, 8, 9] uses observations of the behavior of empires in the fight for spheres of influence. Imperialism is a policy of expanding the government's administrative influence beyond the borders of the country, which is realized both through direct management and indirectly through influence on the markets of food, goods, materials, etc. Thus, all countries are divided into empire and colony. The empires seek to use the resources of other countries or simply to influence their policies by opposing other empires. Regardless of the motivating reasons, empires seek to increase the number of their colonies and extend their influence to the whole world.

The method uses ideas, both evolutionary algorithms, and "swarm intelligence" methods. It begins with the formation of the initial population—countries in the world (solutions on a set of admissible solutions). Some of the best countries (by the size of the objective function) are selected for the role of the imperialist countries, while the rest form colonies. All colonies are assigned to the imperialist states, and their number is determined by the strength of such a state, inversely proportional to the value of the objective function. This is how empires are formed: the imperialist state and its colonies. The largest number of colonies corresponds to the most powerful imperialist state. Then each colony begins to move toward its imperialist state. The strength of the empire is determined by the strength of the imperialist state and its colonies (the share of the average strength of the colonies is added to the strength of the state). The competition between empires leads either to an increase (at least, to non-decreasing) of the strength of the empire or to a decrease in it. Weak empires disappear with time. The described mechanisms should lead to a situation, where there is only one empire in the world, and all other countries are its colonies (this is the condition for the end of the process). The position of the imperialist state is taken as an approximate solution of the problem.

Solution search algorithm. Solution search algorithm involves the following steps.

Step 1. Set the method parameters: the size of population (the number of countries) N_{pop}, the number of imperialist countries N_{imp}, parameters of colony shift β, γ, and colony influence parameter ξ.
Step 2. Generate N_{pop} countries (solutions from set D) using uniform distribution low. Calculate the value of objective function and order solutions (countries) by increasing the objective function: $x^1, x^2, \ldots, x^{N_{pop}}$ (the solution x^1 corresponds to the smallest value of the objective function).
Step 3. Choose imperialist states. Select N_{imp} solutions (countries) from among the first in the list of solutions. N_{imp} corresponds to the best values of the objective

function. Further, they will be called imperialistic. Count the number of colonies $N_{col} = N_{pop} - N_{imp}$.
Step 4. Form empires.

Step 4.1. Calculate the normalized value of each imperialist state:

$$\tilde{f}(x^j_{imp}) = f(x^j_{imp}) - f(x^{N_{pop}}), j = 1, \ldots, N_{imp}. \tag{20.33}$$

Step 4.2. Find the normalized strength of each imperialist state:

$$P^j = \left| \frac{\tilde{f}(x^j_{imp})}{\sum\limits_{j=1}^{N_{imp}} \tilde{f}(x^j_{imp})} \right|, j = 1, \ldots, N_{imp}. \tag{20.34}$$

Step 4.3. Find the number of colonies of each imperialist state:

$$N^j_c \underset{round}{=} \left[P^j \cdot N_{col}\right],\ j = 1, \ldots, N_{imp} - 1, N^{N_{imp}}_c = N_{col} - \sum_{j=1}^{N_{imp}-1} N^j_c, \tag{20.35}$$

where $round[\cdot]$ is the round-off operation.
Step 4.4. For any imperialist state, select N^j_c countries randomly from among the colonies. The imperialist state and chosen colonies form an empire.

Step 5. The shift of the colonies of the empire to the imperialist state (assimilation procedure). For any empire, consistently fulfill, $j = 1, \ldots, N_{imp}$.

Step 5.1. Choose the first colony in the empire with number j randomly.
Step 5.2. Find new colony locations x^{new}:

$$x^{new} = x^{old} + U(0; \beta \cdot d) \cdot V_1 + d \cdot \mathrm{tg}\theta \cdot V_2, \tag{20.36}$$

where β, θ are the parameters, d is the distance from the colony to the imperialist state, $U(a, b)$ is the random variable uniformly distributed on $[a, b]$, V_1 is the identity vector directed from the colony to the imperialist state, V_2 is the random identity vector perpendicular V_1: $\{V_2\} = \{\{rand\} | V_1 \cdot V_2 = 0, \|V_2\| = 1, \|V_1\| = 1\}$, $\theta = U(-\gamma, \gamma), d = \sqrt{\sum\limits_{i=1}^{n} (x^j_{imp,i} - x_{c,i})^2}$, j is the imperialist state number, x^j_{imp} is the position of the imperialist state, x_c is the colony position. Thus, vectors V_1 and V_2 can be found as follows:

$$V_1 = \frac{1}{d}\left(x^j_{imp,1} - x_{c,1}; \ldots; x^j_{imp,n} - x_{c,n}\right), \tag{20.37}$$

$$V_1 \cdot \tilde{V}_2 = V_{1,1} \cdot V_{2,1} + \ldots + V_{1,n} \cdot \tilde{V}_{2,n} = 0 \quad \text{(orthogonality condition)}, \tag{20.38}$$

where components $\tilde{V}_{2,1}, \ldots, \tilde{V}_{2,n-1}$ can be generated randomly on a segment $[-1,\ 1]$, and $\tilde{V}_{2,n}$ find from the orthogonality condition. Next, normalize the vector $\tilde{V}_2 : V_2 = \frac{\tilde{V}_2}{\left\|\tilde{V}_2\right\|}$.

Step 5.3. Calculate the value of the objective function $f(x^{new})$.
If $f(x^{new}) \geq f(x^j_{imp})$, then go to Step 5.4.
If $f(x^{new}) < f(x^j_{imp})$, then let $x_c = x^j_{imp}$, $x^j_{imp} = x^{new}$ (the colony becomes an imperialist state, and the former imperialist state becomes a colony).
Go to Step 5.1.
Step 5.4. If the number of colonies that have changed position has reached N^j_c, then complete the procedure. Otherwise, randomly select the next colony, which has not yet changed position, and go to Step 5.2.

Step 6. Competition between empires.

Step 6.1. Find the total cost of the empire:

$$TC^j = f(x^j_{imp}) + \xi \cdot \frac{\sum_{i=1}^{N^j_c} f(x^i_c)}{N^j_c}, j = 1, \ldots, N, \tag{20.39}$$

where ξ is the positive number less than 1, x^i_c is the position of ith colony of jth empire determined by the position x^j_{imp} of the imperialist state.
Step 6.2. Find normalized total cost of empire:

$$NTC^j = TC^j - \max_{i \in \{1,\ldots,N_{imp}\}}\{TC^i\}. \tag{20.40}$$

Step 6.3. Find the level of influence of each empire:

$$P^j = \left| \frac{NTC^j}{\sum_{i=1}^{N_{imp}} NTC^i} \right|, j = 1, \ldots, N_{imp}. \tag{20.41}$$

Step 6.4. Find the weakest empire with the lowest value $P^j, j \in \{1, \ldots, N_{imp}\}$.
Step 6.5. In the found empire, find the weakest colony with the highest value of the objective function.
Step 6.6. Generate vectors:

$$P = (p^1, p^2, \ldots, p^{N_{imp}}); R = (r^1, r^2, \ldots, r^{N_{imp}}), \tag{20.42}$$

where $r^j = U(0,\ 1)$,

$$D = P - R = (d^1, d^2, \ldots, d^{N_{imp}}) = (p^1 - r^1, \ldots, p^{N_{imp}} - r^{N_{imp}}). \quad (20.43)$$

Step 6.7. The colony defined in Step 6.5 should be included in the empire corresponding to the largest value among the components of the row vector D.

Step 7. Disintegration of the weakest empires.
If there are no colonies in the empire, it will cease to exist (the country is included in the empire, as defined in Step 6.7). There is a new number N_{imp}.
Step 8. If $N_{imp} = 1$, then process is complete. The solution of the problem is to consider the position of the imperialist state. Else, go to Step 5.

Recommendation on the parameters selection. The number of countries N_{pop} determines the number of calculations of the objective function at each iteration. For a problem with a large range of admissible solutions, it is recommended to take a larger parameter value N_{pop}. Recommended parameter values $N_{pop} \in [50,\ 300]$.

The number of imperialist countries N_{imp} determines how long the search for new solutions will continue. At the end of the search, only one empire remains. Recommended values for a standard set of functions $N_{imp} \in [5,\ 30]$.

The colony shift parameters β, γ determine the movement of the colonies to its empire (see Step 5.2). Recommended parameter values $\beta = 2,\ \gamma = \frac{\pi}{4}$.

Colony influence parameter ξ. Recommended parameter value $\xi = 0.1$.

20.2.5 *Hybrid Multi-agent Algorithm*

After analyzing the test function optimization results of the methods described, it can be seen that the result obtained by the imperialist competition algorithm differs from the other two. The members of the population at the last iteration are scattered around the maximum point, while in the other two methods, all the individuals merged into one point, and the result obtained by this method gives a large deviation from the exact solution.

Imperialist competition algorithm does not stop with the achievement of a given number of iterations (like the other two methods), but stops when all empires are dissolved except for one. The position of this empire is the solution of the problem.

As a result, the idea arose to create a hybrid algorithm [3], which includes the imperialist competition algorithm and a method of a krill herd. By combining these two methods, one can achieve a shorter algorithm time and obtain a more accurate result, since at the beginning, the method that imitates imperialist competition will help to reduce the scope of the solution search by obtaining the final population in some neighborhood of the extreme point, and the following method will help clarify the solution in the found area.

20.3 Application of Multi-agent Methods for Optimal Open-Loop Control Problems

In this section, the statement of the optimal open-loop control problem is given in Sect. 20.3.1. In Sect. 20.3.2, optimal control problem solutions are proposed. Section 20.3.3 provides the designed software, while Sect. 20.3.4 describes the statement of optimal open-loop control test problems and their solutions.

20.3.1 Statement of the Problem

Let the behavior of the control object model be described by an ordinary differential equation:

$$\dot{x}(t) = f(t, x(t), u(t)), \tag{20.44}$$

where x is the system state vector, $x = (x_1, \ldots, x_n)^T \in R^n$, u is the control vector, $u = (u_1, \ldots, u_q)^T \in U \subseteq R^q$, U is some given set of admissible control values determined by the direct product of segments $[a_1, b_1] \times \cdots \times [a_q, b_q]$, $t \in T = [t_0, \ t_1]$ is the system time interval, start time t_0 and terminal time t_1 are set, $f(t, x, u)$ is the continuous vector function, R^n is n-dimensional Euclidean space.

The initial condition $x(t_0) = x_0$ sets the initial state of the system.

We define the set of admissible processes $\boldsymbol{D}(t_0, x_0)$ as a set of pairs $d = (x(\cdot), u(\cdot))$ that include trajectory $x(\cdot)$ and control $u(\cdot)$ (where $\forall t \in T : \ x(t) \in R^n, \ \ u(t) \in U$, functions $x(\cdot)$ are continuous and piecewise-differentiable, and $u(\cdot)$ piecewise-continuous), satisfying Eq. 20.44 with the initial condition.

On the set $\boldsymbol{D}(t_0, x_0)$, we define the cost functional

$$I(d) = F(x(t_1)). \tag{20.45}$$

Need to find such a pair $d^* = (x^*(\cdot), u^*(\cdot)) \in \boldsymbol{D}(t_0, x_0)$ that $I(d^*) = \min_{d \in \boldsymbol{D}(t_0, x_0)} I(d)$.

We consider Eq. 20.44 linear in control, which has the form:

$$\dot{x}(t) = A(x(t)) + B(t)u(t), \tag{20.46}$$

where $A(x)$ is the nonlinear function and $B(t)$ is the matrix $n \times q$ depending on time.

In Eq. 20.46, the structure of optimal open-loop control is relay according to the maximum principle. Therefore, it is proposed to look for an approximate solution in a parametric form determined by the number of control switching moments and their values.

20.3.2 Search Algorithm of Optimal Open-Loop Control

Search algorithm includes the following steps.

Step 1. Initialization. Select a method from the group of multi-agent algorithms and set its parameters. Set a switching number $p = 0$ in the control $u(t)$, wherein $t_{\Pi_0} \in \{t_0, t_1\}$.
Step 2. Generate the initial population (controls) of *NP* individuals on the time interval $t \in [t_0,\ t_1]$. The resulting $1, \ldots, NP$ sequences of values are switching points $t_R \in [t_0,\ t_1]$ in the control $u(t)$.
Step 3. Generate control by generated switching point values

$$u_p^j(t) = a_p\chi(t_0) + (a_p - b_p)\sum_{k=0}^{p}(-1)^k\chi(t - t_{R_k}), \qquad (20.47)$$

where $\chi(t) = \begin{cases} 0 & \text{if } t \le 0 \\ 1 & \text{if } t > 0 \end{cases}, j \in \overline{1, NP}, p \in \overline{1, q}, a_p \le u \le b_p.$

Step 4. Integrate *NP* systems of differential equations (Eq. 20.46) with controls $u^1(t), \ldots, u^{NP}(t)$ using the fourth order Runge–Kutta method. For any individual, obtain the corresponding trajectories $x_1^1, \ldots, x_1^{NP}, \ldots, x_n^1, \ldots, x_n^{NP}$ and calculate the values of the cost functional $I^1, \ldots, I^{NP}$.
Step 5. Fulfill the next iteration of the selected method of minimizing Eq. 20.45. Obtain new positions of individuals $1', \ldots, NP'$ (switching point values). Go to Step 3.
Step 6. The loop (Step 3–Step 5) ends, when a certain number of iterations are reached. The best individual is selected (set of control switching points). The corresponding control and trajectory, as well as, the value of the cost functional I_p^*, are taken as an approximate solution of the problem with the switching number equaled to p.
Step 7. If $I_p^* < I_{p-1}^*$ (condition is checked under $p \ge 1$), then let $p = p + 1$ and go to Step 2. If $I_p^* \ge I_{p-1}^*$, then the search procedure for optimal open-loop control is completed, and control with p switching is selected.

20.3.3 Software

Software [3] was developed based on all methods described above. To create it, we used Microsoft Visual Studio development environment, the programming language is C #.

With the help of the developed software, the effectiveness of the described algorithms on a standard set of test examples (for example: Shaffer function, root function, trapfall, and Rosenbrock function.) was explored. Also, the set of problems on finding the optimal open-loop control was solved (Tables 20.1, 20.2, 20.3, 20.4, 20.5 and

Table 20.1 Formulation of the task 1

Parameters	Values
The dimension of the state vector	$n = 2$
Time interval	$t \in [0, 1]$
Control constraint	$-1 \leq u \leq 1$
Initial value	$x(0) = (0, 0)$
System of differential equations	$\begin{cases} \dot{x}_1 = x_2 + \sin x_1 + u \\ \dot{x}_2 = x_1 \cos x_2 u \end{cases}$
Cost functional	$I(u) = x_2(1)$

Table 20.2 The results of solving the task 1

Optimization method	Coordinates of points $(x_1(1),\ x_2(1))$	Switching point coordinates	The value of the functional I
Fish school search	(0.43494, –0.13587)	0.49	–0.13587
Krill herd	(0.42459, –0.13478)	0.48	–0.13571
Imperialist competitive algorithm	(0.44665, –0.13598)	0.5	–0.13598
Known solution [10]	(0.440804, –0.13593)	0.5	–0.13599

Table 20.3 Formulation of the task 2

Parameters	Values
The dimension of the state vector	$n = 2$
Time interval	$t \in [0, 2]$
Control constraint	$-1 \leq u \leq 2$
Initial value	$x(0) = (-1, 0)^T$
System of differential equations	$\begin{cases} \dot{x}_1 = x_2^2 + u \\ \dot{x}_2 = 8 \sin x_1 + x_1 - x_2 - u \end{cases}$
Cost functional	$I(u) = -x_2(2)$

Table 20.4 The results of solving the task 2

Optimization method	Coordinates of points $(x_1(2),\ x_2(2))$	Switching point coordinates	The value of the functional I
Fish school search	(16.50987, 6.37294)	(0.62, 1.52, 1.74, 1.93)	−16.50987
Krill herd	(15.93114, 6.18436)	(0.57, 0.58, 0.58, 1.85)	−15.93114
Imperialist competitive algorithm	(16.67731, 6.55339)	(0.57, 1.35, 1.52, 1.86)	−16.67731
Known solution [10]	(16.76268, 6.35095)	(0.5, 1.25, 1.5, 1.8)	−16.76268

Table 20.5 Formulation of the task 3

Parameters	Values
The dimension of the state vector	$n = 2$
Time interval	$t \in [0; 1, 6]$
Control constraint	$-2 \le u \le 1$
Initial value	$x(0) = (1, 0)^T$
System of differential equations	$\begin{cases} \dot{x}_1 = \dfrac{1}{\cos x_1 + 2} + 3\sin x_2 + u \\ \dot{x}_2 = x_1 + x_2 + u \end{cases}$
Cost functional	$I(u) = -x_1(1, 6) + \frac{1}{2}x_2(1, 6)$

Table 20.6 The results of solving the task 3

Optimization method	Coordinates of points $(x_1(1.6), x_2(1.6))$	Switching point coordinates	The value of the functional I
Fish school search	(3.43034, 12.81994)	1.24	−2.97963
Krill herd	(3.52562, 13.00372)	1.27	−2.97624
Imperialist competitive algorithm	(3.93412, 13.53938)	1.39	−2.83557
Known solution [10]	(3.46114, 12.884)	1.26	−2.98086

20.6).

On the initial form of the software (Fig. 20.1), the user can select a task to find the optimal open-loop control, set the switching number in the control, select the optimization method, and specify its parameters.

The result of the program are the coordinates of the points $x_1(t_1)$, $x_2(t_2)$, the optimal value of the cost functional I, and the coordinates of the switching points. After finding the optimal control, the program displays the graphs of the control function and trajectories.

20.3.4 Solving the Problem of Finding Optimal Open-Loop Control

Task 1. Formulation of the task (Table 20.1) [3].

The best switching number: $p = 1$.

Optimization method and its parameters: fish school search ($NP = 5$, $ITER = 100$, $W_{scale} = 150$, $W = 100$, $step_{vol} = 0.1$, $step_{ind} = 0.01$), krill herd ($NP = 10$,

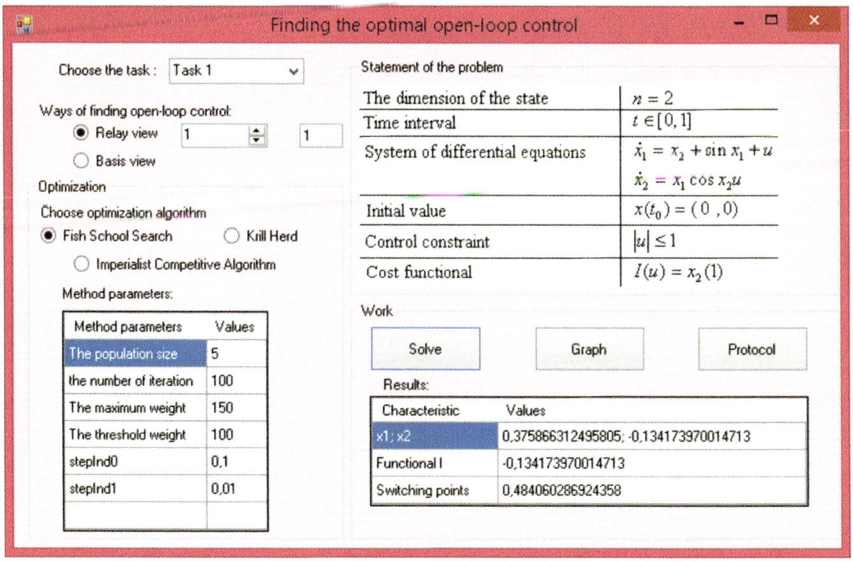

Fig. 20.1 Software interface

$ITER = 100$, $S_{\max} = 0.01$, $mu = 0.5$, $V_f = 0.02$, $D_{\max} = 0.005$, $c_i = 0.2$), and imperialist competitive algorithm ($N_{pop} = 150$, $N_{imp} = 15$, $ITER = 500$, $\beta = 0.2$, $\gamma = 0.02$, $\xi = 0.01$).

The results of solving the task are presented in Table 20.2.

Graphs of optimal trajectories and controls are shown in Fig. 20.2.

Task 2. Formulation of the task (Table 20.3) [3].

The best switching number: $p = 4$.

Optimization method and its parameters: fish school search ($NP = 30$, $ITER = 500$, $W_{scale} = 5000$, $W = 4500$, $step_{vol} = 0.1$, $step_{ind} = 0.01$), krill herd ($NP = 40$, $ITER = 1000$, $S_{\max} = 0.01$, $mu = 0.5$, $V_f = 0.02$, $D_{\max} = 0.005$, $c_i = 0.2$), and imperialist competitive algorithm ($N_{pop} = 150$, $N_{imp} = 15$, $ITER = 500$, $\beta = 0.2$, $\gamma = 0.02$, $\xi = 0.01$).

The results of solving the task are presented in Table 20.4.

Graphs of optimal trajectories and controls are shown in Fig. 20.3.

Task 3. Formulation of the task (Table 20.5) [3].

The best switching number: $p = 1$.

Optimization method and its parameters: fish school search ($NP = 5$, $ITER = 100$, $W_{scale} = 300$, $W = 250$, $step_{vol} = 0.1$, $step_{ind} = 0.01$), krill herd ($NP = 10$, $ITER = 100$, $S_{\max} = 0.01$, $mu = 0.5$, $V_f = 0.02$, $D_{\max} = 0.005$, $c_i = 0.2$), and imperialist competitive algorithm ($N_{pop} = 40$, $N_{imp} = 4$, $ITER = 500$, $\beta = 0.2$, $\gamma = 0.02$, $\xi = 0.01$).

The results of solving the task are presented in Table 20.6.

Graphs of optimal trajectories and controls are shown in Fig. 20.4.

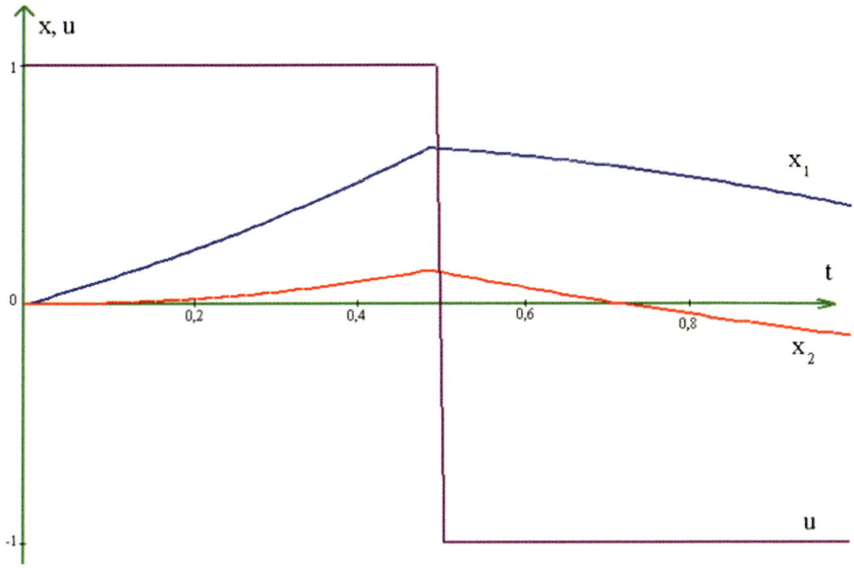

Fig. 20.2 Trajectories and control

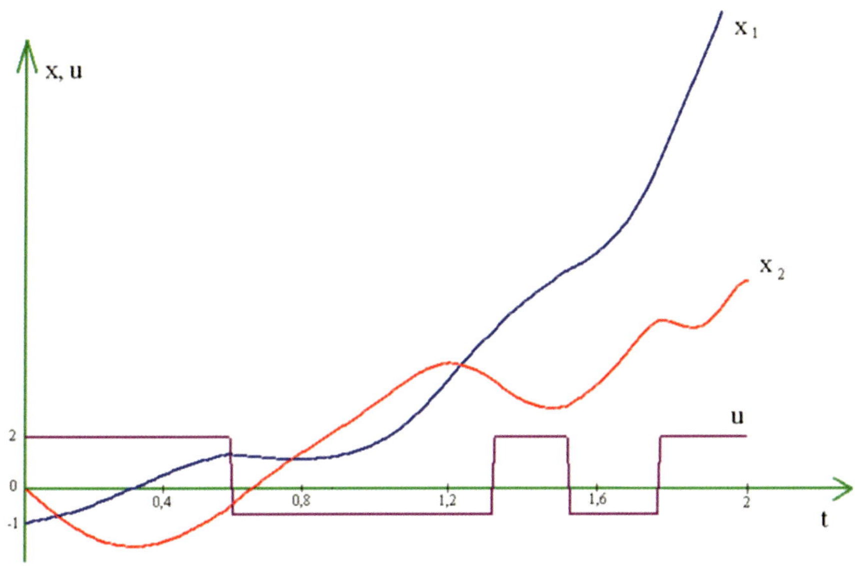

Fig. 20.3 Trajectories and control

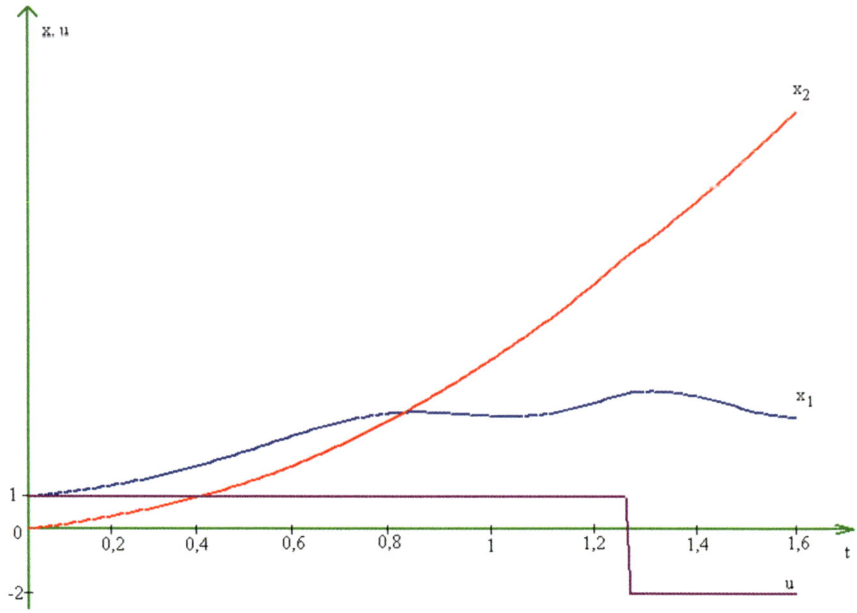

Fig. 20.4 Trajectories and control

20.4 Conclusions

The chapter outlined multi-agent methods for finding a constrained global extremum: fish school search, krill herd, and imperialist competitive algorithm. Based on these methods, an algorithm has been developed for finding the optimal open-loop control of deterministic systems, which is looking for control in a relay mode with a certain switching number. Using the described methods, the software was created that allows to find the solution to the optimal open-loop deterministic control problem. The software draws the graphics of the trajectories and controls and also calculates the optimal value of the criterion and coordinates of the switching points in the control.

References

1. Panteleev, A.V., Metlitskaya, D.V.: An application of genetic algorithms with binary and real coding for approximate synthesis of suboptimal control in deterministic systems. Autom. Remote Control **72**(11), 2328–2338 (2011)
2. Panteleev, A.V., Metlitskaya, D.V.: Using the method of artificial immune systems to seek the suboptimal program control of deterministic systems. Autom. Remote Control **75**(11), 1922–1935 (2014)
3. Karane, M.M.C.: Comparative analysis of multi-agent methods for constrained global optimization. In: IV International Conference on Information Technologies in Engineering Education, pp. 128–133 (2018)

4. Bastos Filho, C.J.A., de Lima Neto, F.B., Lins, A.J.C.C., Nascimento, A.I.S., Lima, M.P.: A novel search algorithm based on fish school behavior. In: 2008 IEEE International Conference on Systems, Man and Cybernetics, pp. 2646–2651 (2008)
5. Bastos Filho, C.J.A., de Lima Neto, F.B., Lins, A.J.C.C., Nascimento, A.I.S., Lima, M.P.: Fish school search: overview. In: Chiong, R. (Ed.) Nature-Inspired Algorithms for Optimisation. SCI, vol. 193, pp. 261–277. Springer, Heidelberg (2009)
6. Bacanin, N., Pelevic, B., Tuba, M.: Krill herd (KH) algorithm for portfolio optimization. In: Mathematics and Computers in Business, Manufacturing and Tourism, pp. 39–44 (2013)
7. Gandomi, A.H., Alavi, A.H.: Krill herd: a new bio-inspired optimization algorithm. Commun. Nonlinear Sci. Numer. Simul. **17**(12), 4831–4845 (2012)
8. Atashpaz-Gargari, E., Lucas, C.: Imperialist competitive algorithm: an algorithm for optimization inspired by imperialist competition. In: IEEE Congress on Evolutionary Computation, pp. 4661–4667 (2007)
9. Kaveh, A., Talatahari, S.: Imperialist competitive algorithm for engineering design problems. Asian J. Civ. Eng. (Build. Hous.) **11**(6), 675–697 (2010)
10. Finkelstein, E.A.: Computational technologies of approximation of the reachable set of a controlled system. Dissertation for the degree of Canadian Technological Sciences. Institute of System Dynamics and Control Theory, Irkutsk (in Russian) (2018)

Chapter 21
Spectral Method for Analysis of Diffusions and Jump Diffusions

Gevorg Y. Baghdasaryan, Marine A. Mikilyan, Andrei V. Panteleev and Konstantin A. Rybakov

Abstract The chapter discusses the use of the spectral form of mathematical description, or the spectral method, for the statistical analysis of stochastic dynamical systems: diffusions and jump diffusions, i.e., for solving Fokker–Planck–Kolmogorov equation and Kolmogorov–Feller equation for the probability density of the state vector for these dynamical systems. The spectral form of mathematical description allows to transform linear partial differential equations or partial integro-differential equations into a system of linear algebraic equations, which determines coefficients according to orthogonal series expansions for the probability density with respect to an arbitrary orthonormal system of functions. As an example for testing, the Dryden wind turbulence model and its modification, allowing to take into account not only continuous random effects but also impulse ones, are considered.

21.1 Introduction

The spectral form of mathematical description, or the spectral method, is used for solving various problems of the control theory [1]. At the beginning, it was applied to solve the output processes analysis problem of linear nonstationary control systems.

G. Y. Baghdasaryan · M. A. Mikilyan
Institute of Mechanics, National Academy of Sciences of Armenia, 24, Marshal Baghramyan Ave., 0019 Yerevan, Armenia
e-mail: gevorg.baghdasaryan@rau.am

M. A. Mikilyan
e-mail: marine.mikilyan@rau.am

Russian-Armenian University, 123, Hovsep Emin St, 0051 Yerevan, Armenia

A. V. Panteleev · K. A. Rybakov (✉)
Moscow Aviation Institute (National Research University), 4, Volokolamskoe Shosse, Moscow 125993, Russian Federation
e-mail: rkoffice@mail.ru

A. V. Panteleev
e-mail: avpanteleev@inbox.ru

L. C. Jain et al. (eds.), *Advances in Theory and Practice of Computational Mechanics*, Smart Innovation, Systems and Technologies 173,
https://doi.org/10.1007/978-981-15-2600-8_21

Further, its scope was significantly expanded [2–6]. Now it is used for the analysis, synthesis, and identification of control systems, which are described by ordinary differential equations and partial differential equations, integro-differential equations, equations with fractional derivatives, etc.

The use of the spectral form of mathematical description implies the representation of input and output signals as sequences of coefficients according to orthogonal series expansions. Such signal characteristics are called the spectral characteristics. The characteristics of linear dynamical systems in the spectral form of mathematical description are the spectral characteristics of linear operators [7]. Elementary and specialized algorithms to use the spectral form of mathematical description are developed for computer algebra systems and applications [8]. These algorithms include subroutines for calculating the spectral characteristics of typical input signals, spectral characteristics of multiplication operators with typical multipliers, differentiation and integration operators, shift operators, and operators of fractional integration and differentiation with respect to various orthonormal systems of functions such as Legendre, Chebyshev, Laguerre, and Hermite polynomials, trigonometric and complex exponential functions, Walsh and Haar functions, functions defined by wavelets or splines, etc.

In this chapter, the output processes analysis problem for nonlinear stochastic dynamical control systems is concerned. The results obtained in [4] for diffusions are expanded to jump diffusions. As an example for testing the spectral method, the Dryden wind turbulence model and its jump diffusion modification are considered.

The remainder of this chapter is organized as follows. In Sect. 21.2, the spectral method formalism is proposed. The spectral method for analysis problem of diffusions is described in Sect. 21.3. Section 21.4 provides the spectral method for analysis problem of jump diffusions. The chapter is summarized in Sect. 21.5.

21.2 Spectral Method Formalism

Hereinafter, the necessary definitions for multidimensional matrices as well spectral characteristics of functions and linear operators are given in Sects. 21.2.1–21.2.2, respectively.

21.2.1 Multidimensional Matrices

Multidimensional matrices are needed for further presentation of main results. Therefore, the necessary definitions for multidimensional matrices are given below. We will denote $(p+q)$-dimensional matrix with entries $a_{i_1\dots i_p j_1\dots j_q}$ by $A(p,q)$, $i_1,\dots,i_p,j_1,\dots,j_q = 0,1,2,\dots$ The separation of indices into two groups $i_1,\dots,i_p$ and $j_1,\dots,j_q$ allows to attribute the structure of a multidimensional matrix [7], it is important to define the product of multidimensional matrices [9].

1. Let $\alpha, \beta \in \mathbb{R}$ and let $A(p, q) = [a_{i_1 \dots i_p j_1 \dots j_q}]$ and $B(p, q) = [b_{i_1 \dots i_p j_1 \dots j_q}]$ be the infinite $(p+q)$-dimensional matrices. The expression $\alpha A(p, q) + \beta B(p, q)$ is the infinite $(p+q)$-dimensional matrix $C(p, q) = [c_{i_1 \dots i_p j_1 \dots j_q}]$ if

$$c_{i_1 \dots i_p j_1 \dots j_q} = \alpha a_{i_1 \dots i_p j_1 \dots j_q} + \beta b_{i_1 \dots i_p j_1 \dots j_q}, \quad i_1, \dots, i_p, j_1, \dots, j_q = 0, 1, 2, \dots$$

2. Let $A(p, r) = [a_{i_1 \dots i_p k_1 \dots k_r}]$ and $B(r, q) = [b_{k_1 \dots k_r j_1 \dots j_q}]$ be the infinite $(p+r)$-dimensional and $(r+q)$-dimensional matrices, respectively. The product $A(p, r) \cdot B(r, q)$ is the infinite $(p+q)$-dimensional matrix $C(p, q) = [c_{i_1 \dots i_p j_1 \dots j_q}]$ if

$$c_{i_1 \dots i_p j_1 \dots j_q} = \sum_{k_1, \dots, k_r = 0}^{\infty} a_{i_1 \dots i_p k_1 \dots k_r} b_{k_1 \dots k_r j_1 \dots j_q} < \infty,$$
$$i_1, \dots, i_p, j_1, \dots, j_q = 0, 1, 2, \dots$$

An infinite $2p$-dimensional matrix $E(p, p)$ is said to be the identity matrix if $A(p, p) \cdot E(p, p) = E(p, p) \cdot A(p, p) = A(p, p)$ for each $2p$-dimensional matrix $A(p, p)$.

3. Let $A(p, p)$ be an infinite $2p$-dimensional matrix. An infinite $2p$-dimensional matrix $B(p, p)$ is said to be the two-sided inverse of $A(p, p)$ if $A(p, p) \cdot B(p, p) = B(p, p) \cdot A(p, p) = E(p, p)$. We will use the notation $A^{-1}(p, p)$ to denote the two-sided inverse of $A(p, p)$.
4. Let $A(p, q) = [a_{i_1 \dots i_p j_1 \dots j_q}]$ and $B(r, s) = [b_{k_1 \dots k_r l_1 \dots l_s}]$ be the infinite $(p+q)$-dimensional and $(r+s)$-dimensional matrices, respectively. The tensor product $A(p, q) \otimes B(r, s)$ is the infinite $(p+r+q+s)$-dimensional matrix $C(p+r, q+s) = [c_{i_1 \dots i_p k_1 \dots k_r j_1 \dots j_q l_1 \dots l_s}]$ if

$$c_{i_1 \dots i_p k_1 \dots k_r j_1 \dots j_q l_1 \dots l_s} = a_{i_1 \dots i_p j_1 \dots j_q} b_{k_1 \dots k_r l_1 \dots l_s},$$
$$i_1, \dots, i_p, k_1, \dots, k_r, j_1, \dots, j_q, l_1, \dots, l_s = 0, 1, 2, \dots$$

5. Let $A(p, q) = [a_{i_1 \dots i_p j_1 \dots j_q}]$ be an infinite $(p+q)$-dimensional matrix. An infinite $(q+p)$-dimensional matrix $B(q, p) = [b_{j_1 \dots j_q i_1 \dots i_p}]$ is said to be the transpose of $A(p, q)$ if

$$b_{j_1 \dots j_q i_1 \dots i_p} = a_{i_1 \dots i_p j_1 \dots j_q}, \quad i_1, \dots, i_p, j_1, \dots, j_q = 0, 1, 2, \dots$$

We will use the notation $A^{\mathrm{T}}(p, q)$ to denote the transpose of $A(p, q)$.

21.2.2 Spectral Characteristics of Functions and Linear Operators

In this section, we introduce necessary definitions and propositions for the spectral characteristics and the spectral transform.

Let $\{q_{i_0}(t)\}_{i_0=0}^{\infty}$ be an orthonormal basis of $L_2(\mathbb{T})$ and let $\{p_{i_1\dots i_n}(x)\}_{i_1,\dots,i_n=0}^{\infty}$ be an orthonormal basis of $L_2(\mathbb{R}^n)$ [10], where $t \in \mathbb{T} = [t_0, T]$ and $x \in \mathbb{R}^n$. Then

$$\{e_{i_0 i_1 \dots i_n}(t, x) = q_{i_0}(t) p_{i_1\dots i_n}(x)\}_{i_0,i_1,\dots,i_n=0}^{\infty} \tag{21.1}$$

is the orthonormal basis of $L_2(\mathbb{T} \times \mathbb{R}^n)$.

Definition 1 An infinite $(n + 1)$-dimensional matrix $H(n + 1, 0) = [h_{i_0 i_1 \dots i_n}]$ is called the spectral characteristic of a square-integrable function $h(t, x)$, i.e., $h(t, x) \in L_2(\mathbb{T} \times \mathbb{R}^n)$, if

$$h_{i_0 i_1 \dots i_n} = (e_{i_0 i_1 \dots i_n}(t, x), h(t, x))_{L_2(\mathbb{T}\times\mathbb{R}^n)} = \int_{\mathbb{T}} \int_{\mathbb{R}^n} e_{i_0 i_1 \dots i_n}(t, x) h(t, x) \mathrm{d}x \mathrm{d}t,$$
$$i_0, i_1, \dots, i_n = 0, 1, 2, \dots,$$

and

$$h(t, x) = \sum_{i_0,i_1,\dots,i_n=0}^{\infty} h_{i_0 i_1 \dots i_n} e_{i_0 i_1 \dots i_n}(t, x), \quad (t, x) \in \mathbb{T} \times \mathbb{R}^n. \tag{21.2}$$

Thus, $H(n + 1, 0) = \mathbb{S}[h(t, x)]$ and $h(t, x) = \mathbb{S}^{-1}[H(n + 1, 0)]$, where $\mathbb{S}$ and $\mathbb{S}^{-1}$ denote the spectral transform and the spectral inversion, respectively.

Similarly, the spectral characteristic of a square-integrable function $h(t) \in L_2(\mathbb{T})$ with respect to the orthonormal basis $\{q_{i_0}(t)\}_{i_0=0}^{\infty}$ and the spectral characteristic of a square-integrable function $h(x) \in L_2(\mathbb{R}^n)$ with respect to the orthonormal basis $\{p_{i_1\dots i_n}(x)\}_{i_1,\dots,i_n=0}^{\infty}$ can be defined. For example, an infinite n-dimensional matrix $H(n, 0) = [h_{i_1\dots i_n}]$ is called the spectral characteristic of a function $h(x) \in L_2(\mathbb{R}^n)$ if

$$h_{i_1\dots i_n} = (p_{i_1\dots i_n}(x), h(x))_{L_2(\mathbb{R}^n)} = \int_{\mathbb{R}^n} p_{i_1\dots i_n}(x) h(x) \mathrm{d}x, \quad i_1, \dots, i_n = 0, 1, 2, \dots$$

and

$$h(x) = \sum_{i_1,\dots,i_n=0}^{\infty} h_{i_1\dots i_n} p_{i_1\dots i_n}(x), \quad x \in \mathbb{R}^n. \tag{21.3}$$

Definition 2 An infinite $2(n + 1)$-dimensional matrix $A(n + 1, n + 1) = [a_{i_0 i_1 \dots i_n j_0 j_1 \dots j_n}]$ is said to be the spectral characteristic of a linear operator $\mathcal{A} : D_{\mathcal{A}} \subseteq$

$L_2(\mathbb{T} \times \mathbb{R}^n) \to L_2(\mathbb{T} \times \mathbb{R}^n)$ if

$$a_{i_0 i_1 \dots i_n j_0 j_1 \dots j_n} = (e_{i_0 i_1 \dots i_n}(t, x), \mathcal{A} e_{j_0 j_1 \dots j_n}(t, x))_{L_2(\mathbb{T} \times \mathbb{R}^n)}$$
$$= \int_{\mathbb{T}} \int_{\mathbb{R}^n} e_{i_0 i_1 \dots i_n}(t, x) \mathcal{A} e_{j_0 j_1 \dots j_n}(t, x) \mathrm{d}x \mathrm{d}t,$$
$$i_0, i_1, \dots, i_n, j_0, j_1, \dots, j_n = 0, 1, 2, \dots$$

The spectral transform of linear operators is also denoted by $\mathbb{S}$, therefore, $\mathbb{S}[\mathcal{A}] = A(n+1, n+1)$.

Basic properties of the spectral transform for functions and linear operators are as follows:

1. For any $h_l(t, x) \in L_2(\mathbb{T} \times \mathbb{R}^n)$ and $\alpha_l \in \mathbb{R}$, $l = 1, 2, \dots, L$, the following relation is satisfied (the linearity property of the spectral transform):

$$\mathbb{S}\left[\sum_{l=1}^{L} \alpha_l h_l(t, x)\right] = \sum_{l=1}^{L} \alpha_l \mathbb{S}[h_l(t, x)].$$

2. If the function $h(t, x) \in L_2(\mathbb{T} \times \mathbb{R}^n)$ such that $h(t^*, x) = h^*(x) \in L_2(\mathbb{R}^n)$, $t^* \in \mathbb{T}$, $q(1, 0; t^*)$ is the infinite column matrix with entries $q_{i_0}(t^*)$, i.e.,

$$q(1, 0; t^*) = [\, q_0(t^*) \;\; q_1(t^*) \;\; q_2(t^*) \;\; \dots \,]^{\mathrm{T}},$$

$H(n+1, 0)$ and $H^*(n, 0)$ are spectral characteristics of $h(t, x)$ and $h^*(x)$, respectively, then

$$(q^{\mathrm{T}}(1, 0; t^*) \otimes E(n, n)) \cdot H(n+1, 0) = H^*(n, 0),$$

where $E(n, n)$ is the $2n$-dimensional identity matrix.

3. If $\mathcal{A} : D_{\mathcal{A}} \subseteq L_2(\mathbb{T} \times \mathbb{R}^n) \to L_2(\mathbb{T} \times \mathbb{R}^n)$ is a linear operator, $h(t, x) \in D_{\mathcal{A}}$, and $A(n+1, n+1)$ is the spectral characteristic of $\mathcal{A}$, then

$$\mathbb{S}[\mathcal{A} h(t, x)] = A(n+1, n+1) \cdot \mathbb{S}[h(t, x)].$$

4. If $\mathcal{A} : D_{\mathcal{A}} \subseteq L_2(\mathbb{T} \times \mathbb{R}^n) \to L_2(\mathbb{T} \times \mathbb{R}^n)$ and $\mathcal{B} : D_{\mathcal{B}} \subseteq L_2(\mathbb{T} \times \mathbb{R}^n) \to R_{\mathcal{B}} \subseteq D_{\mathcal{A}}$ are linear operators, $\mathcal{C} = \mathcal{A} \circ \mathcal{B}$ is a composition of $\mathcal{A}$ and $\mathcal{B}$, $A(n+1, n+1)$, $B(n+1, n+1)$, and $C(n+1, n+1)$ are spectral characteristics of $\mathcal{A}$, $\mathcal{B}$, and $\mathcal{C}$, respectively, then

$$C(n+1, n+1) = A(n+1, n+1) \cdot B(n+1, n+1).$$

Spectral characteristics of functions with a similar properties can also be defined for elements, which do not belong to $L_2(\mathbb{T} \times \mathbb{R}^n)$ (e.g., for elements of $L_p(\mathbb{T} \times \mathbb{R}^n)$, where $p < 2$, or for distributions [1, 11]).

21.3 Spectral Method for Analysis of Diffusions

In this section, the statement of the problem is given in Sect. 21.3.1. Section 21.3.2 provides the detailed description of the spectral method for solving Fokker–Planck–Kolmogorov equation, and Sect. 21.3.3 contains numerical results for the analysis of Dryden wind turbulence model.

21.3.1 Problem Statement

Let $X(t)$ be an $\mathbb{R}^n$-valued random process that satisfies Itô Stochastic Differential Equation (SDE):

$$dX(t) = f(t, X(t))\mathrm{d}t + \sigma(t, X(t))dW(t), \quad X(t_0) = X_0, \tag{21.4}$$

where $t \in \mathbb{T} = [t_0, T]$, $f(t, x){:}\mathbb{T} \times \mathbb{R}^n \to \mathbb{R}^n$ is the n-dimensional function, $\sigma(t, x)$: $\mathbb{T} \times \mathbb{R}^n \to \mathbb{R}^{n\times s}$ is the $(n \times s)$-dimensional matrix function, $W(t)$ is the standard s-dimensional Wiener process, X_0 is the initial state with a given probability density $\varphi_0(x)$ (X_0 and $W(t)$ are independent).

Functions $f(t, x)$ and $\sigma(t, x)$ satisfy the conditions for the existence and uniqueness of the strong or weak solution of SDEs [12], and $\mathrm{E}|X_0|^2 < +\infty$, where E is the expectation or mean.

For any $t \in \mathbb{T}$, the most comprehensive statistical characteristic of $X(t)$ is the probability distribution function $F(t, x) = F(t, x_1, \ldots, x_n)$, $x \in \mathbb{R}^n$:

$$F(t, x) = \Pr\{X_1(t) < x_1, \ldots, X_n(t) < x_n\},$$

where $\Pr\{\,\cdot\,\}$ is the probability. This characteristic can be expressed by an integral of the probability density $\varphi(t, x) = \varphi(t, x_1, \ldots, x_n)$ as follows:

$$F(t, x_1, \ldots, x_n) = \int_{-\infty}^{x_1} \cdots \int_{-\infty}^{x_n} \varphi(t, x_1, \ldots, x_n)\mathrm{d}x_1 \ldots \mathrm{d}x_n.$$

Thus,

$$\varphi(t, x_1, \ldots, x_n) = \frac{\partial^n F(t, x_1, \ldots, x_n)}{\partial x_1 \ldots \partial x_n}.$$

It is known that if the probability density $\varphi(t, x)$ exists then it satisfies Fokker–Planck–Kolmogorov equation or Kolmogorov's forward equation [7, 12–15]:

$$\frac{\partial \varphi(t, x)}{\partial t} = -\sum_{i=1}^{n} \frac{\partial}{\partial x_i}[f_i(t, x)\varphi(t, x)] + \frac{1}{2}\sum_{i=1}^{n}\sum_{j=1}^{n} \frac{\partial^2}{\partial x_i \partial x_j}[g_{ij}(t, x)\varphi(t, x)],$$

$$\varphi(t_0, x) = \varphi_0(x) \tag{21.5}$$

or

$$\frac{\partial \varphi(t,x)}{\partial t} = \mathcal{A}\varphi(t,x), \quad \varphi(t_0, x) = \varphi_0(x), \tag{21.6}$$

where $\mathcal{A}$ is the forward diffusion operator defined by:

$$\mathcal{A}\varphi(t,x) = -\sum_{i=1}^{n} \frac{\partial}{\partial x_i}[f_i(t,x)\varphi(t,x)] + \frac{1}{2}\sum_{i=1}^{n}\sum_{j=1}^{n} \frac{\partial^2}{\partial x_i \partial x_j}[g_{ij}(t,x)\varphi(t,x)]. \tag{21.7}$$

We assume that

(a) There exists the probability density $\varphi(t,x)$ for the random process $X(t)$.
(b) Probability densities $\varphi(t,x)$ and $\varphi_0(x)$ are the square-integrable functions, i.e., $\varphi(t,x) \in L_2(\mathbb{T} \times \mathbb{R}^n)$, $\varphi_0(x) \in L_2(\mathbb{R}^n)$.
(c) For any $\xi(t,x) \in C_0^\infty(\mathbb{T} \times \mathbb{R}^n)$, the following equation is satisfied:

$$\int_{\mathbb{T}}\int_{\mathbb{R}^n} \xi(t,x)\frac{\partial \varphi(t,x)}{\partial t}\mathrm{d}x\mathrm{d}t = \sum_{i=1}^{n}\int_{\mathbb{T}}\int_{\mathbb{R}^n} \frac{\partial \xi(t,x)}{\partial x_i} f_i(t,x)\varphi(t,x)\mathrm{d}x\mathrm{d}t$$
$$+ \frac{1}{2}\sum_{i=1}^{n}\sum_{j=1}^{n}\int_{\mathbb{T}}\int_{\mathbb{R}^n} \frac{\partial^2 \xi(t,x)}{\partial x_i \partial x_j} g_{ij}(t,x)\varphi(t,x)\mathrm{d}x\mathrm{d}t,$$

where $C_0^\infty(\mathbb{T} \times \mathbb{R}^n)$ denotes the space of functions that have the compact support and continuous derivatives of all orders, i.e., there exists a weak solution [16] of the equation for the probability density $\varphi(t,x)$.

In relations given above, $g_{ij}(t,x)$ are entries of the $(n \times n)$-dimensional symmetric matrix function $g(t,x) = \sigma(t,x)\sigma^{\mathrm{T}}(t,x)$, $i, j = 1, 2, \ldots, n$.

Thus, the analysis problem of diffusions (Eq. 21.4) is formulated as follows. Given functions $f(t,x)$, $\sigma(t,x)$, defining the Itô SDE and probability density $\varphi_0(x)$ of the initial state X_0, find the probability density $\varphi(t,x)$.

21.3.2 Spectral Method for Solving Fokker–Planck–Kolmogorov Equation

Apply the spectral transform to the left-hand and right-hand sides of Eq. 21.5 using the linearity property of the spectral transform (see Sect. 21.2.2). Then

$$\mathbb{S}\left[\frac{\partial\varphi(t,x)}{\partial t}\right] = -\sum_{i=1}^{n}\mathbb{S}\left[\frac{\partial}{\partial x_i}[f_i(t,x)\varphi(t,x)]\right] + \frac{1}{2}\sum_{i=1}^{n}\sum_{j=1}^{n}\mathbb{S}\left[\frac{\partial^2}{\partial x_i \partial x_j}[g_{ij}(t,x)\varphi(t,x)]\right]. \qquad (21.8)$$

Further, we will use new notations ($i, j = 1, 2, \ldots, n$):

(i) $\mathcal{P}(n+1, n+1)$ is the spectral characteristic of the differentiation operator $\partial/\partial t$.
(ii) $\mathcal{P}_i(n+1, n+1)$ and $\mathcal{P}_{ij}(n+1, n+1)$ are the spectral characteristics of differentiation operators $\partial/\partial x_i$ and $\partial^2/\partial x_i \partial x_j$, respectively.
(iii) $F_i(n+1, n+1)$ and $G_{ij}(n+1, n+1)$ are the spectral characteristics of multiplication operators with multipliers $f_i(t,x)$ and $g_{ij}(t,x)$, respectively.

Let $\Phi(n+1, 0)$ and $\Phi_0(n, 0)$ be the spectral characteristics of probability densities $\varphi(t,x)$ and $\varphi_0(x)$, respectively, i.e.,

$$\Phi(n+1,0) = [\varphi_{i_0 i_1 \ldots i_n}], \quad \varphi_{i_0 i_1 \ldots i_n} = \int_{\mathbb{T}}\int_{\mathbb{R}^n} \varphi(t,x) e_{i_0 i_1 \ldots i_n}(t,x) \mathrm{d}x \mathrm{d}t, \qquad (21.9)$$

$$\Phi_0(n,0) = [\varphi_{0 i_1 \ldots i_n}], \quad \varphi_{0 i_1 \ldots i_n} = \int_{\mathbb{R}^n} \varphi_0(x) p_{i_1 \ldots i_n}(x) \mathrm{d}x, \qquad (21.10)$$

$$i_0, i_1, \ldots, i_n = 0, 1, 2, \ldots$$

Then we have

$$\mathbb{S}\left[\frac{\partial\varphi(t,x)}{\partial t}\right] = P(n+1,n+1)\cdot\Phi(n+1,0) - q(1,0;t_0)\otimes\Phi_0(n,0),$$

$$\mathbb{S}\left[\frac{\partial}{\partial x_i}[f_i(t,x)\varphi(t,x)]\right] = \mathcal{P}_i(n+1,n+1)\cdot F_i(n+1,n+1)\cdot\Phi(n+1,0),$$

$$\mathbb{S}\left[\frac{\partial^2}{\partial x_i \partial x_j}[g_{ij}(t,x)\varphi(t,x)]\right] = \mathcal{P}_{ij}(n+1,n+1)\cdot G_{ij}(n+1,n+1)\cdot\Phi(n+1,0),$$

where

$$P(n+1,n+1) = \mathcal{P}(n+1,n+1) + (q(1,0;t_0)\cdot q^{\mathrm{T}}(1,0;t_0))\otimes E(n,n), \qquad (21.11)$$

$$\mathcal{P}_{ij}(n+1,n+1) = \mathcal{P}_i(n+1,n+1)\cdot\mathcal{P}_j(n+1,n+1), \qquad (21.12)$$

$$i, j = 1, 2, \ldots, n.$$

To prove this, it is necessary to use properties of the spectral transform for functions and linear operators from Sect. 21.2.2. In particular,

$$\mathbb{S}\left[\frac{\partial\varphi(t,x)}{\partial t}\right] = \mathcal{P}(n+1,n+1)\cdot\Phi(n+1,0),$$

where the spectral characteristic $\mathcal{P}(n+1,n+1)$ can be represented in the form

$$\mathcal{P}(n+1,n+1) = P(n+1,n+1) - (q(1,0;t_0)\cdot q^{\mathrm{T}}(1,0;t_0))\otimes E(n,n)$$

by the property 3. Using the property 2, we obtain

$$\begin{aligned}
&\mathcal{P}(n+1,n+1)\cdot\Phi(n+1,0)\\
&\quad= (P(n+1,n+1) - (q(1,0;t_0)\cdot q^{\mathrm{T}}(1,0;t_0))\otimes E(n,n))\cdot\Phi(n+1,0)\\
&\quad= P(n+1,n+1)\cdot\Phi(n+1,0)\\
&\qquad - (q(1,0;t_0)\otimes E(n,n))\cdot(q^{\mathrm{T}}(1,0;t_0)\otimes E(n,n))\cdot\Phi(n+1,0)\\
&\quad= P(n+1,n+1)\cdot\Phi(n+1,0) - (q(1,0;t_0)\otimes E(n,n))\cdot\Phi_0(n,0)\\
&\quad= P(n+1,n+1)\cdot\Phi(n+1,0) - q(1,0;t_0)\otimes\Phi_0(n,0).
\end{aligned}$$

Properties 3 and 4 imply relations for first-order and second-order derivatives of the probability density $\varphi(t,x)$ with multipliers $f_i(t,x)$ and $g_{ij}(t,x)$.

Thus, Eq. 21.8 reduces to

$$\begin{aligned}
&P(n+1,n+1)\cdot\Phi(n+1,0) - q(1,0;t_0)\otimes\Phi_0(n,0)\\
&\quad= -\sum_{i=1}^{n}\mathcal{P}_i(n+1,n+1)\cdot F_i(n+1,n+1)\cdot\Phi(n+1,0)\\
&\qquad+\frac{1}{2}\sum_{i=1}^{n}\sum_{j=1}^{n}\mathcal{P}_{ij}(n+1,n+1)\cdot G_{ij}(n+1,n+1)\cdot\Phi(n+1,0), \qquad (21.13)
\end{aligned}$$

consequently, the spectral characteristic $\Phi(n+1,0)$ satisfies the equation

$$\begin{aligned}
&P(n+1,n+1)\cdot\Phi(n+1,0) - q(1,0;t_0)\otimes\Phi_0(n,0)\\
&\quad= A(n+1,n+1)\cdot\Phi(n+1,0), \qquad (21.14)
\end{aligned}$$

where

$$\begin{aligned}
A(n+1,n+1) = &-\sum_{i=1}^{n}\mathcal{P}_i(n+1,n+1)\cdot F_i(n+1,n+1)\\
&+\frac{1}{2}\sum_{i=1}^{n}\sum_{j=1}^{n}\mathcal{P}_{ij}(n+1,n+1)\cdot G_{ij}(n+1,n+1). \qquad (21.15)
\end{aligned}$$

In fact, the $2(n+1)$-dimensional matrix $A(n+1,n+1)$ is the spectral characteristic of the forward diffusion operator $\mathcal{A}$ defined by Eq. 21.7. Also note that the spectral characteristic $\Phi(n+1,0)$ is called the generalized characteristic function (the ordinary characteristic function is Fourier transform of the probability density), and Eq. 21.13 is called the generalized characteristic function equation [2, 4, 7].

Thus, the analysis problem of diffusions (Eq. 21.4) is reduced to solving the infinite system of linear algebraic equations (Eq. 21.13) with unknown entries of the spectral characteristic $\Phi(n+1,0)$. Some aspects of solving the infinite system of linear algebraic equations are given in [17, 18]. The exact solution of Eqs. 21.13–21.14 is defined by:

$$\Phi(n+1,0) = (P(n+1,n+1) - A(n+1,n+1))^{-1} \cdot (q(1,0;t_0) \otimes \Phi_0(n,0)). \tag{21.16}$$

Then, the probability density $\varphi(t,x)$ is given by Eq. 21.2:

$$\varphi(t,x) = \mathbb{S}^{-1}[\Phi(n+1,0)] = \sum_{i_0,i_1,\ldots,i_n=0}^{\infty} \varphi_{i_0 i_1 \ldots i_n} e_{i_0 i_1 \ldots i_n}(t,x), \quad (t,x) \in \mathbb{T} \times \mathbb{R}^n, \tag{21.17}$$

where $\varphi_{i_0 i_1 \ldots i_n}$ are entries of the spectral characteristic $\Phi(n+1,0)$.

To find the approximate solution of the analysis problem of diffusions (Eq. 21.4) all spectral characteristics should be truncated on all dimensions. The methodical inaccuracy caused by the spectral characteristic truncation is described in [1, 2]. In this case, we obtain the probability density $\varphi(t,x)$ as a partial sum

$$\varphi(t,x) \approx \sum_{i_0=0}^{L_0-1} \sum_{i_1=0}^{L_1-1} \cdots \sum_{i_n=0}^{L_n-1} \varphi_{i_0 i_1 \ldots i_n} e_{i_0 i_1 \ldots i_n}(t,x), \quad (t,x) \in \mathbb{T} \times \mathbb{R}^n, \tag{21.18}$$

where $L_0, L_1, \ldots, L_n$ are the truncation orders of spectral characteristics.

The algorithm for solving the analysis problem of diffusions (Eq. 21.4) by the spectral method is given below:

1. Specify basis systems $\{q_{i_0}(t)\}_{i_0=0}^{\infty}$ and $\{p_{i_1 \ldots i_n}(x)\}_{i_1,\ldots,i_n=0}^{\infty}$ for $L_2(\mathbb{T})$ and $L_2(\mathbb{R}^n)$, respectively. Form the basis system $\{e_{i_0 i_1 \ldots i_n}(t,x)\}_{i_0,i_1,\ldots,i_n=0}^{\infty}$ for $L_2(\mathbb{T} \times \mathbb{R}^n)$ by Eq. 21.1. Specify truncation orders $L_0, L_1, \ldots, L_n$ for all spectral characteristics.
2. Find the column matrix $q(1,0;t_0)$ with entries $q_{i_0}(t_0)$, i.e.,

$$q(1,0;t_0) = [\; q_0(t_0) \;\; q_1(t_0) \;\; \ldots \;\; q_{L_0-1}(t_0) \;]^{\mathrm{T}}.$$

3. Find spectral characteristics $\mathcal{P}(n+1,n+1)$ and $\mathcal{P}_i(n+1,n+1)$ of differentiation operators $\partial/\partial t$ and $\partial/\partial x_i$, respectively, $i = 1,2,\ldots,n$. Find the matrix $P(n+1,n+1)$ and spectral characteristics $\mathcal{P}_{ij}(n+1,n+1)$ of differentiation operators $\partial^2/\partial x_i \partial x_j$ using Eqs. 21.11–21.12, $i, j = 1,2,\ldots,n$.
4. Find spectral characteristics $F_i(n+1,n+1)$ and $G_{ij}(n+1,n+1)$ of multiplication operators with multipliers $f_i(t,x)$ and $g_{ij}(t,x)$, respectively, $i, j = 1,2,\ldots,n$.

5. Find the spectral characteristic $A(n+1, n+1)$ of the forward diffusion operator $\mathcal{A}$ using Eq. 21.15.
6. Find the spectral characteristic $\Phi_0(n, 0)$ of the probability density $\varphi_0(x)$ for the initial state X_0 using Eq. 21.10.
7. Find the solution $\Phi(n+1, 0)$ of the generalized characteristic function Eq. 21.14 using Eq. 21.16.
8. Find the approximate solution $\varphi(t, x)$ of Fokker–Planck–Kolmogorov equation by Eq. 21.18.

21.3.3 Dryden Wind Turbulence Model

Consider one-dimensional Dryden wind turbulence model that is a mathematical model of continuous gusts [19–22]. It is given by the linear SDE:

$$dX(t) = -\mu X(t)\mathrm{d}t + \sigma dW(t), \quad X(0) = X_0, \quad \mu = \frac{V_t}{L_t}, \quad \sigma = \sqrt{2\mu}\sigma_0, \tag{21.19}$$

where $t \in \mathbb{T} = [0, T]$, V_t is the longitudinal flight velocity, L_t is the turbulence scale length, σ_0 is the standard deviation of the wind velocity, $W(t)$ is the one-dimensional standard Wiener process, i.e.,

$$n = s = 1, \quad t_0 = 0, \quad f(t, x) = -\mu x, \quad \sigma(t, x) = \sigma, \quad g(t, x) = \sigma^2.$$

Here, $X(t)$ is the wind velocity and X_0 is the initial wind velocity. If X_0 is a constant or X_0 is a random variable having a normal distribution (Gaussian distribution), then $X(t)$ is Gaussian process (more precisely, Ornstein–Uhlenbeck process [12]) with the probability density $\varphi(t, x)$ that is defined by the mean $m(t) = \mathrm{E}X(t)$ and the second-order moment $M(t) = \mathrm{E}X^2(t)$ or the mean $m(t) = \mathrm{E}X(t)$ and the variance $D(t) = \mathrm{E}(X(t) - m(t))^2 = M(t) - m^2(t)$:

$$m(t) = m_0\mathrm{e}^{-\mu t} \quad (m_0 = \mathrm{E}X_0), \quad M(t) = \left(M_0 - \frac{\sigma^2}{2\mu}\right)\mathrm{e}^{-2\mu t} + \frac{\sigma^2}{2\mu} \quad (M_0 = \mathrm{E}X_0^2),$$

$$D(t) = \left(D_0 - \frac{\sigma^2}{2\mu}\right)\mathrm{e}^{-2\mu t} + \frac{\sigma^2}{2\mu} \quad (D_0 = \mathrm{E}(X_0 - m_0)^2 = M_0 - m_0^2).$$

Functions $m(t)$ and $M(t)$ satisfy the following Ordinary Differential Equations (ODEs):

$$\dot{m}(t) = -\mu m(t), \quad m(0) = m_0, \tag{21.20}$$

$$\dot{M}(t) = -2\mu M(t) + \sigma^2, \quad M(0) = M_0 = \mathrm{E}X_0^2. \tag{21.21}$$

The mean $m(t)$ and the second-order moment $M(t)$ also satisfy ODEs (Eqs. 21.20–21.21), when X_0 has a non-Gaussian distribution, but in this case $X(t)$ is the non-Gaussian process.

Fokker–Planck–Kolmogorov equation corresponding to SDE (Eq. 21.19) can be written as

$$\frac{\partial\varphi(t,x)}{\partial t} = \mu\frac{\partial}{\partial x}[x\varphi(t,x)] + \frac{\sigma^2}{2}\frac{\partial^2\varphi(t,x)}{\partial x^2}, \quad \varphi(0,x) = \varphi_0(x), \tag{21.22}$$

where $\varphi_0(x)$ is the probability density for the initial wind velocity X_0. Let X_0 be a random variable having a standard normal distribution, i.e.,

$$\varphi_0(x) = \frac{1}{\sqrt{2\pi}}\mathrm{e}^{-\frac{1}{2}x^2}.$$

Further, we apply the algorithm for solving analysis problem of diffusions (Eq. 21.4) by the spectral method.

1. Let $\{q_{i_0}(t)\}_{i_0=0}^{\infty}$ be Legendre polynomials $\{\hat{P}_{i_0}(t)\}_{i_0=0}^{\infty}$ and let $\{p_{i_1}(x)\}_{i_1=0}^{\infty}$ be Hermite functions $\{\hat{\Phi}_{i_1}(x)\}_{i_1=0}^{\infty}$, where $t \in \mathbb{T}$ and $x \in \mathbb{R}$, then $\{e_{i_0 i_1}(t,x) = q_{i_0}(t)p_{i_1}(x) = \hat{P}_{i_0}(t)\hat{\Phi}_{i_1}(x)\}_{i_0,i_1=0}^{\infty}$. Thus,

$$\hat{P}_i(t) = \sqrt{\frac{2i+1}{T}}\sum_{k=0}^{i}(-1)^{i-k}C_{i+k}^{i}C_i^{i-k}\frac{t^k}{T^k}, \quad C_k^i = \frac{k!}{i!\,(k-i)!},$$

$$\hat{\Phi}_i(x) = \sqrt{\frac{i!}{2^i\sqrt{\pi}}}\mathrm{e}^{-x^2/2}\sum_{k=0}^{\lfloor i/2\rfloor}\frac{(-1)^k(2x)^{i-2k}}{k!\,(i-2k)!}, \quad i = 0,1,2,\ldots$$

   Let us assume that $L_0 = L_1 = 32$ (truncation orders).
2. The column matrix $q(1,0;0)$ with entries $q_{i_0}(0)$ is

$$q(1,0;0) = [\,\hat{P}_0(0)\ \ \hat{P}_1(0)\ \ \ldots\ \ \hat{P}_{L_0-1}(0)\,]^{\mathrm{T}}, \quad \hat{P}_i(0) = (-1)^i\sqrt{\frac{2i+1}{T}}.$$

3. Spectral characteristics $\mathcal{P}(2,2)$ and $\mathcal{P}_1(2,2)$ of differentiation operators $\partial/\partial t$ and $\partial/\partial x$ are defined by:

$$\mathcal{P}(2,2) = [p_{i_0 i_1 j_0 j_1}], \quad p_{i_0 i_1 j_0 j_1} = \int_{\mathbb{T}}\int_{\mathbb{R}} e_{i_0 i_1}(t,x)\frac{\partial e_{j_0 j_1}(t,x)}{\partial t}\mathrm{d}x\mathrm{d}t$$

$$= \int_{\mathbb{T}}\hat{P}_{i_0}(t)\frac{d\hat{P}_{j_0}(t)}{\mathrm{d}t}\mathrm{d}t\int_{\mathbb{R}}\hat{\Phi}_{i_1}(x)\hat{\Phi}_{j_1}(x)\mathrm{d}x = \int_{\mathbb{T}}\hat{P}_{i_0}(t)\frac{d\hat{P}_{j_0}(t)}{\mathrm{d}t}\mathrm{d}t\cdot\delta_{i_1 j_1},$$

$$\mathcal{P}_1(2,2) = [p_{1 i_0 i_1 j_0 j_1}], \quad p_{1 i_0 i_1 j_0 j_1} = \int_{\mathbb{T}}\int_{\mathbb{R}} e_{i_0 i_1}(t,x)\frac{\partial e_{j_0 j_1}(t,x)}{\partial x}\mathrm{d}x\mathrm{d}t$$

$$= \int_{\mathbb{T}} \hat{P}_{i_0}(t)\hat{P}_{j_0}(t)\mathrm{d}t \int_{\mathbb{R}} \hat{\Phi}_{i_1}(x)\frac{d\hat{\Phi}_{j_1}(x)}{\mathrm{d}x}\mathrm{d}x = \delta_{i_0 j_0} \cdot \int_{\mathbb{R}} \hat{\Phi}_{i_1}(x)\frac{d\hat{\Phi}_{j_1}(x)}{\mathrm{d}x}\mathrm{d}x,$$

where $\delta_{i_0 j_0}$ and $\delta_{i_1 j_1}$ are the Kronecker deltas. Consequently,

$$\mathcal{P}(2,2) = \mathcal{P}(1,1) \otimes E(1,1), \quad \mathcal{P}_1(2,2) = E(1,1) \otimes \mathcal{P}_1(1,1),$$

and

$$\begin{aligned} P(2,2) &= \mathcal{P}(2,2) + (q(1,0;0) \cdot q^{\mathrm{T}}(1,0;0)) \otimes E(1,1) \\ &= (\mathcal{P}(1,1) + q(1,0;0) \cdot q^{\mathrm{T}}(1,0;0)) \otimes E(1,1), \\ \mathcal{P}_{11}(2,2) &= \mathcal{P}_1^2(2,2) = E(1,1) \otimes \mathcal{P}_1^2(1,1), \end{aligned}$$

where $E(1,1)$ is the two-dimensional identity matrix, $\mathcal{P}(1,1)$ is the matrix with entries

$$p_{ij} = \int_{\mathbb{T}} \hat{P}_i(t)\frac{d\hat{P}_j(t)}{\mathrm{d}t}\mathrm{d}t = \frac{1}{T}\begin{cases} 2\sqrt{(2i+1)(2j+1)} & \text{for } i < j \text{ and } (j-i) \bmod 2 = 1 \\ 0 & \text{otherwise,} \end{cases}$$

and $\mathcal{P}_1(1,1)$ is the matrix with entries

$$p_{1ij} = -p_{1ji} = \int_{\mathbb{R}} \hat{\Phi}_i(x)\frac{d\hat{\Phi}_j(x)}{\mathrm{d}x}\mathrm{d}x = \frac{1}{2}\begin{cases} \sqrt{2j} & \text{for } j - i = 1 \\ -\sqrt{2i} & \text{for } i - j = 1 \\ 0 & \text{otherwise.} \end{cases}$$

The formulas for entries p_{ij} and p_{1ij} have been derived in [1, 7] using definitions and properties for Legendre polynomials and Hermite polynomials [23, 24].

4. Spectral characteristics $F(2,2)$ and $G(2,2)$ of multiplication operators with multipliers $f(t,x)$ and $g(t,x)$ are defined by:

$$\begin{aligned} F(2,2) &= [f_{i_0 i_1 j_0 j_1}], \quad f_{i_0 i_1 j_0 j_1} = \int_{\mathbb{T}}\int_{\mathbb{R}} f(t,x)e_{i_0 i_1}(t,x)e_{j_0 j_1}(t,x)\mathrm{d}x\mathrm{d}t \\ &= -\mu \int_{\mathbb{T}} \hat{P}_{i_0}(t)\hat{P}_{j_0}(t)\mathrm{d}t \int_{\mathbb{R}} x\hat{\Phi}_{i_1}(x)\hat{\Phi}_{j_1}(x)\mathrm{d}x = -\mu\delta_{i_0 j_0} \cdot \int_{\mathbb{R}} x\hat{\Phi}_{i_1}(x)\hat{\Phi}_{j_1}(x)\mathrm{d}x, \end{aligned}$$

$$\begin{aligned} G(2,2) &= [g_{i_0 i_1 j_0 j_1}], \quad g_{i_0 i_1 j_0 j_1} = \int_{\mathbb{T}}\int_{\mathbb{R}} g(t,x)e_{i_0 i_1}(t,x)e_{j_0 j_1}(t,x)\mathrm{d}x\mathrm{d}t \\ &= \sigma^2 \int_{\mathbb{T}} \hat{P}_{i_0}(t)\hat{P}_{j_0}(t)\mathrm{d}t \int_{\mathbb{R}} \hat{\Phi}_{i_1}(x)\hat{\Phi}_{j_1}(x)\mathrm{d}x = \sigma^2\delta_{i_0 j_0}\delta_{i_1 j_1}. \end{aligned}$$

Consequently,

$$\begin{aligned} F(2,2) &= \mu \cdot E(1,1) \otimes X(1,1), \\ G(2,2) &= \sigma^2 \cdot E(1,1) \otimes E(1,1) = \sigma^2 \cdot E(2,2), \end{aligned}$$

where $F(2,2)$ is the four-dimensional identity matrix, $X(1,1)$ is the matrix with entries

$$\chi_{ij} = \chi_{ji} = \int_{\mathbb{R}} x\hat{\Phi}_i(x)\hat{\Phi}_j(x)\mathrm{d}x = \frac{1}{\sqrt{2}}\begin{cases} \sqrt{j+1} & \text{for } j-i=1 \\ \sqrt{i+1} & \text{for } i-j=1 \\ 0 & \text{otherwise.} \end{cases}$$

This relation for entries χ_{ij} has been derived in [7] using definitions and properties for Hermite polynomials [23, 24].

5. The spectral characteristic $A(2,2)$ can be expressed as

$$A(2,2) = -P_1(2,2)\cdot F(2,2) + \frac{\sigma^2}{2}\cdot P_1^2(2,2).$$

6. The spectral characteristic $\Phi_0(1,0)$ of the probability density $\varphi_0(x)$ for the initial wind velocity X_0 is defined by

$$\Phi_0(1,0) = [\varphi_{0i}], \quad \varphi_{0i} = \int_{\mathbb{R}} \hat{\Phi}_i(x)\varphi_0(x)dx = \frac{1}{\sqrt{2\sqrt{\pi}}}\begin{cases} 1 & \text{for } i=0 \\ 0 & \text{otherwise} \end{cases}$$

because $\varphi_0(x) = (1/\sqrt{2\sqrt{\pi}})\hat{\Phi}_0(x)$.

7. The solution $\Phi(2,0)$ of the generalized characteristic function equation has the following form:

$$\Phi(2,0) = (P(2,2) - A(2,2))^{-1}\cdot(q(1,0;0)\otimes\Phi_0(1,0)).$$

8. The approximate solution $\varphi(t,x)$ of Fokker–Planck–Kolmogorov equation is defined by:

$$\varphi(t,x) \approx \sum_{i_0=0}^{L_0-1}\sum_{i_1=0}^{L_1-1} \varphi_{i_0i_1}\cdot\hat{P}_{i_0}(t)\cdot\hat{\Phi}_{i_1}(x), \quad (t,x)\in\mathbb{T}\times\mathbb{R},$$

where $\varphi_{i_0i_1}$ are entries of the matrix $\Phi(2,0)$.

The approximate solution of the analysis problem for $T = 10$ s, $V_t = 60$ m/s, $L_t = 1000$ m, and $\sigma_0 = 1.5$ m/s is presented in Fig. 21.1.

21.4 Spectral Method for Analysis of Jump Diffusions

This section structure is similar to previous Sect. 21.3, i.e., it includes Sect. 21.4.1 with the statement of the problem, Sect. 21.4.2 with detailed description of the spectral method for solving Kolmogorov–Feller equation, and Sect. 21.4.3 with numerical results for the analysis of Dryden wind turbulence model with jumps.

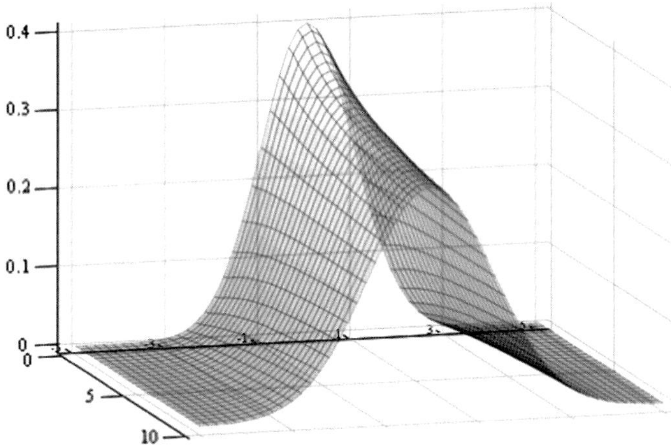

Fig. 21.1 The probability density $\varphi(t, x)$ of the wind velocity $X(t)$

21.4.1 Problem Statement

Let $X(t)$ be an $\mathbb{R}^n$-valued random process that satisfies Itô SDE with a compound Poisson process:

$$dX(t) = f(t, X(t))\mathrm{d}t + \sigma(t, X(t))dW(t) + dP^c(t), \quad X(t_0) = X_0, \qquad (21.23)$$

where all notations except $P^c(t)$ have been introduced in Sect. 21.3.1.

In Eq. 21.23, $P^c(t)$ is the compound Poisson process, which can be defined in different ways [14, 25, 26]. Let $\lambda(t, x)$:$\mathbb{T} \times \mathbb{R}^n \to \mathbb{R}_+$ denote the jump rate (or intensity) and let $\rho(t, \delta)$ denote the probability density for jumps (random increments of $X(t)$). These two functions specify Poisson process $P(t)$ so that

$$\Pr(P(t + \Delta t) - P(t) = 1|X(t) = x) = \lambda(t, x)\Delta t + o(\Delta t)$$

for small Δt, and $X(\tau_j) = X(\tau_j^-) + \Delta_j$, where jumps $\Delta_j \in \mathbb{R}^n$ are random vectors distributed with probability density $\rho(\tau_j, \delta)$, $j = 1, 2, \ldots$, and $\{\tau_j\}$ are points of the Poisson process $P(t)$, $\tau_0 = t_0$, i.e.,

$$P^c(t) = \sum_{j=1}^{P(t)} \Delta_j.$$

Functions $f(t, x)$, $\sigma(t, x)$, $\lambda(t, x)$, $\rho(t, \delta)$ satisfy the conditions for the existence and uniqueness of the strong or weak solution of SDEs with a compound Poisson process [25], and $\mathrm{E}|X_0|^2 < +\infty$.

The probability density $\varphi(t,x)$ satisfies Kolmogorov–Feller equation or Kolmogorov's forward equation [25, 26]:

$$\frac{\partial\varphi(t,x)}{\partial t} = -\sum_{i=1}^{n}\frac{\partial}{\partial x_i}[f_i(t,x)\varphi(t,x)] + \frac{1}{2}\sum_{i=1}^{n}\sum_{j=1}^{n}\frac{\partial^2}{\partial x_i\partial x_j}[g_{ij}(t,x)\varphi(t,x)]$$
$$-\lambda(t,x)\varphi(t,x) + \int_{\mathbb{R}^n}\lambda(t,\xi)\rho(t,x-\xi)\varphi(t,\xi)d\xi, \quad \varphi(t_0,x) = \varphi_0(x). \tag{21.24}$$

Equation 21.24 can be expressed in the form of Eq. 21.6 for the forward jump diffusion operator $\mathcal{A}$ defined by:

$$\mathcal{A}\varphi(t,x) = -\sum_{i=1}^{n}\frac{\partial}{\partial x_i}[f_i(t,x)\varphi(t,x)] + \frac{1}{2}\sum_{i=1}^{n}\sum_{j=1}^{n}\frac{\partial^2}{\partial x_i\partial x_j}[g_{ij}(t,x)\varphi(t,x)]$$
$$-\lambda(t,x)\varphi(t,x) + \int_{\mathbb{R}^n}\lambda(t,\xi)\rho(t,x-\xi)\varphi(t,\xi)d\xi. \tag{21.25}$$

Here, we will suppose that assumptions (a) and (b) from Sect. 21.3.1 are satisfied, and the assumption similar to (c) with respect to Eq. 21.24 is also satisfied.

Thus, the analysis problem of jump diffusions (Eq. 21.23) is formulated as follows. Given functions $f(t,x)$, $\sigma(t,x)$, $\lambda(t,x)$, $\rho(t,\delta)$, defining Itô SDE with a compound Poisson process, and the probability density $\varphi_0(x)$ of the initial state X_0, find the probability density $\varphi(t,x)$.

21.4.2 Spectral Method for Solving Kolmogorov–Feller Equation

Apply the spectral transform to the left-hand and right-hand sides of Eq. 21.24 using the linearity property of the spectral transform (see Sect. 21.2.2). Then

$$\mathbb{S}\left[\frac{\partial\varphi(t,x)}{\partial t}\right]$$
$$= -\sum_{i=1}^{n}\mathbb{S}\left[\frac{\partial}{\partial x_i}[f_i(t,x)\varphi(t,x)]\right]$$
$$+\frac{1}{2}\sum_{i=1}^{n}\sum_{j=1}^{n}\mathbb{S}\left[\frac{\partial^2}{\partial x_i\partial x_j}[g_{ij}(t,x)\varphi(t,x)]\right]$$
$$-\mathbb{S}[\lambda(t,x)\varphi(t,x)] + \mathbb{S}\left[\int_{\mathbb{R}^n}\lambda(t,\xi)\rho(t,x-\xi)\varphi(t,\xi)d\xi\right]. \tag{21.26}$$

We will use notations (i)–(iii) from Sect. 21.3.2 as well new notations:

(iv) $\Lambda(n+1, n+1)$ is the spectral characteristic of the multiplication operator with the multiplier $\lambda(t, x)$.

(v) $R(n+1, n+1)$ is the spectral characteristic of the linear integral operator $\mathcal{R}$ with the kernel $\rho(t, x-\xi)$, i.e.,

$$\mathcal{R}\psi(t, x) = \int_{\mathbb{R}^n} \rho(t, x-\xi)\psi(t, \xi)d\xi, \quad \psi(t, x) : \mathbb{T} \times \mathbb{R}^n \to \mathbb{R}.$$

Therefore,

$$\begin{aligned} &\mathbb{S}[\lambda(t, x)\varphi(t, x)] \\ &\quad = \Lambda(n+1, n+1) \cdot \Phi(n+1, 0), \\ &\mathbb{S}\left[\int_{\mathbb{R}^n} \lambda(t, \xi)\rho(t, x-\xi)\varphi(t, \xi)d\xi\right] \\ &\quad = R(n+1, n+1) \cdot \Lambda(n+1, n+1) \cdot \Phi(n+1, 0), \end{aligned}$$

where spectral characteristics $\Phi(n+1, 0)$ and $\Phi_0(n, 0)$ have been defined by Eqs. 21.9–21.10.

Thus, Eq. 21.26 reduces to

$$\begin{aligned} &P(n+1, n+1) \cdot \Phi(n+1, 0) - q(1, 0; t_0) \otimes \Phi_0(n, 0) \\ &\quad = -\sum_{i=1}^{n} \mathcal{P}_i(n+1, n+1) \cdot F_i(n+1, n+1) \cdot \Phi(n+1, 0) \\ &\qquad + \frac{1}{2}\sum_{i=1}^{n}\sum_{j=1}^{n} \mathcal{P}_{ij}(n+1, n+1) \cdot G_{ij}(n+1, n+1) \cdot \Phi(n+1, 0) \\ &\qquad - \Lambda(n+1, n+1) \cdot \Phi(n+1, 0) + R(n+1, n+1) \cdot \Lambda(n+1, n+1) \cdot \Phi(n+1, 0), \end{aligned} \tag{21.27}$$

consequently, the spectral characteristic $\Phi(n+1, 0)$ satisfies the equation that coincides with Eq. 21.14, but in this case $A(n+1, n+1)$ is the spectral characteristic of the forward jump diffusion operator $\mathcal{A}$ defined by Eq. 21.25:

$$\begin{aligned} A(n+1, n+1) = &-\sum_{i=1}^{n} \mathcal{P}_i(n+1, n+1) \cdot F_i(n+1, n+1) \\ &+ \frac{1}{2}\sum_{i=1}^{n}\sum_{j=1}^{n} \mathcal{P}_{ij}(n+1, n+1) \cdot G_{ij}(n+1, n+1) \\ &- \Lambda(n+1, n+1) + R(n+1, n+1) \cdot \Lambda(n+1, n+1). \end{aligned} \tag{21.28}$$

The spectral characteristic $\Phi(n+1, 0)$ is also called the generalized characteristic function, it satisfies the generalized characteristic function equation (Eq. 21.27).

Thus, the analysis problem of jump diffusions (Eq. 21.23) is reduced to solving the infinite system of linear algebraic system (Eq. 21.27) with unknown entries of the spectral characteristic $\Phi(n+1, 0)$. The exact solution of Eq. 21.27 is defined by Eq. 21.16, the probability density $\varphi(t, x)$ is given by Eq. 21.17 or Eq. 21.18 for the approximate solution of the analysis problem of jump diffusions.

The algorithm for solving analysis problem of jump diffusions (Eq. 21.23) by the spectral method is given below (items 1–4 are identical to the algorithm for solving analysis problem of diffusions (Eq. 21.4) in Sect. 21.3.2):

5. Find the spectral characteristic $\Lambda(n+1, n+1)$ of the multiplication operator with the multiplier $\lambda(t, x)$ and the spectral characteristic $R(n+1, n+1)$ of the linear integral operator $\mathcal{R}$ with the kernel $\rho(t, x-\xi)$.
6. Find the spectral characteristic $A(n+1, n+1)$ of the forward jump diffusion operator $\mathcal{A}$ using Eq. 21.28.
7. Find the spectral characteristic $\Phi_0(n, 0)$ of the probability density $\varphi_0(x)$ for the initial state X_0 using Eq. 21.10.
8. Find the solution $\Phi(n+1, 0)$ of the generalized characteristic function equation (Eq. 21.14) using Eq. 21.16.
9. Find the approximate solution $\varphi(t, x)$ of Kolmogorov–Feller equation by Eq. 21.18.

21.4.3 Dryden Wind Turbulence Model with Jumps

Consider the modified one-dimensional Dryden wind turbulence model with jumps. It is given by the linear SDE with a compound Poisson process:

$$dX(t) = -\mu_* X(t)\mathrm{d}t + \sigma_* dW(t) + dP^c(t), \quad X(0) = X_0, \tag{21.29}$$

where $t \in \mathbb{T} = [0, T]$, μ_* and σ_* are constants that will be specified below.

The compound Poisson process $P^c(t)$ is defined by the constant jump rate λ and the probability density for jumps Δ_j (see Sect. 21.4.1) that equals the probability density for the initial wind velocity X_0 given in Sect. 21.3.3, i.e.,

$$n = s = 1, \quad t_0 = 0, \quad f(t, x) = -\mu_* x, \quad \sigma(t, x) = \sigma_*, \quad g(t, x) = \sigma_*^2,$$
$$\lambda(t, x) = \lambda, \quad \rho(t, \delta) = \varphi_0(\delta).$$

The mean $m(t) = \mathrm{E}X(t)$ and the second-order moment $M(t) = \mathrm{E}X^2(t)$ satisfy the following ODEs:

$$\dot{m}(t) = -\mu_* m(t) + \lambda m^\Delta, \quad m(0) = m_0, \tag{21.30}$$

$$\dot{M}(t) = -2\mu_* M(t) + \sigma_*^2 + 2\lambda m^\Delta m(t) + \lambda M^\Delta, \quad M(0) = M_0 = \mathrm{E}X_0^2, \quad (21.31)$$

where m^Δ and M^Δ are the mean and the second-order moment of jumps. Thus, $m^\Delta = 0$ and $M^\Delta = 1$, because jumps have a standard normal distribution. To compensate the compound Poisson process with respect to the mean and the second-order moment the constants μ_* and σ_* should be specified as follows:

$$\mu_* = \mu, \quad \sigma_*^2 + \lambda M^\Delta = \sigma^2 \quad (\sigma_* = \sqrt{2\mu\sigma_0^2 - \lambda}, \quad 2\mu\sigma_0^2 \geq \lambda),$$

where μ , σ and σ_0 have been defined in Sect. 21.3.3.

Hence, Eqs. 21.30–21.31 coincide with Eqs. 21.20–21.21, respectively.

Kolmogorov–Feller equation corresponding to SDE (Eq. 21.29) can be written as:

$$\frac{\partial \varphi(t,x)}{\partial t} = \mu_* \frac{\partial}{\partial x}[x\varphi(t,x)] + \frac{\sigma_*^2}{2}\frac{\partial^2 \varphi(t,x)}{\partial x^2}$$
$$- \lambda\varphi(t,x) + \lambda \int_{\mathbb{R}} \varphi_0(x-\xi)\varphi(t,\xi)d\xi, \quad \varphi(t_0,x) = \varphi_0(x). \quad (21.32)$$

Further, we apply the algorithm for solving analysis problem of jump diffusions (Eq. 21.23) by the spectral method (items 1–4 should be used from the solving analysis problem for Dryden turbulence wind model (Eq. 21.19) in Sect. 21.3.3).

5. The spectral characteristic $\Lambda(2,2)$ of the multiplication operator with the multiplier $\lambda(t,x)$ is defined by:

$$\Lambda(2,2) = [\lambda_{i_0 i_1 j_0 j_1}], \quad \lambda_{i_0 i_1 j_0 j_1} = \int_{\mathbb{T}}\int_{\mathbb{R}} \lambda(t,x) e_{i_0 i_1}(t,x) e_{j_0 j_1}(t,x)\mathrm{d}x\mathrm{d}t$$
$$= \lambda \int_{\mathbb{T}} \hat{P}_{i_0}(t)\hat{P}_{j_0}(t)\mathrm{d}t \int_{\mathbb{R}} \hat{\Phi}_{i_1}(x)\hat{\Phi}_{j_1}(x)\mathrm{d}x = \lambda\delta_{i_0 j_0}\delta_{i_1 j_1},$$

where $\delta_{i_0 j_0}$ and $\delta_{i_1 j_1}$ are the Kronecker deltas. Consequently,

$$\Lambda(2,2) = \lambda \cdot E(1,1) \otimes E(1,1) = \lambda \cdot E(2,2),$$

where $E(1,1)$ and $E(2,2)$ are two-dimensional and four-dimensional identity matrices, respectively.

The spectral characteristic $R(2,2)$ of the linear integral operator $\mathcal{R}$ with the kernel $\varphi_0(x-\xi)$ satisfies the following relations:

$$R(2,2) = [r_{i_0 i_1 j_0 j_1}],$$
$$r_{i_0 i_1 j_0 j_1} = \int_{\mathbb{T}}\int_{\mathbb{R}} e_{i_0 i_1}(t,x) \int_{\mathbb{R}} e_{j_0 j_1}(t,\xi)\varphi_0(x-\xi)d\xi\mathrm{d}x\mathrm{d}t$$
$$= \int_{\mathbb{T}} \hat{P}_{i_0}(t)\hat{P}_{j_0}(t)\mathrm{d}t \int_{\mathbb{R}} \hat{\Phi}_{i_1}(x) \int_{\mathbb{R}} \hat{\Phi}_{j_1}(\xi)\varphi_0(x-\xi)d\xi\mathrm{d}x$$

$$= \delta_{i_0 j_0} \int_{\mathbb{R}} \hat{\Phi}_{i_1}(x)[\hat{\Phi}_{j_1}(x) * \varphi_0(x)]\mathrm{d}x,$$

where $\hat{\Phi}_{j_1}(x) * \varphi_0(x)$ is the convolution of $\Phi_{j_1}(x)$ and $\varphi_0(x)$. Therefore,

$$R(2, 2) = E(1, 1) \otimes R(1, 1),$$

where entries

$$\varrho_{ij} = \int_{\mathbb{R}} \hat{\Phi}_i(x)[\hat{\Phi}_j(x) * \varphi_0(x)]\mathrm{d}x$$

of the matrix $R(1, 1)$ can be calculated by a recurrence relation based on properties for the Hermite polynomials [23, 24]. Here, they are calculated numerically.

6. The spectral characteristic $A(2, 2)$ is expressed as:

$$A(2, 2) = -\mathcal{P}_1(2, 2) \cdot F(2, 2) + \frac{\sigma^2}{2} \cdot \mathcal{P}_1^2(2, 2) - \Lambda(2, 2) + \Lambda(2, 2) \cdot R(2, 2).$$

The spectral characteristic $\Phi_0(1, 0)$ of the probability density $\varphi_0(x)$ for the initial wind velocity X_0, the relation for the solution $\Phi(2, 0)$ of the generalized characteristic function equation and the approximate solution $\varphi(t, x)$ of Kolmogorov–Feller equation are similar to ones given in Sect. 21.3.3 for the solving analysis problem for Dryden turbulence wind model (Eq. 21.19) (see items 6–8).

The approximate solution of the analysis problem for $T = 10$ s, $V_t = 60$ m/s, $L_t = 1000$ m, $\sigma_0 = 1.5$ m/s, and $\lambda = 0.2$ is presented in Fig. 21.2. In fact, this solution has minimal differences compared with the approximate solution of the analysis

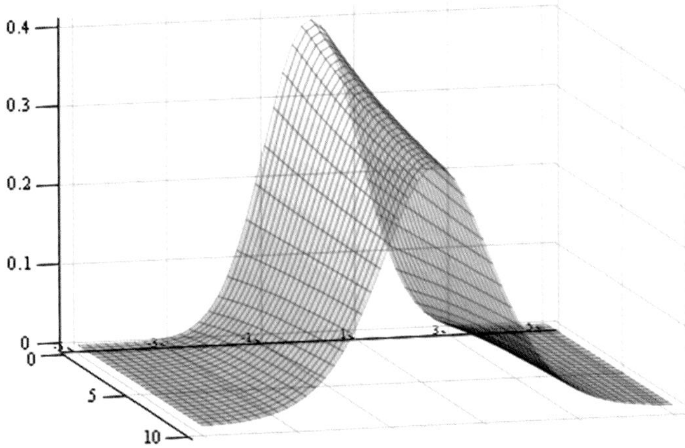

Fig. 21.2 The probability density $\varphi(t, x)$ of the wind velocity $X(t)$

problem from Sect. 21.3.3. This is a consequence of the selecting of compound Poisson process parameters so that wind velocities $X(t)$ defined by linear SDEs (Eqs. 21.19 and 21.29) have the zero means and the equal second-order moments.

21.5 Conclusions

In this chapter, the output processes analysis problem for nonlinear stochastic dynamical control systems is concerned. The results from [4] for diffusions (the spectral method for solving Fokker–Planck–Kolmogorov equation) are expanded to jump diffusions (the spectral method for solving Kolmogorov–Feller equation). A detailed description of proposed methods is supplemented by step-by-step algorithms for solving analysis problem and numerical experiments. The spectral method can also be used for other problems if mathematical model includes the partial differential equations [16, 24, 27, 28].

References

1. Solodovnikov, V.V., Semenov, V.V., Peschel, M., Nedo, D.: Design of Control Systems on Digital Computers: Spectral and Interpolational Methods. Mashinostroenie, Moscow (in Russian), Verlag Technik, Berlin (in German) (1979)
2. Semenov, V.V., Sotskova, I.L.: The spectral method for solving Fokker–Planck–Kolmogorov equation for stochastic control system analysis. In: 2nd IFAC Symposium on Stochastic Control, pp. 503–508. Pergamon Press, Oxford (1987)
3. Rybakov, K.A., Sotskova, I.L.: Spectral method for analysis of switching diffusions. IEEE Trans. Autom. Control **52**(7), 1320–1325 (2007)
4. Panteleev, A.V., Rybakov, K.A.: Analyzing nonlinear stochastic control systems in the class of generalized characteristic functions. Autom. Remote Control **72**(2), 393–404 (2011)
5. Rybin, V.V.: Modeling of Nonstationary Integer-Order and Fractional-Order Control Systems by Grid-Projection Spectral Method. MAI Publication, Moscow (in Russian) (2013)
6. Kozhevnikov, A.S., Rybakov, K.A.: Analysis of nonlinear stochastic systems with jumps generated by Erlang flow of events. Open J. Appl. Sci. **3**(1), 1–7 (2013)
7. Panteleev, A.V., Rybakov, K.A., Sotskova, I.L.: Spectral Method of Nonlinear Stochastic Control System Analysis. Vuzovskaya kniga, Moscow (in Russian) (2015)
8. Rybakov, K.A., Rybin, V.V.: Algorithms and software for calculating automated control systems in the spectral form of mathematical description. In: Modern Science: Theoretical, Practical and Innovative Aspects of Progress, vol. 2, pp. 171–199. Scientific Cooperation Publication, Rostov-on-Don (in Russian) (2018)
9. Sokolov, N.P.: Operations on multidimensional matrices. Sov. Math., Dokl. **6**, 1115–1118 (1965)
10. Hutson, V., Pym, J.S., Cloud, M.J.: Applications of Functional Analysis and Operator Theory. Elsevier, Amsterdam (2005)
11. Antosik, P., Mikusiński, J., Sikorski, R.: Theory of Distributions. The Sequential Approach. Elsevier Scientific, Amsterdam (1973)
12. Øksendal, B.: Stochastic Differential Equations. An Introduction with Applications. Springer, Berlin (2000)

13. Cerrai, S.: Second order PDE's in Finite and Infinite Dimension. A Probabilistic Approach. Springer, Berlin (2001)
14. Pugachev, V.S., Sinitsyn, I.N.: Stochastic Systems: Theory and Applications. World Scientific, Singapore (2002)
15. Risken, H.: The Fokker-Planck Equation: Methods of Solution and Applications. Springer, Berlin (1996)
16. Ladyženskaja, O.A., Solonnikov, V.A., Ural'ceva, N.N.: Linear and Quasi-Linear Equations of Parabolic Type. AMS, Providence (1968)
17. Cooke, R.G.: Infinite Matrices and Sequence Spaces. MacMillan & Co., London (1950)
18. Lindner, M.: Infinite Matrices and Their Finite Sections: An Introduction to the Limit Operator Method. Birkhäuser, Basel (2006)
19. Dryden, H.L.: A review of the statistical theory of turbulence. Quart. Appl. Math. **1**(1), 7–42 (1943)
20. Dobrolensky, Y.: Flight Dynamics in Turbulent Atmosphere. Mashinostroenie, Moscow (in Russian) (1969)
21. Flying qualities of piloted aircraft: MIL-HDBK-1797. U.S. Department of Defense, Washington (1997)
22. Kulikov, V.E.: Forming filter for differentiable turbulent wind simulation. In: Proceedings of the Moscow Institute of Electromechanics and Automation, vol. 7, pp. 36–42 (in Russian) (2013)
23. Bateman, H., Erdélyi, A.: Higher Transcendental Functions, vol. 2. McGraw-Hill Book Company, New York (1953)
24. Korn, G.A., Korn, T.M.: Mathematical Handbook for Scientists and Engineers. Dover Publication, New York (2000)
25. Øksendal, B., Sulem, A.: Applied Stochastic Control of Jump Diffusions. Springer, Cham (2005)
26. Hanson, F.B.: Applied Stochastic Processes and Control for Jump-Diffusions: Modeling, Analysis, and Computation. SIAM, Philadelphia (2007)
27. Baghdasaryan, G., Mikilyan, M.: Effects of Magnetoelastic Interactions in Conductive Plates and Shells. Springer, Cham (2016)
28. Baghdasaryan, G., Danoyan, Z.: Magnetoelastic Waves. Springer, Singapore (2018)

Chapter 22
Long-Period Lunar Perturbations in Earth Pole Oscillatory Process: Theory and Observations

Sergej S. Krylov, Vadim V. Perepelkin and Alexandra S. Filippova

Abstract In this chapter, the dynamic effects of the Earth pole motion in the celestial-mechanical problem statement as the "deformable Earth–Moon problem in the gravitational field of the Sun" are discussed. The orbits of the Moon and the Earth–Moon barycenter are assumed as known and given ones. Combination harmonics in the Earth pole motion are found and their connection with perturbations caused by the Moon's orbit precession is shown. Applying a numerical–analytical approach, the additional components of the Earth pole motion model were determined in an explicit form.

22.1 Introduction

Creating dynamic models of the Earth pole motion, which allow to determine its position on the Earth surface with high accuracy is fundamental in solving a number of problems in astrometry, navigation, and geophysics.

It is known [1] that the two main components of the Earth pole motion are two harmonics with periods of 365 and 433 days, respectively, with relatively slowly varying parameters. Also in the polar motion, a trend and high-frequency oscillations can be identified that have mostly irregular nature. Oscillations with periods of 365 and 433 days are mainly due to the orbital motion, deformability of the Earth's figure, and the influence of mobile geomedia, leading to variations in the Earth dynamic characteristics—its figure and the vector of the Earth's own angular momentum.

S. S. Krylov · V. V. Perepelkin (✉) · A. S. Filippova
Moscow Aviation Institute (National Research University), 4, Volokolamskoe Shosse, Moscow 125993, Russian Federation
e-mail: vadimkin1@yandex.ru

S. S. Krylov
e-mail: krylov@mai.ru

A. S. Filippova
e-mail: filippova.alex@gmail.com

L. C. Jain et al. (eds.), *Advances in Theory and Practice of Computational Mechanics*, Smart Innovation, Systems and Technologies 173,
https://doi.org/10.1007/978-981-15-2600-8_22

Considering the celestial-mechanical problem statement as a basis for building a more complex model of the Earth's rotational motion that includes mobile media, it is of a significant interest to take into account the influence of the perturbations from the Earth–Moon system spatial motion on the Earth pole oscillatory process.

High-precision astrometric measurements of the Earth Orientation Parameters (EOP) [2, 3], such as variations in the Earth pole coordinates, allow us to conclude that there are interrelated dynamic processes in the Earth–Moon–Sun system, the description and explanation of which in the scientific literature is mostly missing. For example, a transformation of the coordinate system can be found, depending on the annual Earth motion around the Sun and the Chandler wobble (associated with the Earth's internal structure and large-scale geophysical phenomena), that result in the detection of the Earth pole oscillatory process synchronous with the precession motion of the lunar orbit. This leads to the need for the development in the existing theory of the deformable Earth motion relative to its center of mass and for the construction of a refined mathematical model of EOP forecast. The latter can be implemented by an adequate choice of the model complexity and disturbing factors that are taken into account. The complexity of the model must correspond to the measurement accuracy and the duration of the data processing interval [4]. The latter is achieved by extensively analyzing the basic functions composition, their number and by adjusting of parameters. The justification of the developed model is carried out by performing numerical experiments using the least squares method and spectral correlation analysis.

The aim of this research is to study the effect of lunar–solar long-period disturbances on the Earth pole motion. A mathematical description of the Earth pole motion model and gravitational-tidal lunar–solar disturbances is given in Sect. 22.2. In Sect. 22.3, the gravitational-tidal mechanism of the Earth pole motion with a frequency close to the frequency of the lunar orbit precession is discussed. Section 22.4 concludes the chapter.

22.2 Mathematical Description of the Earth Pole Oscillatory Processes

In Sect. 22.2.1, a model of the pole motion is considered that is based on the dynamic Euler–Liouville equations and takes into account the gravitational-tidal disturbing moments and angular momentum of the atmosphere. In Sect. 22.2.2, the tidal lunar–solar potential is described. Variations in geopotential coefficients that appear due to the gravitational-tidal perturbations are discussed in Sect. 22.2.3.

22.2.1 Celestial-Mechanical Model of the Earth Pole Motion

Using the developed in [4] dynamic model of the Earth pole oscillations for a long-term forecast, it is possible to achieve high accuracy of motion approximation up to 10–12 years (for ~10^3–10^4 measurements quantity) and make a forecast within 6 years that correspond to observation data.

According to [4], the differential equations for the model of Earth pole motion can be obtained from the dynamic Euler–Liouville equations with a variable inertia tensor:

$$\frac{\mathrm{d}}{\mathrm{d}t}J\boldsymbol{\omega}+\boldsymbol{\omega}\times J\boldsymbol{\omega}=\mathbf{M},\quad \boldsymbol{\omega}=(p,q,r)^T,\quad J=J^*+\delta J,\quad J^*=\text{const},\tag{22.1}$$

$$J^*=\text{diag}(A^*,B^*,C^*),\quad \delta J=\delta J(t),\quad \|\delta J\|\ll\left\|J^*\right\|,\tag{22.2}$$

$$\mathbf{M}=\mathbf{M}^S+\mathbf{M}^L-\dot{\mathbf{h}}-\boldsymbol{\omega}\times\mathbf{h}.\tag{22.3}$$

Here, J is the matrix of the variable inertia tensor, $\boldsymbol{\omega}$ is the angular velocity vector in the Earthbound coordinate system [1, 4], which is qualitatively and quantitatively corresponds with the International Terrestrial Reference Frame (ITRF) [1]. The axes of the chosen coordinate system approximately coincide with the main central axes of inertia of the "frozen" Earth figure taking into account "equatorial bulge". It is assumed that small variations of the inertia tensor δJ include various harmonic components that resulted from the regular perturbation effects of gravitational diurnal tides from the Sun and the Moon and, possibly, from other perturbations, for example, annual, monthly, etc. $\mathbf{M}^S$ and $\mathbf{M}^L$ are the disturbing moments from the Sun and the Moon, respectively, and $\mathbf{h}$ is the vector of the angular momentum of the atmosphere.

Due to the smallness of $p,q\quad(p,q\ll r)$, in the first approximation in p, q from Eqs. 22.1–22.3 after averaging over the spin angle we obtain the system of equations:

$$\dot{p}+\frac{C-B}{A}rq=\frac{M_p}{A}-\frac{J_{qr}}{A}r^2,\tag{22.4}$$

$$\dot{q}-\frac{C-A}{B}rp=\frac{M_q}{B}+\frac{J_{pr}}{B}r^2,\tag{22.5}$$

$$\dot{r}+\frac{B-A}{C}pq=\frac{M_r}{C},\tag{22.6}$$

where M_p, M_q, M_r are the disturbing moments of forces, J_{pr}, J_{qr} are the Earth centrifugal moments of inertia. Here, it is assumed that

$$\begin{gathered}A=A^*(1+\alpha_p),\,B=B^*(1+\alpha_q),\,C=C^*(1+\alpha_r),\\ \delta A(\varphi_2)=A^*\alpha_p,\,\delta B(\varphi_2)=B^*\alpha_q,\,\delta C(\varphi_2)=C^*\alpha_r,\end{gathered}\tag{22.7}$$

where $\alpha_{p,q,r}$ describes daily tidal "bulge", φ_2 is the spin angle.

Then from Eqs. 22.4–22.6 with $r = r_0 = \text{const}$, we get the equations for the components p, q of the Earth pole motion:

$$\dot{p} + \frac{C^* - B^*}{A^*} rq = \left(-\frac{C^*}{A^*}\alpha_r + \frac{B^*}{A^*}\alpha_q + \frac{C^* - B^*}{A^*}\alpha_p \right) rq + \frac{-J_{qr} r^2 + M_p}{A^*}, \tag{22.8}$$

$$\dot{q} - \frac{C^* - A^*}{B^*} rp = \left(\frac{C^*}{B^*}\alpha_r - \frac{A^*}{B^*}\alpha_p - \frac{C^* - A^*}{B^*}\alpha_q \right) rp + \frac{J_{pr} r^2 + M_q}{B^*}. \tag{22.9}$$

Since the daily tidal bulges have a daily mean equal to zero, after averaging over φ_2 the equations take the form:

$$\begin{gathered} \dot{p} + N_p q = j^0_{qr} r^2 + \mu_p, \quad p(t_0) = p_0, \\ \dot{q} - N_q p = -j^0_{pr} r^2 + \mu_q, \quad q(t_0) = q_0, \\ N_p = \tfrac{C^* - B^*}{A^*} r_0, \quad N_q = \tfrac{C^* - A^*}{B^*} r_0, \quad r_0 = 7.29 \times 10^{-5}\ \text{rad/s}, \end{gathered} \tag{22.10}$$

where μ_p, μ_q are the perturbing torque–weight ratios of forces, $j^0_{pr} = -\langle J_{pr}/B^* \rangle_\varphi \neq 0$, $j^0_{qr} = -\langle J_{qr}/A^* \rangle_\varphi \neq 0$ are the tidal "bulges", and $N = \sqrt{N_p N_q} \cong 0.84 \div 0.85$ cycles per year is the Chandler frequency.

The achievement of a prediction high accuracy of the Earth pole motion is connected on the one hand with taking into account various disturbing factors, and, on the other hand, with the construction of a generalizing dynamic model that allows one to analyze subtle effects in the Earth pole oscillatory process on a qualitative level.

22.2.2 Lunar–Solar Perturbations in the Model of Earth Pole Motion

Lunar–solar gravitational-tidal forces have the potential, which is represented as a series of spherical functions [5]. For example, in this case, when the Moon is assumed as a gravitating point-like body or a sphere, the Moon tidal potential U_M can be represented using Eq. 22.11.

$$U_M = \sum_{n=2}^{\infty} U_{Mn} \tag{22.11}$$

The first term of the series (for $n = 2$) corresponds to zonal, tesseral, and sectorial lunar tides on the Earth's surface. The expression U_{M2} in the tidal potential series can be represented as:

$$U_{M2} = P_{20}(\cos\theta)a_{20}(t) + P_{21}(\cos\theta)(a_{21}(t)\cos\varphi + b_{21}(t)\sin\varphi) + P_{22}(\cos\theta)(a_{22}(t)\cos 2\varphi + b_{22}(t)\sin 2\varphi), \tag{22.12}$$

where $P_{2m}(\cos\theta)$ are the associated Legendre functions, θ and φ are the geographic coordinates of a point, r is the distance between a point and the Earth's center of mass. The coefficients $a_{2m}(t)$, $b_{2m}(t)$ depend on time and are determined by the Moon position relative to the Earth. Denoting θ_M, φ_M as the latitude and longitude of the Moon, respectively, the decomposition coefficients of Eq. 22.12 can be expressed in the following form:

$$\begin{aligned}
a_{20} &= \tfrac{1}{2} g\,\xi^3 \tfrac{r^2}{R_E} \tfrac{R_E}{\bar{R}_{EM}} \tfrac{m_M}{m_E} (3\cos^2\theta_M - 1),\\
a_{21} &= -3\tfrac{8\pi}{5} N_{21}^2\, g\,\xi^3 \tfrac{r^2}{R_E} \left(\tfrac{R_E}{\bar{R}_{EM}}\right)^3 \tfrac{m_M}{m_E} \cos\theta_M \sin\theta_M \cos\varphi_M,\\
b_{21} &= -3\tfrac{8\pi}{5} N_{21}^2\, g\,\xi^3 \tfrac{r^2}{R_E} \left(\tfrac{R_E}{\bar{R}_{EM}}\right)^3 \tfrac{m_M}{m_E} \cos\theta_M \sin\theta_M \sin\varphi_M,\\
a_{22} &= -3\tfrac{8\pi}{5} N_{22}^2\, g\,\xi^3 \tfrac{r^2}{R_E} \left(\tfrac{R_E}{\bar{R}_{EM}}\right)^3 \tfrac{m_M}{m_E} \sin^2\theta_M \cos 2\varphi_M,\\
b_{22} &= -3\tfrac{8\pi}{5} N_{22}^2\, g\,\xi^3 \tfrac{r^2}{R_E} \left(\tfrac{R_E}{\bar{R}_{EM}}\right)^3 \tfrac{m_M}{m_E} \sin^2\theta_M \sin 2\varphi_M,\\
\xi &= \tfrac{\bar{R}_{EM}}{R_{EM}}, \quad N_{2m} = (-1)^m \sqrt{\tfrac{5}{4\pi}\tfrac{(2-m)!}{(2+m)!}},
\end{aligned} \tag{22.13}$$

where $\bar{R}_{EM}$ is the average distance between the Earth's center of mass and the Moon, R_{EM} is the current distance between the Earth's center of mass and the Moon, R_E is the average Earth radius ($R_E \cong 6.38 \times 10^6$m), m_E is the mass of the Earth, m_M is the mass of the Moon, g is the gravitational acceleration.

The classical theory of the lunar motion for the problem "Earth–Moon system in the gravitational field of the Sun" allows one to take into account a number of dynamic effects of variations in the Earth axial rotation velocity including also the main inequalities of the Moon's motion [5, 6].

For the spatial version of the bounded three-body problem Earth–Moon–Sun, the perturbed motion equation for the lunar orbit node Ω_M and inclination of I of the lunar orbit plane to the ecliptic have the form [7]:

$$\dot{\Omega}_M = -\frac{3}{4}\frac{n_S^2}{n_M}[1 - \cos 2(l_M - \Omega_M) - \cos 2(l_S - \Omega_M) + \cos 2\lambda], \tag{22.14}$$

$$\dot{I} = -\frac{3}{4}\frac{n_S^2}{n_M} \sin I[\sin 2(l_S - \Omega_M) - \sin 2(l_M - \Omega_M) + \sin 2\lambda]. \tag{22.15}$$

Here, n_M, n_S are the sidereal mean motion of the Moon and the Sun, respectively, fluctuations of the angle I occur with the lunar orbit node period 18.61 years, l_M, l_S are the mean longitude of the Moon and the Sun, respectively, a_M is the semi-major axis of the Moon orbit, $(l_M - \Omega_M)$ is the angle between the Moon and the ascending node of the lunar orbit, and $\lambda \cong (n_M - n_S)t + \lambda_0$.

The tidal potential contains harmonics with different periods [1]. The main components (with the largest amplitudes) have periods of a year, half a year, 13.66 and 26.73 days, as well as, less significant 9.1 days and a third of the year. Along with the main frequencies of the Moon's motion, there is a stable harmonic with an argument of 2λ and the period of half of the synodic month, and a high frequency harmonic with an argument of $(2\lambda + M)$ and the period of 9.56 days, where M is the mean Moon anomaly, which is affected by changes in the mean longitude and perigee displacement. The lunar inequality associated with the argument $(2\lambda - M)$ is an eviction with a period of 31.81 days. The presence of these components is due to the perturbation corresponding to the second zonal harmonic of the tidal potential.

A quasiperiodic lunar perturbation (the precessional motion of the lunar orbit and small variations in the inclination of its plane) corresponding to Eqs. 22.14–22.15 is included in the additional tidal potential.

The tidal potential U_{M2} is expressed by the sum of harmonic terms with combination frequencies [7]:

$$\begin{aligned} U_{M2} = &-\frac{1}{4}\kappa g\frac{r^2}{R_E}(1-3\cos^2\theta)\sum_i^{n_0} A_i\cos(\nu_i t+\psi_i^0) \\ &-\frac{1}{2}\kappa g\frac{r^2}{R_E}\sin 2\theta\sum_i^{n_1} B_i\cos(\nu_i t+\nu_\varphi t+\psi_i^1) \\ &-\frac{1}{2}\kappa g\frac{r^2}{R_E}\sin^2\theta\sum_i^{n_2} C_i\cos(\nu_i t+2\nu_\varphi t+\psi_i^2), \\ \kappa = &\frac{3}{2}\frac{m_M}{m_E}\left(\frac{R_E}{R_{EM}}\right)^3 = 0.843\times10^{-7},\quad 0\le r<R_E. \end{aligned} \tag{22.16}$$

Here, t is the Greenwich mean solar time, n_0, n_1, n_2 are the number of zonal, tesseral, and sectorial harmonics, respectively, and coefficients A_i, B_i, C_i are their amplitudes. The values of $\nu_i = a_i\dot{l}_M + b_i\dot{l}_S + c_i\dot{p}_M + d_i\dot{p}_S + e_i\dot{\Omega}_M$ are linear combinations of the angular parameters $l_{M,S}$, $p_{M,S}$, Ω_M derivations with integer coefficients. The parameters l_M and l_S are the mean longitudes of the Moon and the Sun, respectively, with periods of 27.55 and 365.25 sidereal days. The value of p_M is the mean longitude of the Moon perigee varying with a period of 8.85 years, and p_S is the mean longitude of the Sun perigee changing with a period of 25,700 years. The parameter Ω_M determines the longitude of the Moon ascending node and it varies with a period of 18.61 years.

22.2.3 Variations of the Geopotential Coefficients

Due to the presence of perturbations of the tidal potential for the figure of the Earth, an additional perturbing potential $\delta W(t)$ arises, which depends on time. With the expansion of the potential δW, the largest term in terms of magnitude is the perturbation from the second harmonic δW_2:

$$\delta W_2 = \frac{f m_E R_E^2}{r^3} \Delta Y_2(\theta, \varphi), \tag{22.17}$$

where $\Delta Y_2(\theta, \varphi)$ is the change in the normalized spherical function, and is expressed in terms of the coefficients of the second order of the geopotential series [8]:

$$\begin{aligned} \Delta Y_2(\theta, \varphi) &= \delta c_{20} P_{20}(\cos\theta) + [\delta c_{21} \cos\varphi + \delta s_{21} \sin\varphi] P_{21}(\cos\theta) \\ &\quad + [\delta c_{22} \cos 2\varphi + \delta s_{22} \sin 2\varphi] P_{22}(\cos\theta). \end{aligned} \tag{22.18}$$

The variations of the geopotential coefficients caused by the perturbations from the Sun and the Moon are obtained from Eq. 22.16 and have the following form:

$$\begin{gathered} \delta c_{20} = \varkappa \sum_{j=1}^{n_0} a_{0j} \cos\Theta_{0j}, \quad \delta c_{21} = \varkappa \sum_{j=1}^{n_1} a_{1j} \cos\Theta_{1j}, \\ \delta s_{21} = \varkappa \sum_{j=1}^{n_1} a_{1j} \sin\Theta_{1j}, \quad \delta c_{22} = \varkappa \sum_{j=1}^{n_2} a_{2j} \cos\Theta_{2j}, \\ \delta s_{22} = \varkappa \sum_{j=1}^{n_2} a_{2j} \sin\Theta_{2j}, \\ \varkappa = \chi \zeta g \left(\frac{R_E}{\overline{R}_{EM}} \right)^3 \frac{R_E^2 m_M}{f m_E}, \end{gathered} \tag{22.19}$$

where $\Theta_{ij} = \{\Theta_{ij}^M, \Theta_{ij}^S\}$ are linear combinations of the angles τ_0, $l_{M,S}$, $p_{M,S}$, Ω_M due to the Moon and the Sun perturbations, respectively.

The harmonics $\cos(\Theta_{ij})$, when i equals to 0, 1, and 2, correspond to zonal, tesseral, and sectorial tides, respectively. The value of $\tau_0 = t - l_M - l_S$ is the Greenwich mean lunar time, ζ is the ratio of the vertical tidal displacement of the Earth's surface to the displacement of the equipotential surface of the tidal potential, the coefficient χ equals to 1 for a model of the homogeneous Earth and 0.843 for a real Earth.

The variations of the coefficients of the second zonal harmonic of the geopotential can be expressed in terms of the variations of the inertia tensor components

$$\delta J = \begin{pmatrix} \delta A & -\delta J_{pq} & -\delta J_{pr} \\ -\delta J_{pq} & \delta B & -\delta J_{qr} \\ -\delta J_{pr} & -\delta J_{qr} & \delta C \end{pmatrix} \tag{22.20}$$

in the following way:

$$\delta c_{20} = \frac{\delta A + \delta B - 2\delta C}{2m_E R_E^2}, \quad \delta c_{21} = \frac{\delta J_{pr}}{m_E R_E^2}, \quad \delta s_{21} = \frac{\delta J_{qr}}{m_E R_E^2},$$
$$\delta c_{22} = \frac{\delta B - \delta A}{4m_E R_E^2}, \quad \delta s_{22} = \frac{\delta J_{pq}}{2m_E R_E^2}. \tag{22.21}$$

To further describe the variations of the inertia tensor and geopotential coefficients, it is convenient to adopt the following notation in the expansion of the inertia tensor δJ_{ij}:

$$\delta J_{ij} = \delta J_{ij}^{(t)} + \delta J_{ij}^{(\varphi)} + \delta J_{ij}^{(2\varphi)},$$

where $\delta J_{ij}^{(t)}$ are the intra-annual and interannual variations, $\delta J_{ij}^{(\varphi)}$, $\delta J_{ij}^{(2\varphi)}$ are daily and semidiurnal variations, respectively.

Intra-day variations $\delta J_{ij}^{(\varphi)}$, $\delta J_{ij}^{(2\varphi)}$ contain oscillation components with combinational frequencies of the ν_i spatial variant of the problem “deformable Earth–Moon in the gravitational field of the Sun”.

22.3 Gravitational and Tidal Perturbations in the Model of the Earth Pole Motion

In this section, expressions for lunar–solar gravitational-tidal moments are given in Sect. 22.3.1. In Sect. 22.3.2, more subtle effects of the disturbed oscillatory process of the Earth pole are considered in the canonical action-angle variables.

22.3.1 Gravitational-Tidal Lunar–Solar Moment of Forces

Considering the gravitational-tidal perturbations from the Moon and the Sun, we consider the moments of gravitational forces structure that are included in the right-hand side of Eqs. 22.1–22.3. The expressions for the components of the moments of gravitational forces, for example, from the Sun, were obtained in [4, 9–11] and have the following form:

$$
\begin{aligned}
M_p^S &= 3\omega^2[(C^* + \delta C - (B^* + \delta B))\gamma_q\gamma_r + \delta J_{qr}(\gamma_q^2 - \gamma_r^2) + \delta J_{pq}\gamma_p\gamma_r - \delta J_{pr}\gamma_p\gamma_q],\\
M_q^S &= 3\omega^2[(A^* + \delta A - (C^* + \delta C))\gamma_p\gamma_r + \delta J_{pr}(\gamma_r^2 - \gamma_p^2) + \delta J_{qr}\gamma_p\gamma_q - \delta J_{pq}\gamma_q\gamma_r],\\
M_r^S &= 3\omega^2[(B^* + \delta B - (A^* + \delta A))\gamma_p\gamma_q + \delta J_{pq}(\gamma_p^2 - \gamma_q^2) + \delta J_{qr}\gamma_p\gamma_r - \delta J_{pr}\gamma_q\gamma_r],\\
&\omega = \omega_*(1 + e_S \cos\nu_S)^{3/2},\\
&\gamma_p = \sin\theta_S \sin\varphi_S, \quad \gamma_q = \sin\theta_S\cos\varphi_S, \quad \gamma_r = \cos\theta_S,
\end{aligned}
\tag{22.22}
$$

where ω_* is a constant determined by gravitational and focal parameters, ν_S is a true anomaly of the Earth when moving in an elliptical orbit with eccentricity e_S, γ_p, γ_q, γ_r are the direction cosines of the radius vector in the bound coordinate system, $\psi_S, \theta_S, \varphi_S$ are Euler angles defining the orientation of that coordinate system relative to the orbital one, where z axis is directed to the attracting center, namely, the Sun, A^*, B^*, C^* are the effective main central moments of inertia with regard to the "frozen" Earth deformations. They can be calculated with sufficient accuracy. The coefficients $\delta A, \delta B, \delta C, \delta J_{pq}, \delta J_{qr}$, and δJ_{pr} are due to tidal diurnal and semidiurnal gravitational influences of the Moon and the Sun.

After averaging over the fast variable φ_S (φ_S is the angle of Earth's spin) for $M_{p,q,r}^S$ simple expressions are obtained:

$$
\begin{aligned}
M_p^S &= 3\omega_0^2\Big[\chi_{1p}^S \sin^2\theta_S + \chi_{2p}^S \sin\theta_S\cos\theta_S - \langle\delta J_{qr}\rangle_{\varphi_S}\cos 2\theta_S\Big],\\
M_q^S &= 3\omega_0^2\Big[\chi_{1q}^S \sin^2\theta_S + \chi_{2q}^S \sin\theta_S\cos\theta_S + \langle\delta J_{pr}\rangle_{\varphi_S}\cos 2\theta_S\Big],\\
M_r^S &= 3\omega_0^2\big[\chi_{1r}^S \sin^2\theta_S + \chi_{2r}^S \sin\theta_S\cos\theta_S\big].
\end{aligned}
\tag{22.23}
$$

The values of $\chi_{1p,1q,1r}^S$, $\chi_{2p,2q,2r}^S$ are the tidal coefficients due to semidiurnal and diurnal tides. They are obtained by averaging coefficients at $\sin^2\theta_S$ and $\sin\theta_S\cos\theta_S$ in the components of the gravitational moment from the Sun by φ_S:

$$
\begin{aligned}
\chi_{1p}^S &= -\langle\delta J_{qr}\sin^2\varphi_S\rangle_{\varphi_S} - \tfrac{1}{2}\langle\delta J_{pr}\sin 2\varphi_S\rangle_{\varphi_S},\\
\chi_{2p}^S &= \tfrac{1}{2}\langle(\delta C - \delta B)\cos\varphi_S\rangle_{\varphi_S} + \langle\delta J_{pq}\sin\varphi_S\rangle_{\varphi_S},\\
\chi_{1q}^S &= \langle\delta J_{pr}\cos^2\varphi_S\rangle_{\varphi_S} + \tfrac{1}{2}\langle\delta J_{qr}\sin 2\varphi_S\rangle_{\varphi_S},\\
\chi_{2q}^S &= \tfrac{1}{2}\langle(\delta A - \delta C)\sin\varphi_S\rangle_{\varphi_S} - \langle\delta J_{pq}\cos\varphi_S\rangle_{\varphi_S},\\
\chi_{1r}^S &= \tfrac{1}{2}\langle(\delta B - \delta A)\sin 2\varphi_S\rangle_{\varphi_S} - \langle\delta J_{pq}\cos 2\varphi_S\rangle_{\varphi_S},\\
\chi_{1r}^S &= \tfrac{1}{2}\langle\delta J_{qr}\sin\varphi_S\rangle_{\varphi_S} - \langle\delta J_{pr}\cos\varphi_S\rangle_{\varphi_S}.
\end{aligned}
\tag{22.24}
$$

The values of the coefficients $\chi_{1p,1q,1r}^S$, $\chi_{2p,2q,2r}^S$ are small values that can be determined based on observational data.

The expressions of the direction cosines of the Sun's radius vector are written using Euler kinematic equations, which specify the orientation of the related axes relative to the orbital coordinate system [4]:

$$
p = \dot{\psi}_S \sin\theta_S \sin\varphi_S + \dot{\theta}_S\cos\varphi_S + \omega_0(\nu_S)(\sin\psi_S\cos\varphi_S
$$

$$
\begin{aligned}
&+\cos\psi_S \sin\varphi_S \cos\theta_S),\\
q &= \dot{\psi}_S \sin\theta_S \cos(\varphi_S) - \dot{\theta}_S \sin\varphi_S + \omega_0(\nu_S)(-\sin\psi_S \sin\varphi_S +\\
&+\cos\psi_S \cos\varphi_S \cos\theta_S),\\
r &= \dot{\varphi}_S + \dot{\psi}_S \cos\theta_S - \omega_0(\nu_S)\cos\psi_S \sin\theta_S,\\
\omega_0(\nu_S) &= \dot{\nu}_S = \omega_*(1 + e_S \cos\nu_S)^2,
\end{aligned}
\tag{22.25}
$$

where $e_S = 0.0167$ is the orbit eccentricity, $\nu_S(t)$ is the true anomaly. Conducting some transformations in Eq. 22.23 and integrating them in the first approximation, we obtain the following expressions:

$$
\begin{aligned}
r = r^0, \quad \varphi_S \approx r^0 t + \varphi^0, \quad \nu_S = \omega_* t + \nu_S^0, \quad \cos\theta_S(\nu_S) = a\cos\nu_S,\\
\sin\theta_S \cos\theta_S = b\cos\nu_S + d\cos 3\nu_S + \ldots,
\end{aligned}
\tag{22.26}
$$

where the initial values are $\theta_S^0 = 66^\circ 33'$, $\psi_S^0 \in [0, 2\pi]$, coefficients for cos ν_S are in the intervals $a \in [0.4, 1]$, $b \in \left[0.4, \frac{4}{3\pi}\right]$ and depend on the initial values θ_S^0, ψ_S^0, and the coefficient d is much less than one [4].

In a simplified version of the problem for the stationary lunar orbit (Keplerian orbit) the disturbing moment from the Moon has a structure similar to Eq. 22.22. The perturbing moment from the Moon $\mathbf{M}^L$ leads to small-scale tidal changes in the speed of the Earth axial rotation and the position of the Earth pole at relatively short time intervals. In particular, in order to select a component with a period of 9.1 days in the expansion of the lunar moment $\mathbf{M}^L$ it is necessary to keep the third harmonic in the expression $\sin\theta_M \cos\theta_M$, i.e.,

$$
\sin\theta_M \cos\theta_M = b(\theta_M^0, \psi_M^0)\cos\nu_M + d\cos 3\nu_M + \ldots,
$$

where ν_M is true anomaly of the Moon.

In the general case, when the Moon orbit performs a known precessional-nutation motion, the perturbing moment from the Moon is determined by the expression:

$$
\mathbf{M}^L = -m_M \mathbf{R}_{EM} \times \nabla W. \tag{22.27}
$$

Here, $\mathbf{R}_{EM}$ is the radius is the vector of the center of mass of the Moon relative to the Earth's center of mass, W is the Earth external gravitational potential. The decomposition of a geopotential in a series of spherical functions has the form [8]:

$$
W = \frac{f m_E}{R_E} \sum_{n=0}^{\infty} \sum_{m=0}^{n} \left(\frac{R_E}{r}\right)^{(n+1)} P_{nm}(\cos\theta_M)(c_{nm}\cos m\varphi_M + s_{nm}\sin m\varphi_M).
\tag{22.28}
$$

As noted above, the largest harmonic in the geopotential decomposition is the second one. Then the expressions for the components of the gravitational-tidal moment take the form:

$$
\begin{aligned}
M_p^L &= \frac{3}{2}\frac{f m_E m_M R_E^2}{r^3}\big(-c_{20}\sin 2\theta_M \sin\varphi_M + s_{21}(3\cos^2\theta_M - 1)\\
&\quad + \sin^2\theta_M[s_{21}\cos 2\varphi_M - c_{21}\sin 2\varphi_M]\\
&\quad + 2\sin 2\theta_M[s_{22}\cos\varphi_M - c_{22}\sin 2\varphi_M]\big),\\
M_q^L &= \frac{3}{2}\frac{f m_E m_M R_E^2}{r^3}\big(c_{20}\sin 2\theta_M \cos\varphi_M - c_{21}(3\cos^2\theta_M - 1)\\
&\quad + \sin^2\theta_M[c_{21}\cos 2\varphi_M + s_{21}\sin 2\varphi_M]\\
&\quad -2\sin 2\theta_M[c_{22}\cos\varphi_M + s_{22}\sin 2\varphi_M]\big),\\
M_r^L &= \frac{3}{2}\frac{f m_E m_M R_E^2}{r^3}(\sin 2\theta_M[c_{21}\sin\varphi_M - s_{21}\cos\varphi_M]\\
&\quad + 4\sin^2\theta_M[c_{22}\sin 2\varphi_M - s_{22}\cos 2\varphi_M]\big).
\end{aligned}
\tag{22.29}
$$

Coefficients of the geopotential c_{2m}, s_{2m} consist of constant coefficients c_{2m}^*, s_{2m}^* and periodic tidal variations of δc_{2m}, δs_{2m}. After multiplying the tidal components in $M_{p,q,r}^L$, the expressions containing long-period terms and terms resulting in diurnal librations, are obtained.

Expressions of the components of the $\delta M_{p,q,r}^L$ additional gravitational-tidal moment from the Moon are

$$
\begin{aligned}
\delta M_p^L &= -2\delta c_{20}\sum_{j=1}^{n1} b_{0j}^M \sin\Theta_{1j}^M + \delta s_{21}\sum_{j=1}^{n0} b_{0j}^M\cos\Theta_{0j}^M + \delta s_{21}\sum_{j=1}^{n2} b_{2j}^M\cos\Theta_{2j}^M\\
&\quad - \delta c_{21}\sum_{j=1}^{n2} b_{2j}^M\sin\Theta_{2j}^M - 4\delta c_{22}\sum_{j=1}^{n1} b_{1j}^M\sin\Theta_{1j}^M + 4\delta s_{22}\sum_{j=1}^{n1} b_{1j}^M\cos\Theta_{1j}^M,\\
\delta M_q^L &= 2\delta c_{20}\sum_{j=1}^{n1} b_{0j}^M\cos\Theta_{1j}^M - \delta c_{21}\sum_{j=1}^{n0} b_{0j}^M\cos\Theta_{0j}^M + \delta s_{21}\sum_{j=1}^{n2} b_{2j}^M\sin\Theta_{2j}^M\\
&\quad + \delta c_{21}\sum_{j=1}^{n2} b_{2j}^M\cos\Theta_{2j}^M - 4\delta c_{22}\sum_{j=1}^{n1} b_{1j}^M\cos\Theta_{1j}^M - 4\delta s_{22}\sum_{j=1}^{n1} b_{1j}^M\sin\Theta_{1j}^M,\\
\delta M_r^L &= \delta c_{21}\sum_{j=1}^{n1} b_{1j}^M\sin\Theta_{1j}^M + \delta s_{21}\sum_{j=1}^{n1} b_{1j}^M\cos\Theta_{1j}^M\\
&\quad + \delta c_{22}\sum_{j=1}^{n2} b_{2j}^M\sin\Theta_{2j}^M + \delta s_{21}\sum_{j=1}^{n2} b_{2j}^M\cos\Theta_{2j}^M,
\end{aligned}
\tag{22.30}
$$

which can be decomposed into the sum of the combinational harmonics of the spatial variant of the problem “deformable Earth–Moon in the gravitational field of the Sun”.

Now in the expressions of the gravitational moment from the Moon (Eq. 22.30) let us take into account variations of the geopotential decomposition coefficients caused only by solar disturbances, and vice versa: in expressions of the gravitational moment from the Sun—the variations of the geopotential decomposition coefficients caused only by lunar disturbances. For example, substituting the variations δc_{ij}^{M} caused by the Moon influence in the expressions of additional moment of gravitational forces from the Moon will result in $\delta M_{p,q,r}^{L} = 0$ for the elastic Earth model, and for real Earth $\delta M_{p,q,r}^{L}$ will be a small value.

The total lunar–solar additional gravitational-tidal moment, for example, for $\delta M_{q}^{SL} = \delta M_{q}^{S} + \delta M_{q}^{L}$ will be

$$\begin{aligned}\delta M_{q}^{SL} &= 2\delta c_{20}^{S}\sum_{j=1}^{n1} b_{0j}^{M}\cos\Theta_{1j}^{M} - \delta c_{21}^{S}\sum_{j=1}^{n0} b_{0j}^{M}\cos\Theta_{0j}^{M} + 2\delta c_{20}^{M}\sum_{j=1}^{n1} b_{0j}^{S}\cos\Theta_{1j}^{S}\\ &\quad - \delta c_{21}^{M}\sum_{j=1}^{n0} b_{0j}^{S}\cos\Theta_{0j}^{S} + \delta s_{21}^{S}\sum_{j=1}^{n2} b_{2j}^{M}\sin\Theta_{2j}^{M} + \delta c_{21}^{S}\sum_{j=1}^{n2} b_{2j}^{M}\cos\Theta_{2j}^{M}\\ &\quad + \delta s_{21}^{M}\sum_{j=1}^{n2} b_{2j}^{S}\sin\Theta_{2j}^{S} + \delta c_{21}^{M}\sum_{j=1}^{n2} b_{2j}^{S}\cos\Theta_{2j}^{S} - 4\delta c_{22}^{S}\sum_{j=1}^{n1} b_{1j}^{M}\cos\Theta_{1j}^{M}\\ &\quad - 4\delta s_{22}^{S}\sum_{j=1}^{n1} b_{1j}^{M}\sin\Theta_{1j}^{M} - 4\delta c_{22}^{M}\sum_{j=1}^{n1} b_{1j}^{S}\cos\Theta_{1j}^{S} - 4\delta s_{22}^{M}\sum_{j=1}^{n1} b_{1j}^{S}\sin\Theta_{1j}^{S},\end{aligned} \tag{22.31}$$

and in the case of the elastic Earth $\delta M_{p,q,r}^{SL} = 0$.

Taking into account the mantle viscosity the tidal bulge from the attracting body in its diurnal motion will be "ahead" of the tidal bulge phase for the absolutely elastic Earth model, which will lead not to the zero value, but to the negative one for the additional gravitational-tidal moment, and to the secular slowing down of the Earth's rotation [12, 13]. In this case, the total moment (Eq. 22.31) will be nonzero. The latter will lead to small-scale librations with combination harmonics. Among these harmonics those associated with the spatial motion of the Earth–Moon system can be distinguished.

22.3.2 The Oscillatory Process of the Earth Pole at the Frequency of the Moon's Orbit Precession

The actual problem of studying the irregular behavior of the main components of the Earth pole oscillations is currently poorly understood. Of a significant interest, there are studies aimed at establishing the geophysical and celestial-mechanical reasons for such behavior of the Chandler and 1-year components, and the construction of refined prediction models for EOP required for solving high-precision satellite navigation problems [8].

In modern works, for example, [14, 15] the problem of the how astronomical factors influence on the planetary scaled geophysical processes and fluctuations of EOP is considered. In particular, the spectral analysis results of EOP measurement data show the presence of harmonics with frequencies close to those of the Earth–Moon system motion. It should be noted that astronomical factors affecting geophysical processes can be identified within the framework of a generalizing celestial-mechanical model of the Earth motion relative to its center of mass.

Now let us assume that the perturbing factors are caused by the deformations of the viscoelastic mantle (the outer deformable layer) due to gravitational perturbations caused by the influence of the Moon and the Sun. The main goal of constructing a model of the Earth pole perturbed motion is to identify the parameters Chandler wobble and to predict its trajectory. It is of interest to study the motion at various time intervals available in observations from intra-year intervals to the period of pole beats.

The perturbed Chandler wobble dynamics of the instantaneous axis is associated, in particular, with a change in the angle δ_2 [4], which determines the change in the amplitude of Chandler wobble. The angular variable δ_2 is the angle between the Earth figure axis and the angular momentum vector.

For a more detailed analysis of the perturbed Chandler wobble of the asymmetric Earth pole, it is convenient to use the canonical action-angle variables [4]. The perturbed Routh functional of the problem in question can be represented using Eq. 22.32 as in [3].

$$R = R_0 + \varepsilon R_1(I_1, I_2, I_3, w_1, w_2, w_3, \mathbf{u}, \dot{\mathbf{u}}) + \varepsilon^2 \ldots \tag{22.32}$$

Here, εR_1 is the perturbing functional due to gravitational tides, $\mathbf{u}$ is the displacement vector of the viscoelastic medium points in the Earth's mantle, I_1, I_2, I_3 are the action variables (I_2 is the value of the deformable Earth angular momentum, and I_1, I_3 are its projections on the axis close to the figure axis of the terrestrial and celestial geocentric coordinate systems, respectively), and angles w_1, w_2, w_3 are the canonically conjugate variables corresponding to the phase of the Earth pole motion, the Earth's own rotation angle, and the precession angle, respectively, $\varepsilon > 0$ is a small dimensionless parameter characterizing the relative magnitude of the disturbing factors in Eq. 22.32.

It can be shown [3] that the equation for δ_2 in the action-angle variables will be

$$\dot{\delta}_2 = \varepsilon \frac{1}{I_2}\left[-\frac{1}{\sin\delta_2} \frac{1+\kappa^2 sn^2(w_1,\lambda)}{\kappa_* dn(w_1,\lambda)} \frac{\partial R_1}{\partial w_1} - \delta_2 \frac{\partial R_1}{\partial w_2} \right]. \tag{22.33}$$

Here, I_2 is the magnitude of the Earth own angular momentum, the angular variable w_1 determines the phase of the Earth pole motion, so that $\dot{w}_1$ has the meaning of the Earth pole instant frequency turning around the "mean pole" (the "mean pole" determines the long-period motion of the Earth pole), the phase angle w_2 is associated with the Earth axial rotation, sn, dn are Jacobi elliptic functions, $\lambda \sim 10^{-2}$ is a small parameter, the values of κ, κ_* are the determined by the inertia moments of the Earth:

$$\kappa^2 = \frac{C^*(A^* - B^*)}{A^*(B^* - C^*)}, \quad \kappa_* = \sqrt{1 + \kappa^2}.$$

From Eq. 22.32, we obtain the equation for δ_2:

$$\dot{\delta}_2^2 \approx 2\left(\frac{\lambda}{\kappa}\right)^2 \left[\frac{2 + \kappa^2}{I_2 \kappa_*} \sin 2w_1 + \frac{4}{I_2} \cos 2w_1\right] \cdot \big[a_1(\delta_1, i, \sigma_*) \cos \Omega_M + a_2^c(w_3, \delta_1, i, \sigma_*) \cos 2\Omega_M + a_2^s(w_3, \delta_1, i, \sigma_*) \sin 2\Omega_M\big], \tag{22.34}$$

where a_1, a_2^s, a_2^c are unknown coefficients determined by the parameters of the Earth deformable layer, the values of which can be estimated from EOP astrometric measurements data, δ_1 is the angle between the angular momentum vector and the axis of the geocentric celestial coordinate system that is orthogonal to the ecliptic (its change is related to the precession and nutation of the Earth), w_3 is the Earth precession angle, i is the angle of inclination of the Moon's orbit plane to the ecliptic, Ω_M is the longitude of the ascending node of the Moon's orbit, that is defined by Eq. 22.14, σ_* is a small dissipation coefficient of the viscoelastic layer of the Earth's mantle.

Equation 22.34 leads to the presence in the oscillatory process of the Earth pole of combinational harmonics associated with the precession of the Moon's orbit. Let us denote $a(t)$ as the variable amplitude of the pole motion, which varies with a period of 6.45 years. Then the equations of motion of the Earth pole in the first approximation in σ_* can be represented as

$$\begin{aligned} x_p &= a(t) \cos w_1 + \delta_2 \cos w_1, \\ y_p &= a(t) \sin w_1 + \delta_2 \sin w_1. \end{aligned} \tag{22.35}$$

The additional terms in Eq. 22.35 are the Chandler wobble modulated by harmonics with the frequency of the precession of the lunar orbit.

Now, having the obtained decomposition of the gravitational-tidal moment, we will seek a solution to Eq. 22.10 in the form of Eq. 22.35 with known perturbations of μ_p, μ_q. To do this, they can be determined using the measurement data from International Earth Rotation and Reference Systems Service (IERS) of the trajectory of the Earth pole from Eq. 22.10. Putting them into Fourier series, we leave only the considered combination harmonics $w_1 + \Omega_M$, $w_1 - \Omega_M$ and define the corresponding pole oscillations in an explicit form. In Fig. 22.1, the additional components of model (Eq. 22.35) are shown.

Next, perform the transformation of the coordinates of the Earth pole:

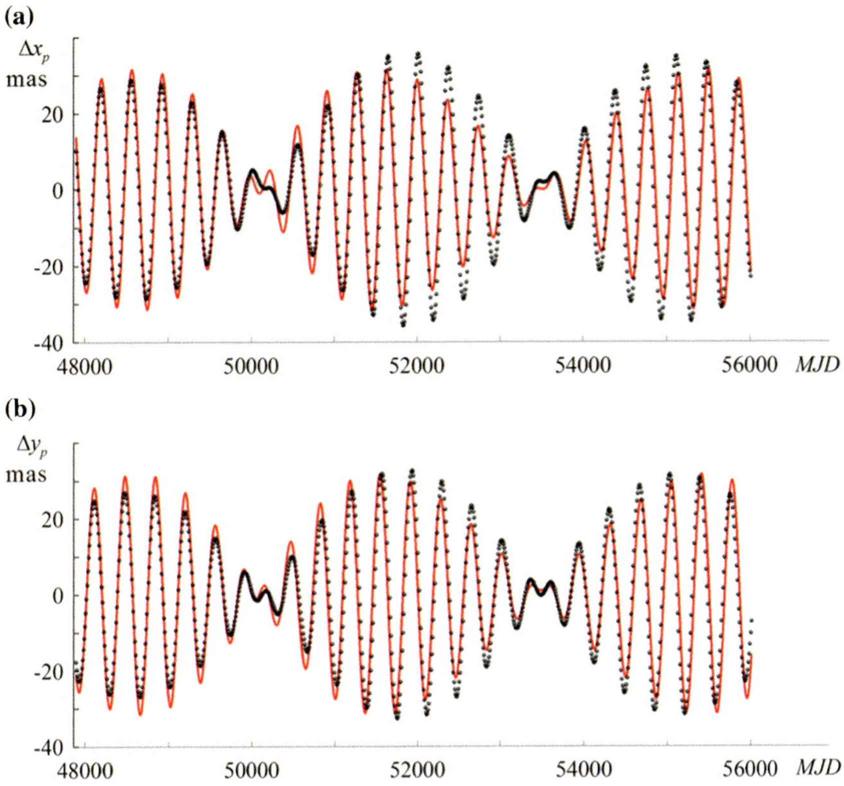

Fig. 22.1 Comparison of Δx_p, Δy_p obtained by processing IERS observations data (discrete points) with a curve obtained using lunar ephemeris: **a** Δx_p values, **b** Δy_p values

$$\delta\varphi = \arctan\left(\frac{\eta_p}{\xi_p + \sqrt{\xi_p^2 + \eta_p^2}}\right), \quad \begin{pmatrix} \xi_p \\ \eta_p \end{pmatrix} = \Pi(w_h - w_1)\,\Pi(w_1)\begin{pmatrix} \Delta x_p \\ \Delta y_p \end{pmatrix}. \tag{22.36}$$

The following notation is introduced into Eq. 22.36: Π is the plane rotation matrix, a_0 is the average value of the amplitude of the pole oscillations around the "mean pole" (that is without the trend component), c_x, c_y are the set the position of the "mean pole" and contain constants, secular terms and variations with periods longer than 6 years, $\dot{w}_h - \dot{w}_1 = \nu_T$ is the frequency of the 6-year amplitude modulation of the Earth pole oscillatory motion, Δx_p, Δy_p are the second terms on the right-hand side of Eq. 22.35 for x and y, respectively.

In Fig. 22.2, the polar angle $\delta\varphi$ is compared with the oscillations of the intersection point between the equator and the lunar orbit. The units of the oscillation amplitudes are radians, and τ is the time in standard years. In new coordinate system, it is possible to illustrate synchronous oscillations of the Earth pole with the precessional motion

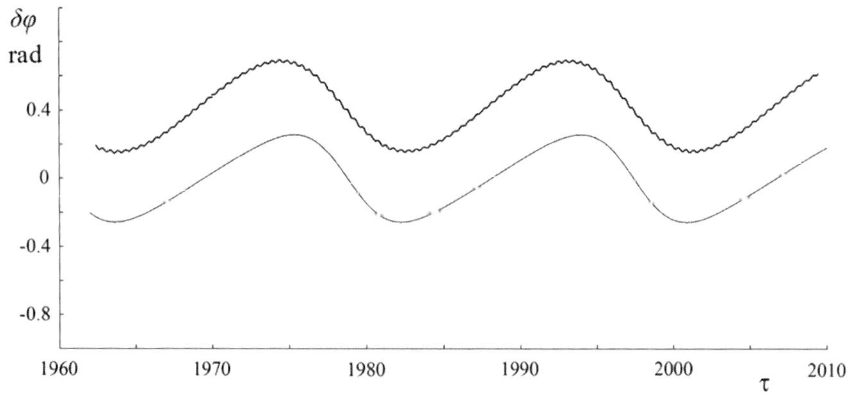

Fig. 22.2 Comparison of the polar angle variations $\delta\varphi$ (bottom line) with the oscillations along the equator of the point of intersection of the lunar orbit and the equator (upper line), constructed using the lunar ephemeris

of the lunar orbit and to determine the regular component of $\delta\varphi$ phase variation, which makes it possible to use lunar ephemeris when predicting additional terms in the model of Earth pole motion.

To do this, making the inverse transformation

$$\begin{pmatrix} \Delta x_p \\ \Delta y_p \end{pmatrix} = \sqrt{\xi_p^2 + \eta_p^2}\, \Pi^{-1}(w_1)\, \Pi^{-1}(w_h - w_1) \begin{pmatrix} \cos\delta\varphi - 1 \\ \sin\delta\varphi \end{pmatrix}, \tag{22.37}$$

we get additional terms in the oscillations of the Earth pole leading to amplitude modulation of its main motion. In Fig. 22.1, the calculated curves are compared with those extracted from the data from IERS. In this case, expressions depending on lunar ephemeris are used in the obtained numerical-analytical model.

22.4 Conclusions

The developed celestial-mechanical model that takes into account the gravitational-tidal lunar–solar perturbations allows to assume the presence of a specific oscillatory process in the Earth pole motion. The perturbation from the Moon leads to the additional combinational harmonics modulated by a harmonic at a frequency close to the lunar orbit precession frequency, and the perturbation from the Sun makes this process nonstationary. More precisely, it can be said that the process will be quasi-stationary until the ratio of the amplitudes of the Chandler and annual components goes through 1 (becomes more than one or, on the contrary, less). Using the data processing of IERS observations over the Earth pole trajectory, a method is proposed for identifying the discovered oscillations.

Acknowledgements This work was carried out within the state task no. 9.7555.2017/BCh.

References

1. International Earth Rotation and Reference Systems Service—IERS Annual Reports, http://www.iers.org. Last accessed 09 Oct 2019
2. Perepelkin, V.V., Rykhlova, L.V., Filippova, A.S.: Long-period variations in oscillations of the Earth's pole due to lunar perturbations. Astron. Rep. **63**(3), 238–247 (2019)
3. Markov, YuG, Perepelkin, V.V., Filippova, A.S.: Analysis of the perturbed Chandler wobble of the Earth pole. Dokl. Phys. **62**(6), 318–322 (2017)
4. Akulenko, L.D., Markov, YuG, Rykhlova, L.V.: Motion of the Earth's poles under the action of gravitational tides in the deformable Earth model. Dokl. Phys. **46**(4), 261–263 (2001)
5. Munk, W.H., MacDonald, G.J.F.: The Rotation of the Earth. Cambridge University Press, New York (1961)
6. Sidorenkov, N.S.: The Interaction Between Earth's Rotation and Geophysical Processes. Wiley-VCH Verlag GmbH and Co. KGaA (2009)
7. Schubert, G.: Treatise on Geophysics, vol. 3. Elsevier, Geodesy (2007)
8. Xu, G.: Sciences of Geodesy—I: Advances and Future Directions. Springer, Berlin, Germany (2010)
9. Kumakshev, S.A.: Gravitational-tidal model of oscillations of Earth's poles. Mech. Solids **53**(2), 159–163 (2018)
10. Kumakshev, S.A.: Model of oscillations of Earth's poles based on gravitational tides. In: Karev, V., Klimov, D., Pokazeev, K. (eds.) Physical and Mathematical Modeling of Earth and Environment Processes. PMMEEP, pp. 157–163 (2017)
11. Klimov, D.M., Akulenko, L.D., Kumakshev, S.A.: The main properties and peculiarities of the Earth's motion relative to the center of mass. Dokl. Phys. **59**(10), 472–475 (2014)
12. Zlenko, A.A.: A celestial-mechanical model for the tidal evolution of the Earth–Moon system treated as a double planet. Astron. Rep. **59**(1), 72–87 (2015)
13. Zlenko, A.A.: The force function of two rigid celestial bodies in Delaunay–Andoyer variables. Astron. Rep. **60**(1), 174–181 (2016)
14. Bizouard, C., Remus, F., Lambert, S., Seoane, L., Gambis, D.: The Earth's Variable Chandler Wobble Astronomy and Astrophysics, vol. 526, pp. A106.1–A106.4 (2011)
15. Zotov, L., Bizouard, C., Shum, C.K.: A possible interrelation between Earth rotation and climatic variability at decadal time-scale. Geodesy Geodyn. **7**(3), 216–222 (2016)

Chapter 23
Application of Modified Fireworks Algorithm for Multiobjective Optimization of Satellite Control Law

Andrei V. Panteleev and Alexander Yu. Kryuchkov

Abstract Application of modified metaheuristic global optimization algorithm "fireworks" to solve problems of multiobjective optimization is discussed. All objectives are numeric and have the same importance. A possible solution to the problem is a vector of real numbers. Generally, the solution to the problem is a set of Pareto optimal possible solutions. The application of the algorithm of finding an effective programmed control in the problem of stabilizing a satellite in a circular orbit is considered. We study a problem with the fixed right end and known finite time, where control is a function in the class of piecewise constant functions and satisfies the constraints. The control must minimize the values of the quadratic objective and the objective of describing fuel consumption simultaneously. Searching of control is divided into two stages. At the first stage, the optimization problem is solved for each of the objectives with penalties. The values of the penalties are selected to satisfy the terminal constraints. At the second stage, the penalties found are used to solve a multiobjective optimization problem.

23.1 Introduction

In the modern world, designing new technical systems is becoming more complex. Requirements for systems in various areas are growing and many different factors have to be taken into account [1, 2]. These factors may reflect the opposite goals. In such circumstances, the search for an acceptable solution becomes a difficult task. As a rule, the solution should be a compromise between conflicting requirements.

A. V. Panteleev (✉) · A. Yu. Kryuchkov
Moscow Aviation Institute (National Research University), 4, Volokolamskoe Shosse, Moscow 125993, Russian Federation
e-mail: avpanteleev@inbox.ru

A. Yu. Kryuchkov
e-mail: a.kryuchkov@phygitalism.com

L. C. Jain et al. (eds.), *Advances in Theory and Practice of Computational Mechanics*, Smart Innovation, Systems and Technologies 173,
https://doi.org/10.1007/978-981-15-2600-8_23

When designing various systems, there is always a question about the choice of parameters that affect the final result. To select the optimal parameters, we can formulate an optimization problem. Each solution to the problem is a set of parameters most suitable from the point of view of the selected objectives. Comparison of solutions is based on the values of the objectives, and each objective can correspond to a particular factor and reflects the degree of influence of the solution on this factor. For example, the smaller objective value corresponds to higher savings in energy consumption. If one objective is selected, then one objective optimization problem is obtained. Many different algorithms have been developed for this problem solution [3]. When there are several objectives, the problem becomes more complicated. In a problem with one objective, an optimal solution is a solution that minimizes or maximizes the objective value. In a multiobjective problem, it is possible to give several concepts of the optimal solution. For example, if the objectives are ordered by importance, then to compare solutions, we can compare the values of objectives based on the lexicographic ordering. The solution will be optimal if it gives the minimum value to the most important objective. Optimal solutions can be infinitely many. There are definitions of optimality according to Geoffrion [4] and Pareto [5]. When choosing the definition of optimality, a specific system of preferences is also fixed. Of course, one can try to reflect the influence of a solution on various factors in one objective, for example, using the scalarization of objectives. It simplifies the optimization problem because it becomes a single-objective optimization problem. There is a difficulty: it is hard to describe the influence of a decision on all factors within one objective, since it is possible to lose some of the information about the links between factors describing different goals. If a multiobjective problem reduced to a single-objective, one has certain disadvantages, for example, there are Pareto-optimal solutions that cannot be found through the scalarization of objectives [6].

In control theory, when finding the optimal control, one objective is usually considered [7, 8]. In modern conditions, it is necessary to solve problems when one objective is not enough to describe all the requirements. Thus, a problem with several objectives appears. It can be solved as a problem with one objective if we combine several objectives into one. The process of combining the objectives itself is not unambiguous. Therefore, it is necessary to develop algorithms for solving multiobjective optimization problems, which allow to find a set of Pareto-optimal solutions, but not requiring scalarization. Some ideas for solving problems in control theory with several objectives were considered in [9].

In the general case, it is impossible to describe Pareto optimal solutions analytically. Therefore, the developing of algorithms for finding an approximate solution is needed. An approximate solution is a finite set of solutions close to the exact solution. Various algorithms based on different ideas were proposed to solve the problems.

In the chapter, a modification of the one-objective optimization “fireworks” algorithm and its application to finding programmed control, which stabilizes a satellite, is considered. The algorithm belongs to metaheuristic algorithms and does not guarantee to find an exact solution. The control is sought in the class of piecewise constant functions, due to the linearity of the dynamic system. The initial problem is reduced to the parametric optimization problem, in which the solution is a vector of real numbers. Numerical global optimization algorithms may be required to solve the problem.

A new problem was formed with modified initial objectives to solve the multiobjective problem and find the control law. Penalties were added to each objective in the new problem. They ought to be found when solving optimization problems for each objective. The next step is to solve the multiobjective problem with modified objectives, but with fixed penalties. Since the problem is solved numerically, therefore, the control is only suboptimal. The solution can be considered acceptable depending on the requirements in applied problems, where finding the exact analytical solution is impossible.

The chapter is organized as follows. Section 23.2 provides the model of a dynamic system. Section 23.3 discusses the idea of solving a multiobjective control problem. Section 23.4 considers a modification of the "fireworks" algorithm. Section 23.5 presents the results of numerical experiments, and Sect. 23.6 gives the conclusions.

23.2 Dynamic System Model

Consider the problem of the stability of the stationary motion of a satellite relative to an axis orthogonal to its circular orbit plane [10]. If we neglect the gravitational moments in comparison with the control moments, then Euler equations are replaced by the following:

$$\begin{cases} \dot{\omega}_x = a \times \omega_z + u_1 \\ \dot{\omega}_z = -a \times \omega_x + u_2 \\ \omega_x(t_0) = \omega_{x0} \\ \omega_z(t_0) = \omega_{z0} \end{cases}. \tag{23.1}$$

where ω_x, ω_z are the angular rotation velocity of the satellite, $a = \text{const}$, t_0 is the start time. The controls u_1, u_2 equal to the ratio of the corresponding torque relative to the axis C_x, C_z by the moment of inertia $I = I_x = I_z$, C_x, C_z are the axes associated with the center of mass of the satellite, where I_x, I_z are the moments of inertia about the axis C_x, C_z. Jet engines control rotation, so the control is constrained: $|u_i| \le h_i$, $i = 1, 2$.

The control ought to minimize two objectives:

$$I_1 = \int_{t_0}^{T} \left[u_1^2(t) + u_2^2(t)\right] dt \to \min_{u_1 \in U_N^1, u_2 \in U_N^2},$$

$$I_2 = \int_{t_0}^{T} [|u_1(t)| + |u_2(t)|] dt \to \min_{u_1 \in U_N^1, u_2 \in U_N^2},$$

where T is the known terminal time, $U_N^j, j = 1, 2$ are the sets of admissible controls:

$$U_N^j = \left\{ \sum_{i=0}^{N-1} b_i \times \chi(t - t_i) : |b_i| \le h_j \right\},$$

where $t_0 < t_1 < \cdots < t_{N-1} < T$ are the switching times, N is the number of switchings, $\chi(\cdot)$ is the Heaviside step function. Terminal constraints $\omega_x(T) = 0,\ \omega_z(T) = 0$ should be satisfied. Minimizing the objective I_2 corresponds to the task of optimizing fuel consumption.

23.3 Sketch of Solution

It is necessary to formulate a multiobjective optimization problem to find a control that minimizes two objectives simultaneously. Since the problem is solved numerically, not analytically, the following approach is proposed:

1. Modify the definitions of the objectives to take into account the terminal constraints by introducing penalties $\lambda_i > 0,\ i = 1, \ldots, 4$:

$$J_1 = \int_{t_0}^{T} \left[u_1^2(t) + u_2^2(t)\right] dt + \lambda_1 \times \omega_x^2(T) + \lambda_2 \times \omega_z^2(T) \to \min_{u_1 \in U_N^1,\ u_2 \in U_N^2},$$

$$J_2 = \int_{t_0}^{T} [|u_1(t)| + |u_2(t)|] dt + \lambda_3 \times \omega_x^2(T) + \lambda_4 \times \omega_z^2(T) \to \min_{u_1 \in U_N^1,\ u_2 \in U_N^2}.$$

2. Select the number of control switchings N as the parameter describing the control laws $u_1(t),\ u_2(t)$.
3. Find penalties values $\lambda_i > 0,\ i = 1, \ldots, 4$ and controls $u_1(t),\ u_2(t)$. The terminal conditions must be satisfied with certain accuracy: $\omega_x(T) \approx 0,\ \omega_z(T) \approx 0$. The solution is encoded as a numerical vector $x \in \mathbb{R}^{2N+4}$: the control $u_1(t)$ is encoded first, then $u_2(t)$ and the penalties $\lambda_i, i = 1, \ldots, 4$. We use the assumption that control takes constant values over equal lengths of time:

$$u_1(t) = \begin{cases} x_i, t \in [h \times (i-1); h \times i), \\ x_N, t \in [h \times (N-1); h \times N] \end{cases}$$

$$u_2(t) = \begin{cases} x_j, t \in \left[h \times (j - N - 1); h \times (j - N)\right), \\ x_{2N}, t \in [h \times (N-1); h \times N] \end{cases}$$

where $x_i \in [-h_1; h_1], i = 1, \ldots, N-1,\ x_j \in [-h_2; h_2], j = N+1, \ldots, 2N-1, h = T/N$. The penalties $\lambda_i, i = 1, \ldots, 4$ are the last two components of the vector x:$x_{2N+i} = \lambda_i$ for the first objective J_1 and $x_{2N+i} = \lambda_{i+2}$ for the second objective $J_2,\ i = 1, 2$. The optimization problems are

$$J_1(x) \to \min_{x \in D}, \tag{23.2}$$

$$J_2(x) \to \min_{x \in D}, \tag{23.3}$$

where $D = \{x \in \mathbb{R}^{2N+2} | -h_1 \le x_i \le h_1, i = 1, \ldots, N, -h_2 \le x_i \le h_2, i = N+1, \ldots, 2N, a_1 \le x_{2N+1} \le b_1, a_2 \le x_{2N+2} \le b_2\}$.

4. After finding the penalty parameters for each objective, they are fixed, and the multiobjectives optimization problem is solved:

$$J(x) = \begin{pmatrix} J_1(x) \\ J_2(x) \end{pmatrix} \to \min_{x \in D^1}, \tag{23.4}$$

where $D^1 = \{x \in \mathbb{R}^{2N} | -h_1 \le x_i \le h_1, i = 1, \ldots, N, -h_2 \le x_i \le h_2, i = N+1, \ldots, 2N\}$.

23.4 Solution of Multiobjective Problem

Hereinafter, Sect. 23.4.1 includes a mathematical formulation of the problem. Section 23.4.2 contains a general description of algorithm. Section 23.4.3 provides a detailed description.

23.4.1 Multiobjective Optimization Problem

The multiobjective optimization problem with constraints is considered. It is assumed that all objectives have the same importance and reducing the value of one objective with fixed values of the other objectives is preferable. A vector of real numbers represents the solution (all objectives are numeric):

$$F(x) = \begin{pmatrix} f_1(x) \\ \vdots \\ f_m(x) \end{pmatrix} \to \min_{x \in D}, \tag{23.5}$$

where $m \ge 2$ is the number of objectives, D is the set of acceptable solutions, $f_j : D \to \mathbb{R},\ j = 1, \ldots, m$:

$$D = \left\{x \in \mathbb{R}^n | a_i \le x_i \le b_i, a_i < b_i, i = 1, \ldots, n\right\}.$$

Need to find an approximation of the set of Pareto optimal solutions. We will need to give several definitions for the further description of the algorithm.

Definition 1 Let $F(x) \in \mathbb{R}^m : F(x) = (f_1(x), \ldots, f_m(x))^T$. It is a vector of objectives of a solution $x \in D$.

Definition 2 Let $F^1 = F\left(x^1\right), F^2 = F\left(x^2\right)$. They are vectors of objectives of $x^1 \in D,\ x^2 \in D$. F^1 dominates $F^2 : F^1 \prec F^2$ if $\forall i \in \{1, \ldots, m\},\ F_i^1 \leq F_i^2$ and $\exists j \in \{1, \ldots, m\} : F_j^1 < F_j^2$.

Definition 3 A solution $x^1 \in D$ is preferable than $x^2 \in D : x^1 \prec x^2 \Leftrightarrow F(x^1) \prec F(x^2)$.

Definition 4 A set $P = \{x \in D | \nexists x' \in D : F\left(x'\right) \prec F(x)\}$ is a set of Pareto optimal solutions.

Definition 5 A set $F(P) = \{F(x) | x \in P\}$ is Pareto front.

The result of solving the problem expressed by Eq. 23.5 will be a finite set of solutions, in which each element is close to some element of P.

23.4.2 Modification of Multiobjective Fireworks Algorithm

The modification of "fireworks" algorithm [11] is based on an imitation of the fireworks process. The firework is accompanied by a cloud of luminous fragments filling the vicinity of an exploding charge. This process is associated with a local search procedure in optimization problems.

Each firework volley determines the transition from one iteration of the search to another (from one generation of solutions to another). Points (decisions) in the set of feasible solutions D are determined for the first volley in an amount NP. An explosion occurs, generating debris scattering from the points of the explosion. The radius of the explosion is determined for each point separately.

Next is the process of forming a new generation of solutions. Non-dominated sorting of solutions is performed based on their vectors of objectives. Let $I = \{x^p | x^p \in D, p = 1, \ldots, NP\}$ be a set of solutions on the current iteration, where $NP \geq 1$. The result of the non-dominated sorting is a partition of the set I into k disjoint subsets $\mathcal{F}_i, i = 1, \ldots, k, 1 \leq k \leq NP, k$ is the number of the last subset in the partition mentioned below.

$$I = \bigcup_{i=1}^{k} \mathcal{F}_i,\ \ \mathcal{F}_i \cap \mathcal{F}_j = \emptyset, i \neq j$$

$$\mathcal{F}_1 = \left\{x \in I | \nexists x' \in I : F\left(x'\right) \prec F(x)\right\}$$

$$\vdots$$

$$\mathcal{F}_l = \left\{ x \in I \backslash \bigcup_{i=1}^{l-1} \mathcal{F}_i \middle| \nexists x' \in I \backslash \bigcup_{i=1}^{l-1} \mathcal{F}_i : F\left(x'\right) \prec F(x) \right\}$$

$$\vdots$$

$$\mathcal{F}_k = I \setminus \bigcup_{i=1}^{k-1} \mathcal{F}_i$$

In other words, non-dominated sorting is a repeating procedure to take the preferred solutions. In the first step, the preferred solutions are selected from I. Further, these preferred solutions are removed from I, and the procedure is repeated to the remainder.

The solutions corresponding to the points of the explosion and the resulting debris are selected from the sets $\mathcal{F}_1, \ \mathcal{F}_2, \ldots, \ \mathcal{F}_k$ until the number of selected solutions is equal to NP. If in some subset of solutions there are more than necessary to take, then part of the solutions are chosen randomly based on the distances to other solutions in this subset in the objective space.

The search process stops when a specified number of iterations is reached.

Fast non-dominated sorting algorithms were proposed in [12, 13]. The algorithm having complexity $O(n \ \log^{m-1} n)$ in the worst case is presented in [13].

Software implementations of the non-dominated sorting algorithm are available on GitHub for the following programming languages: C# [14], Python [15], and Java [16].

23.4.3 Algorithm

Step 1. Choose parameters:

- Number of charges at each iteration $NP \in \mathbb{N}$.
- Parameter $m > 0$ controlling the number of debris.
- Parameters $s_{\min}, \ s_{\max} \in \mathbb{N}, s_{\min} \leq s_{\max}$ are the minimum and maximum number of debris for each charge.
- Maximum explosion amplitude $A_{\max} > 0$.
- Maximum number of iterations $iter_{\max} \in \mathbb{N}$.

Step 2. Let $iter = 1$. Generate NP solutions in the set of feasible solutions $D, I^{iter} = \{x^{1,1}, \ldots, x^{NP,1}\}$:

$$x_i^{p,1} = a_i + Urand(0; 1) \times (b_i - a_i),$$

where $i = 1, \ldots, n, p = 1, \ldots, NP, Urand(0, 1)$ is continuous uniform random variable on $[0, \ 1]$.

Step 3. Non-dominated sorting of I^{iter}. Partition I^{iter} into $\mathcal{F}_1, \ldots, \mathcal{F}_l, 1 \leq l \leq NP$.

Step 4. Explosion and debris generation.

Step 4.1. For each $p = 1, \ldots, NP$ calculate:

1. Subset number q: $x^{p,iter} \in \mathcal{F}_q$.

2. Number of debris:

$$s^{p,iter} = m \times \log_2\left(1 + \frac{l}{q}\right) \times \left(1 - \frac{|\mathcal{F}_q|}{NP}\right),$$

$$\hat{s}^{p,iter} = \begin{cases} s_{\min} & \text{if}\left[s^{p,iter}\right] \le s_{\min} \\ s_{\max} & \text{if}\left[s^{p,iter}\right] \ge s_{\max} \\ \left[s^{p,iter}\right] & \text{otherwise} \end{cases},$$

where $[\cdot]$ is the integer part, l is the number of the last subset in the partition, $\hat{s}^{p,iter}$ is the number of debris generated by the explosion at $x^{p,iter}$.

Step 4.2. Determine the position of debris. For each $p = 1, \ldots, N$, find the position of debris with numbers $s = 1, \ldots, \hat{s}^{p,iter}$:

1. Find $q : x^{p,iter} \in \mathcal{F}_q$.
2. Let $\hat{x}^{p,iter,s} = x^{p,iter}$.
3. For each debris with number s :

   a. Let $\xi = Urand(0,\ 1)$.
   b. Randomly select dimensions:

$$\hat{n} = [n \times \xi].$$

   c. If $\xi < 0.5$, then apply the first algorithm for determining the position of the debris:
      (1) Calculate the magnitude of the explosion:

$$A^{p,iter} = A_{\max} \times \log_2\left(1 + \frac{q}{l}\right) \times \frac{|\mathcal{F}_q|}{NP}.$$

      (2) For each coordinate number i selected from $\hat{n}$:
         i. Calculate the displacement:

$$h_i^s = A^{p,iter} \times Urand(-1, 1).$$

         ii. Calculate the debris coordinate value:
         iii. $h_i^s = A^{p,iter} \times Urand(-1, 1)$
   d. If $\xi \ge 0.5$, then apply the second algorithm for determining the position of the debris:
      (1) For each coordinate number i selected from $\hat{n}$:

$$\hat{x}_i^{p,iter,s} = x_i^{p,iter} \times Nrand(1, 1),$$

where $Nrand(1;\ 1)$ is the normal random variable, 1 is the mean, and 1 is the standard deviation.

Step 4.3. Checking the boundary of the set of acceptable solutions D. For each $p = 1, \ldots, NP$:

1. For each $s = 1, \ldots, \hat{s}^{p,iter}$ check:

   If $\hat{x}_i^{p,iter,s} \notin [a_i; b_i], i = 1, \ldots, n$, then:

$$\hat{x}_i^{p,iter,s} = \begin{cases} Urand(a_i; 0.5 \times (a_i + b_i)) \text{if } \hat{x}_i^{p,iter,s} < a_i \\ Urand(0.5 \times (a_i + b_i), b_i) \text{otherwise} \end{cases}.$$

2. Append $\hat{x}^{p,iter,s}$ to the I^{iter} : $I^{iter} = I^{iter} \cup \left\{\hat{x}^{p,iter,s}\right\}$.

Step 5. Creating a new population.

Step 5.1. Non-dominated sorting I^{iter}: $\mathcal{F}_1, \ldots, \mathcal{F}_l$, where l is the number of the last subset in the partition I^{iter}. Let $iter = iter + 1,\ I^{iter} = \emptyset$.

Step 5.2. Find $u_{\min}$:

$$u_{\min} = \min_{1 \le u \le l} \left\{ u : \left| \bigcup_{i=1}^{u} \mathcal{F}_i \right| \ge NP \right\}.$$

If $\left| \bigcup_{i=1}^{u_{\min}} \mathcal{F}_i \right| = NP$, then $I^{iter} = \bigcup_{i=1}^{u_{\min}} \mathcal{F}_i$, $K = \emptyset$, otherwise $I^{iter} = \bigcup_{i=1}^{u_{\min}-1} \mathcal{F}_i$, $K = \mathcal{F}_{u_{\min}}$. If $u_{\min} = 1$ and $\left|\mathcal{F}_{u_{\min}}\right| > NP$, then $\bigcup_{i=1}^{u_{\min}-1} \mathcal{F}_i$ is empty set. All solutions fall into the set K.

Step 5.3. For each point $x^w \in K$, calculate $R(x^w)$ that is the sum of distances to other points and $p(x^w)$ that is the probability of explosion:

$$R(x^w) = \sum_{x^q \in K} \rho(F(x^w), F(x^q)),$$
$$p(x^w) = \frac{R(x^w)}{\sum_{x^q \in K} R(x^q)},$$

where $\rho(\cdot, \cdot)$ is Euclidean distance.

Step 5.4. Randomly choose from K the set of points (solutions) in the number of $NP - \left|I^{iter}\right|$ based on probability $p(x^w)$ and append it to the set I^{iter}.

Step 5.5. If $iter \le Iter_{\max}$, then go to Step 3, otherwise I^{iter} is an approximate solution.

It is not necessary to perform non-dominated sorting on Step 3 after the first iteration. Information about the partition can be taken after Step 5.1. Such optimization can reduce computational costs.

23.5 Numerical Experiments

The following parameters were selected in the numerical experiments: $t_0 = 0,\ \omega_x(t_0) = 0.01,\ \omega_z(t_0) = 0.1,\ a = 0.00007292123518 \times \sqrt{3},\ T = 1.5,\ h_1 =$

$h_2 = 10$, $a_1 = a_2 = 5000$, $b_1 = b_2 = 30,000$ for the problem (Eq. 23.2), $a_1 = a_2 = 7000$, $b_1 = b_2 = 50,000$ for the problem (Eq. 23.3).

The problem was solved 5 times for each objective using three algorithms: "big bang-big crunch", "fireworks", "grenade explosion method" [17–19]. The values of the penalties λ_i, $i = 1, \ldots, 4$ were chosen based on the results of the solutions. The five best solution results by objectives are listed in tables mentioned below.

The best results were selected using the non-dominated sorting. The selection were based on the three objectives: the final value of the objective in the problem and absolute values $|\omega_x(T)|$, $|\omega_z(T)|$. The links between the parameters of the algorithm and the results obtained are made through the Id column.

First, the problem expressed by Eq. 23.2 was solved. The number of switches $N = 8, 10, 15$. The parameters of the algorithms are listed in Tables 23.1, 23.2, and 23.3. They were the same for all values of N. Table 23.4 shows the results of the problem solution expressed by Eq. 23.2, while Table 23.5 shows the results of the problem solution provided by Eq. 23.3. BBBC name means the "big bang-big crunch" algorithm, FW is "fireworks", GEM is "grenade explosion method" in Tables 23.4 and 23.5. Parameters α and β are given in the "fireworks" algorithm [19] instead of S_{min} and S_{max}. Last of them are used for the algorithm of "fireworks" in the latest version of the software that implements all of the above algorithms [20]. The purpose of the parameters remains the same, but the range has changed. If α and β limited the number of debris by multiplying by m, then parameters S_{min}, S_{max} take values from the set of the natural numbers and limit the number of debris explicitly.

The system of differential equations (Eq. 23.1) was integrated using the fourth order Runge–Kutta algorithm with a step by time is equal 0.006. The results of the

Table 23.1 Parameters of BBBC algorithm

Id	$iter_{max}$	NP	α	β
1	500	900	0.4	0.5
2	400	300	0.2	0.1
3	600	200	0.5	0.5
4	400	300	0.5	0.9
5	1000	200	0.5	0.5

Table 23.2 Parameters of FW algorithm

Id	$iter_{max}$	NP	S_{min}	S_{max}	A_{max}	m
1	300	100	10	30	2.5	20
2	900	60	10	50	1	20
3	150	20	10	50	2	20
4	130	30	10	50	9	20
5	200	200	10	50	0.9	2

Table 23.3 Parameters of GEM algorithm

Id	N_{gr}	N_q	$iter_{\max}$	$R_{t-initial}$	$m_{\min}$	$m_{\max}$	P_{sin}	p_{ts}	R_{rd}	N_{dm}
1	1	100	500	8.485	0.1	0.9	5	0.8	10	1
2	1	100	500	8.485	0.1	0.9	5	0.8	150	1
3	2	75	250	3	0.1	0.9	5	0.8	10	1
4	2	50	900	3	0.1	0.9	5	0.8	150	1
5	2	100	600	3	0.1	0.9	5	0.8	200	1

Table 23.4 Results of solution problem expressed by Eq. 23.2

Id	Algorithm	N	J_1	$\omega_x(T)$	$\omega_z(T)$	λ_1	λ_2
1	BBBC	8	0.007604	−5.8801E−5	−5.0031E−5	6020.491	13228.306
1	BBBC	8	0.007686	−4.7398E−5	1.2918E−4	9156.242	10734.697
5	BBBC	8	0.007785	7.9434E−5	3.3302E−5	7995.033	8813.139
2	FW	8	0.015567	−2.0242E−5	−9.1201E−5	11831.137	15142.621
1	FW	8	0.024222	1.6105E−5	−4.2268E−4	9151.26	7584.314
1	FW	8	0.024643	−1.0386E−4	2.7879E−5	13696.774	28017.862
4	GEM	8	0.556426	−1.3949E−4	9.907E−5	6736.098	5083.258
2	GEM	8	0.612563	5.9419E−5	−2.7177E−4	7162.864	5493.837
2	GEM	8	0.638406	5.0606E−5	−1.4113E−5	5179.473	6840.771
1	BBBC	10	0.008995	−8.3586E−6	3.4592E−5	21240.517	5288.833
3	BBBC	10	0.011183	2.475E−4	2.7719E−6	5115.562	5502.813
5	BBBC	10	0.009019	1.0354E−4	−4.8657E−5	7962.194	7662.075
2	FW	10	0.015147	8.9975E−5	6.9862E−5	15727.856	10001.139
2	FW	10	0.016794	3.6849E−5	−1.6649E−4	6051.207	22030.654
2	FW	10	0.016905	1.8238E−4	6.1224E−5	10031.998	15250.165
2	GEM	10	0.736677	−5.123E−4	2.1993E−4	6161.452	5426.359
2	GEM	10	1.440087	−6.0246E−4	−1.5134E−4	5054.838	9103.089
2	GEM	10	1.531033	−3.3369E−5	−9.9396E−5	9821.737	5487.826
5	BBBC	15	0.00811	−5.7092E−6	−4.5947E−5	7372.927	5518.607
5	BBBC	15	0.075976	7.9654E−5	1.0107E−5	14552.243	18685.553
1	BBBC	15	0.009598	−5.5698E−5	−1.6512E−4	18727.757	5001.869
2	FW	15	0.033248	7.2419E−5	−1.3879E−4	24315.069	5036.152
2	FW	15	0.034547	−8.2631E−5	−6.5246E−5	15625.11	11576.285
2	FW	15	0.0392	−5.9139E−5	−5.1834E−4	29219.92	9410.349
2	GEM	15	1.384027	−1.2177E−3	−6.8241E−6	9554.58	5195.936
5	GEM	15	1.464559	−3.8043E−5	5.2589E−4	5557.229	6332.396
2	GEM	15	1.764458	−1.4677E−4	−2.849E−4	5904.814	7288.434

solution for different values of the number of control switchings $N = 8,\ 10,\ 15$ are listed below in Tables 23.1, 23.2, and 23.3.

Values of the control for $N = 8$ and the problem (Eq. 23.2) are presented in Table 23.6.

Values of the control for $N = 8$ and the problem (Eq. 23.3) are presented in Table 23.7.

The following values of the penalties were selected based on results of problems solutions expressed by Eqs. 23.2 and 23.3. They are listed in Table 23.8.

Table 23.5 Results of solution problem expressed by Eq. 23.3

Id	Algorithm	N	J_2	$\omega_x(T)$	$\omega_z(T)$	λ_3	λ_4
1	BBBC	8	5.160511	1.2955E−4	−2.9862E−4	8617.706	30659.494
5	BBBC	8	7.793863	6.4492E−4	−2.808E−4	14617.899	12032.31
1	BBBC	8	9.523325	6.8663E−4	−1.2991E−5	25216.057	49486.752
2	FW	8	0.110096	4.4293E−5	3.5374E−5	7398.211	23228.186
2	FW	8	0.110418	1.7397E−5	−1.8073E−5	11301.078	26449.15
2	FW	8	0.110385	−5.1218E−5	3.8691E−5	23538.786	7480.885
1	GEM	8	4.272012	3.2021E−3	6.9987E−3	12832.101	15461.899
1	GEM	8	4.274881	5.0938E−3	−5.7929E−4	7186.127	13161.372
5	GEM	8	4.432162	1.3039E−5	−3.0173E−5	10185.876	7386.642
5	BBBC	10	4.714524	−1.8E−4	−3.7626E−4	8597.463	12434.566
1	BBBC	10	6.252822	−1.0633E−4	−2.3524E−4	8816.932	10224.87
1	BBBC	10	6.468312	−2.3239E−6	−6.1775E−4	11334.467	8479.597
2	FW	10	0.110101	3.0433E−5	2.4713E−5	7544.905	49805.834
2	FW	10	0.110188	2.3023E−6	8.403E−6	15014.967	12347.239
2	FW	10	0.110671	5.8278E−5	−5.6617E−6	7719.76	9349.415
1	GEM	10	3.836151	−2.6624E−3	1.428E−3	20703.729	11021.126
5	GEM	10	4.11288	−4.4488E−6	−2.0417E−4	8004.571	8102.188
2	GEM	10	4.280879	2.9121E−4	7.374E−5	7000.455	13175.199
1	BBBC	15	5.827723	2.4443E−4	−2.5118E−4	22853.125	9440.198
1	BBBC	15	7.344124	−7.1157E−4	6.4369E−5	11760.751	31003.365
1	BBBC	15	7.746715	−1.9718E−4	−4.4188E−4	26858.854	13654.348
2	FW	15	0.10872	5.7673E−5	6.8015E−5	11968.315	8067.865
2	FW	15	0.112241	1.2268E−5	−5.6859E−6	17727.378	46942.4
2	FW	15	0.112687	1.4441E−4	−4.8097E−5	10930.355	29076.976
2	GEM	15	3.409267	−4.4071E−4	1.4896E−4	9431.92	7369.179
2	GEM	15	4.315806	2.599E−4	−5.5759E−7	7721.063	7441.143
4	GEM	15	4.417192	−9.7125E−6	−2.986E−5	7574.725	8838.705

Table 23.6 Values of the control

t	$u_1(t)$	$u_2(t)$
0	−0.00151397	−0.02574828
0.1875	−0.00132441	−0.09630083
0.375	−0.00467101	−0.08455635
0.5625	−0.01891687	−0.06688008
0.75	0.00663547	−0.06081657
0.9375	−0.0212785	−0.06917927
1.125	−0.00204489	−0.04615376
1.3125	−0.01087784	−0.08420601

Table 23.7 Values of control

t	$u_1(t)$	$u_2(t)$
0	−0.00014781	−0.50271218
0.1875	−9.58977344E−5	−0.01173894
0.375	−0.05076043	−0.00054373
0.5625	−0.00075133	2.06553592E−9
0.75	−2.08503817E−6	−0.01919405
0.9375	3.78612438E−7	5.06952171E−7
1.125	−0.0006604	−0.00032599
1.3125	−9.60528929E−6	1.27306602E−6

Table 23.8 The penalties for various values of the number of switchings

Penalty	$N = 8$	$N = 10$	$N = 15$
λ_1	6020.49142775647	21240.5168329197	7372.92658580968
λ_2	13228.3055394025	5288.83298590383	5518.60746104122
λ_3	7398.210744558931	15014.9665908026	17727.3778533999
λ_4	23228.1859177857	12347.2391412813	46942.3996244755

The problem expressed by Eq. 23.4 was solved with fixed penalties from Table 23.8 and different values of the number of switchings. The algorithm parameters are presented in Table 23.9.

The values of the objectives are presented in Fig. 23.1 for $N = 8$.

The values of the objectives are presented in Fig. 23.2 for $N = 10$.

The values of the objectives are presented in Fig. 23.3 for $N = 15$.

Table 23.9 Parameters of the modified “fireworks” algorithm

Id	$Iter_{max}$	NP	S_{min}	S_{max}	A_{max}	m
1	900	300	5	20	0.25	5
2	500	500	5	20	0.9	15
3	600	600	10	30	0.8	10
4	1000	200	5	20	0.2	10
5	2000	300	5	20	0.1	4
6	2000	300	5	20	0.05	4

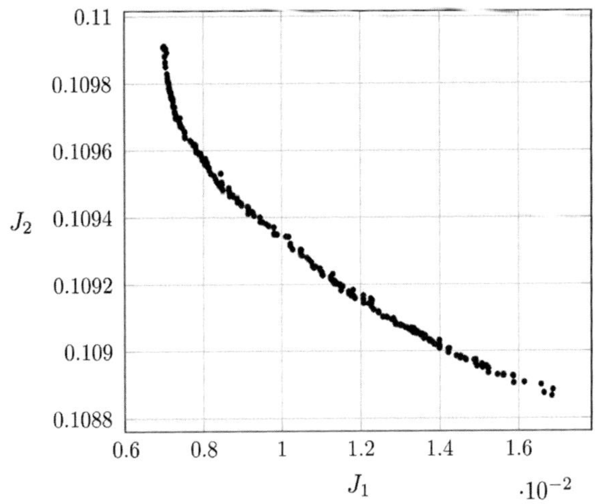

Fig. 23.1 The values of the objectives for $N = 8$

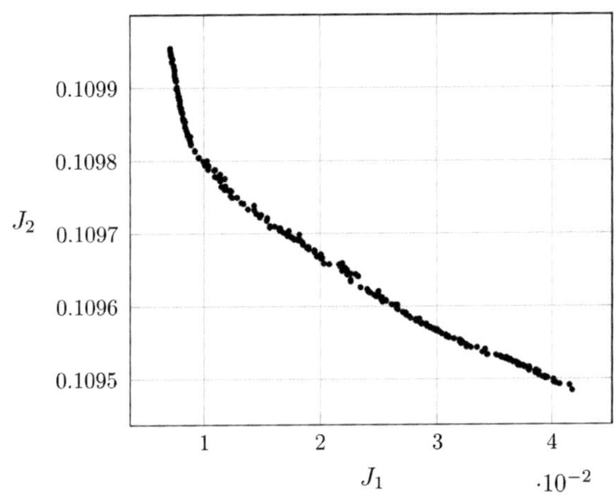

Fig. 23.2 The values of the objectives for $N = 10$

23.6 Conclusions

The numerical algorithm for solving multiobjective optimization problems is proposed. The main difficulties in developing multiobjective optimization algorithms are the convergence and uniformity of solution distribution along Pareto front. Pareto dominance is based on a comparison of vectors of objectives. In practice, it can cause difficulties. Rounding errors complicate the comparison. For example, let two vectors of objectives are equal, except for two components differ from each other by a small value. Does one vector of objectives dominate another? We can get different results depending on the choice: the approximate Pareto front, where it is not there, or not to find part of Pareto front. The solution may be the application

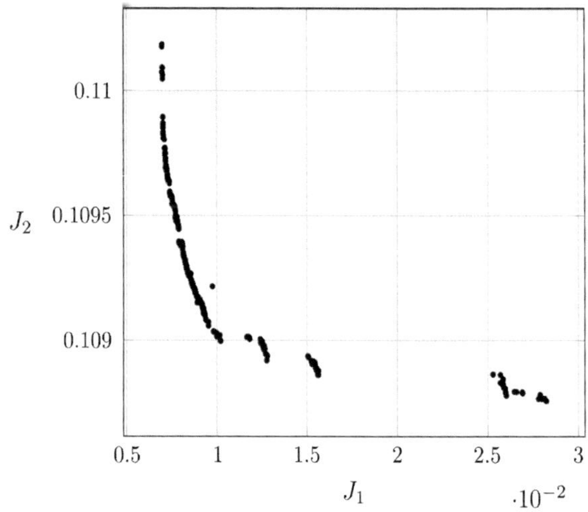

Fig. 23.3 The values of the objectives for $N = 15$

of dominance rules based on fuzzy logic. For example, an approach for solving multiobjective optimization problems based on fuzzy logic is proposed in [21].

An essential challenge is the solution of multiobjective optimization problems with a large number of objectives (three or more). NSGA-3 algorithm for solving problems with a large number of objectives (from three to fifteen) is demonstrated in [22]. However, the authors did not provide a software implementation of their algorithm. The algorithm has different implementations. Comparison between NSGA-2 and NSGA-3 algorithms is presented [23]. It has been shown that NSGA-3 is not always better than NSGA-2.

Nevertheless, the advances in solving multiobjective optimization problems allow us to consider applied problems. For example, the problem of finding programmed control was considered. The control is sought in the class of piecewise constant functions. The initial problem became the parametric optimization problem.

A new problem was formed with modified initial objectives to solve the multiobjective problem. The penalties were added to each objective. The algorithm for solving multiobjective optimization problems is proposed. It is numerical and does not guarantee finding the exact solution. The found control is only suboptimal.

References

1. Arias-Montano, A., Coello, C.A.C., Mezura-Montes, E.: Multiobjective evolutionary algorithms in aeronautical and aerospace engineering. IEEE Trans. Evol. Comput. **16**(5), 662 (2012)
2. Rangaiah, G.: Multi-Objective Optimization. World Scientific, Singapore (2017)
3. Nocedal, J., Wright, S.: Numerical Optimization. Springer, New York (2006)
4. Geoffrion, A.M.: Proper efficiency and the theory of vector maximization. J. Math. Anal. Appl. **22**, 618–630 (1968)

5. Collette, Y., Siarry, P.: Multiobjective Optimization. Springer, Berlin Heidelberg, Berlin (2004)
6. Talbi, E.-G.: Metaheuristics: From Design to Implementation. Wiley, Hoboken, NJ, USA (2009)
7. Panovskiy, V., Panteleev, A.: Meta-heuristic interval methods of search of optimal in average control of nonlinear determinate systems with incomplete information about its parameters. J. Comput. Syst. Sci. Int. **56**(1), 52–63 (2017)
8. Panteleev, A., Pis'mennaya, V.: Application of a memetic algorithm for the optimal control of bunches of trajectories of nonlinear deterministic systems with incomplete feedback. J. Comput. Syst. Sci. Int. **57**(1), 25–36 (2018)
9. Prasad, U., Sarma, I.: Multiobjective optimal control problems: game cooperative solution according to Nash-Harsani. Autom. Remote Control **6**, 99–105 (1975)
10. Babadzanjanz, L.K., Pototskaya, I.Y.: Control in Mechanical Systems with Expenditure Criteria. Izd-vo SPbSU, Saint-Petersburg (in Russian) (2003)
11. Tan, Y., Zhu, Y.: Fireworks algorithm for optimization. In: Tan, Y., Shi, Y., Tan, K.C. (eds.) Advances in Swarm Intelligence, pp. 355–364. Springer, Berlin Heidelberg, Berlin, Heidelberg (2010)
12. Fortin, F.-A., Grenier, S., Parizeau, M.: Generalizing the improved run-time complexity algorithm for non-dominated sorting. In: Proceeding of the Fifteenth Annual Conference on Genetic and Evolutionary Computation Conference. ACM Press, New York, USA (2013)
13. Buzdalov, M., Shalyto, A.: A provably asymptotically fast version of the generalized jensen algorithm for non-dominated sorting. In: Bartz-Beielstein, T., Branke, J., Filipič, B., and Smith, J. (eds.) 13th International Conference on Parallel Problem Solving from Nature—PPSN XIII, pp. 528–537. Springer International Publishing, Cham (2014)
14. KernelA/nds. C# implementation of the non-dominated sorting. https://github.com/KernelA/nds. Last accessed 17 Oct 2019
15. KernelA/nds-py. A Python implementation of the non-dominated sorting. https://github.com/KernelA/nds-py. Last accessed 17 Oct 2019
16. Mbuzdalov/non-dominated-sorting. This repo contains implementations of algorithms for non-dominated sorting and a benchmarking suite. https://github.com/mbuzdalov/non-dominated-sorting. Last accessed 17 Oct 2019
17. Ahrari, A., Atai, A.: Grenade explosion method—a novel tool for optimization of multimodal functions. Appl. Soft Comput. **10**(4), 1132–1140 (2010)
18. Erol, O., Eksin, I.: A new optimization method: big bang-big crunch. Adv. Eng. Softw. **37**(2), 106–111 (2006)
19. Panteleev, A., Kryuchkov, A.: Metaheuristic optimization methods for parameters estimation of dynamic systems. Civ. Aviat. High Technol. **20**(2), 37–45 (2017)
20. KernelA/eoptimization-library. The library for constrained optimization. Implemented three algorithms: big bung-big crunch, Fireworks, Grenade explosion. https://github.com/KernelA/eoptimization-library. Last accessed 17 Oct 2019
21. Jamwal, P., Abdikenov, B., Hussain, S.: Evolutionary optimization using equitable fuzzy sorting genetic algorithm (EFSGA). IEEE Access. **7**, 8111–8126 (2019)
22. Deb, K., Jain, H.: An evolutionary many-objective optimization algorithm using reference-point-based nondominated sorting approach. Part I: solving problems with box constraints. IEEE Trans. Evol. Comput. **18**(4), 577–601 (2014)
23. Ishibuchi, H., Imada, R., Setoguchi, Y., Nojima, Y.: Performance comparison of NSGA-II and NSGA-III on various many-objective test problems. In: 2016 IEEE Congress on Evolutionary Computation, pp. 3045–3052 (2016)

Chapter 24
Approximate Filtering Methods in Continuous-Time Stochastic Systems

Konstantin N. Chugai, Ivan M. Kosachev and Konstantin A. Rybakov

Abstract The goal of this chapter is to consider algorithms based on the particle method for solving the optimal filtering problem for nonlinear continuous-time stochastic observation systems not only by the minimum mean squared error estimate, but also by the maximum a posteriori estimate. Particle filters are proposed on the basis of Duncan–Mortensen–Zakai equation, as well as, on the basis of robust Duncan–Mortensen–Zakai equation. To find the mode of the conditional distribution approximately, Edgeworth series is used for the conditional probability density expansion. This approach allows to reduce significantly the computation time in contrast to finding the mode by estimating the conditional probability density, for example, the histogram or kernel estimations.

24.1 Introduction

Methods and algorithms for solving the optimal filtering problem have a lot of practical applications [1–9], such as the radio signal detection from a noise, navigation, and telemetry information processing for moving objects, parameter identification, etc.

We suggest filtering algorithms based on the particle method for nonlinear continuous-time stochastic observation systems for the unbiased estimate with a

---

K. N. Chugai
Research Institute of the Belarusian Armed Forces, Minsk 220103, Republic of Belarus
e-mail: konstantin.ch40@gmail.com

I. M. Kosachev
Military Academy of the Republic of Belarus, 220, pr. Nezavisimosti, Minsk 220057, Republic of Belarus
e-mail: kosachev1301@mail.ru

K. A. Rybakov (✉)
Moscow Aviation Institute (National Research University), 4, Volokolamskoe shosse, 125993 Moscow, Russian Federation
e-mail: rkoffice@mail.ru

L. C. Jain et al. (eds.), *Advances in Theory and Practice of Computational Mechanics*, Smart Innovation, Systems and Technologies 173,
https://doi.org/10.1007/978-981-15-2600-8_24

minimum mean squared error and the maximum a posteriori estimate. For obtaining the maximum a posteriori estimate, we apply Edgeworth series [10] for the expansion of marginal conditional probability densities. This approach allows to estimate the mode of the conditional distribution approximately, since a partial sum of Edgeworth series is used for marginal conditional probability densities (conditional probability densities for each component of the state) instead of the conditional probability density of the state vector. However, this approach has an important advantage such that it significantly reduces the computation time in contrast to finding the mode by consistently estimating the conditional probability density.

Traditionally, the particle method is associated with Duncan–Mortensen–Zakai equation (the equation for the unnormalized conditional probability density of the state) [11]. Here, we use not only this equation, but also robust Duncan–Mortensen–Zakai equation (the equation for the unnormalized conditional probability density of the special vector state) [12]. As a rule, the robust Duncan–Mortensen–Zakai equation has been studied for stationary observation models [12–15]. In this chapter, the robust Duncan–Mortensen–Zakai equation for nonstationary observation models is concerned. Some preliminaries have been obtained in [16]. According to the proposed form in contrast [17], the optimal filtering problem for the nonstationary case can be solved by applying the particle method [18–21].

The rest of this chapter is structured as follows. Section 24.2 provides the statement of the optimal filtering problem. Equations for the conditional probability density and conditional central moments of the state are given in Sect. 24.3. New filtering algorithms based on the particle method for nonlinear continuous-time stochastic observation systems are described in Sect. 24.4. Section 24.5 presents the conclusions for this chapter.

24.2 Optimal Filtering Problem

Consider a signal observation model described by Itô Stochastic Differential Equations (SDEs) [22, 23]:

$$dX(t) = f(t, X(t))dt + \sigma(t, X(t))dW(t), \quad X(t_0) = X_0, \tag{24.1}$$

$$dY(t) = c(t, X(t))dt + \zeta(t)dV(t), \quad Y(t_0) = Y_0 = 0, \tag{24.2}$$

where $X \in \mathbb{R}^n$ is the state, $Y \in \mathbb{R}^m$ is the observation, $t \in \mathbb{T} = [t_0, T]$, $f(t, x) : \mathbb{T} \times \mathbb{R}^n \to \mathbb{R}^n$ is the n-dimensional function, $\sigma(t, x) : \mathbb{T} \times \mathbb{R}^n \to \mathbb{R}^{n\times s}$ is the $(n \times s)$-dimensional matrix function, $c(t, x) : \mathbb{T} \times \mathbb{R}^n \to \mathbb{R}^m$ is the m-dimensional function, $\zeta(t) : \mathbb{T} \to \mathbb{R}^{m\times d}$ is the $(m \times d)$-dimensional matrix function such that the symmetric matrix $\eta(t) = \zeta(t)\zeta^{\mathrm{T}}(t)$ is nonsingular, i.e., $\det \eta(t) \neq 0$ for any $t \in \mathbb{T}$. Then, $W(t)$ and $V(t)$ are the standard s-dimensional and d-dimensional Wiener processes, respectively, X_0 is the initial state with a probability density $\varphi_0(x)$ such that $\mathrm{E}|X_0|^2 < \infty$, where E is the expectation or mean. The initial state X_0 and Wiener processes $W(t)$, $V(t)$ are independent.

Functions $f(t,x)$, $\sigma(t,x)$, $c(t,x)$, and $\zeta(t)$ are given, they satisfy the conditions on the existence and uniqueness of the solution of SDEs [22], i.e., there exist two positive constants c_1 and c_2 such that for all (t,x), $(t,x') \in \mathbb{T} \times \mathbb{R}^n$

$$|f(t,x)|^2 + |\sigma(t,x)|^2 + |c(t,x)|^2 \leqslant \mathrm{c}_1(1+|x|^2),$$
$$|f(t,x) - f(t,x')|^2 + |\sigma(t,x) - \sigma(t,x')|^2 + |c(t,x) - c(t,x')|^2 \leqslant \mathrm{c}_2|x-x'|^2,$$

where

$$|x|^2 = \sum_{i=1}^{n} x_i^2, \quad |f(t,x)|^2 = \sum_{i=1}^{n} f_i^2(t,x),$$
$$|\sigma(t,x)|^2 = \sum_{i=1}^{n}\sum_{l=1}^{s} \sigma_{il}^2(t,x), \quad |c(t,x)|^2 = \sum_{j=1}^{m} c_j^2(t,x).$$

Such conditions can be weakened, especially for functions $f(t,x)$ and $c(t,x)$. In fact, it is sufficient to use weakened existence and uniqueness conditions for SDEs solution from [24, 25].

The optimal filtering problem is to find an estimate $\hat{X}(t)$ given the observations $Y_0^t = \{Y(\tau), \tau \in [t_0, t]\}$ so that $\hat{X}(t) = \psi(t, Y_0^t)$, where the function $\psi(t,\cdot)$ satisfies the following condition:

$$\mathrm{E}\Pi(\mathcal{E}(t)) \to \min_{\psi(t,\cdot)} \quad \forall t \in \mathbb{T}, \tag{24.3}$$

in which $\mathcal{E}(t) = X(t) - \hat{X}(t)$ is the estimation error and $\Pi(\varepsilon)$ is the loss function [26].

If $\Pi(\varepsilon) = \varepsilon^{\mathrm{T}} L \varepsilon$, where L is the $(n \times n)$-dimensional positive semidefinite matrix, i.e., $\Pi(\varepsilon)$ is the quadratic loss function, then

$$\hat{X}^{\mathrm{MMSE}}(t) = \psi^{\mathrm{MMSE}}(t, Y_0^t) = \mathrm{E}[X(t)|Y_0^t] = \int_{\mathbb{R}^n} x p(t, x|Y_0^t) dx,$$

and, if $\Pi(\varepsilon) = 1 - \delta(\varepsilon)$, where $\delta(\varepsilon)$ is the Dirac delta function, i.e., $\Pi(\varepsilon)$ is the simple loss function, then

$$\hat{X}^{\mathrm{MAP}}(t) = \psi^{\mathrm{MAP}}(t, Y_0^t) = \arg\max_{x \in \mathbb{R}^n} p(t, x|Y_0^t). \tag{24.4}$$

In the relations given above, $p(t, x|Y_0^t)$ is the conditional probability density of the state X [2, 19, 23]. Equation 24.3 defines the unbiased estimate $\hat{X}^{\mathrm{MMSE}}(t)$ with a minimum mean squared error, and Eq. 24.4 defines the maximum a posteriori estimate $\hat{X}^{\mathrm{MAP}}(t)$, i.e., the mode of the conditional distribution with the probability density $p(t, x|Y_0^t)$.

24.3 Equations for Conditional Probability Density

The optimal estimate $\hat{X}(t)$ defined by Eqs. 24.3–24.4 is expressed in terms of the conditional probability density $p(t, x|Y_0^t)$. This density satisfies Stratonovich–Kushner equation [27, 28]:

$$\frac{\partial p(t, x|Y_0^t)}{\partial t} = \mathcal{A}p(t, x|Y_0^t) + \left[\lambda(t, x, \dot{Y}(t)) - \left\langle\lambda(t, X(t), \dot{Y}(t))\right\rangle\right]p(t, x|Y_0^t),$$
$$p(t_0, x) = \varphi_0(x), \tag{24.5}$$

where $\mathcal{A}$ is the forward diffusion operator [22]:

$$\begin{aligned}\mathcal{A}p(t, x|Y_0^t) = &- \sum_{i=1}^{n} \frac{\partial}{\partial x_i}[f_i(t, x)p(t, x|Y_0^t)]\\ &+ \frac{1}{2}\sum_{i=1}^{n}\sum_{j=1}^{n} \frac{\partial^2}{\partial x_i \partial x_j}[g_{ij}(t, x)p(t, x|Y_0^t)]\\ = &- \nabla^{\mathrm{T}}(f(t, x)p(t, x|Y_0^t)) + \frac{1}{2}\mathrm{tr}[\nabla\nabla^{\mathrm{T}}(g(t, x)p(t, x|Y_0^t))],\end{aligned}$$

$g(t, x)$ is the $(n \times n)$-dimensional symmetric matrix function $\sigma(t, x)\sigma^{\mathrm{T}}(t, x)$, $\mathrm{tr}[\cdot]$ is the trace of the matrix, and the function $\lambda(t, x, z)$ is specified as follows:

$$\begin{aligned}\lambda(t, x, z) = &\sum_{k=1}^{m}\sum_{r=1}^{m} c_k(t, x)q_{kr}(t)\left(z_r - \frac{1}{2}c_r(t, x)\right)\\ =&c^{\mathrm{T}}(t, x)q(t)\left(z - \frac{1}{2}c(t, x)\right), \quad q(t) = \eta^{-1}(t),\end{aligned} \tag{24.6}$$

and

$$\left\langle\lambda(t, X(t), \dot{Y}(t))\right\rangle = \mathrm{E}[\lambda(t, X(t), \dot{Y}(t))|Y_0^t] = \int_{\mathbb{R}^n} \lambda(t, x, \dot{Y}(t))p(t, x|Y_0^t)dx.$$

Equation 24.5 is the nonlinear Stochastic Partial Differential Equation (SPDE) in Stratonovich interpretation. It can be used to derive equations for conditional moments of the state X.

Let $\mu_{i_1 \dots i_R}(t)$ be the Rth conditional central moment of the state X, i.e.,

$$\mu_{i_1 \dots i_R}(t) = \left\langle \overset{\circ}{X}_{i_1}(t) \dots \overset{\circ}{X}_{i_R}(t) \right\rangle,$$

where $\overset{\circ}{X}_i(t) = X_i(t) - \hat{X}_i^{\text{MMSE}}(t)$, $i = 1, 2, \dots, n$. Then differential equations for conditional central moments are as follows [29]:

$$\begin{aligned}\dot{\mu}_{i_1 \dots i_R}(t) = &\sum_{q=1}^{R} \left\langle f_{i_q}(t, X(t))(\overset{\circ}{X}_{i_1 \dots i_{q-1} i_{q+1} \dots i_R}(t) - \mu_{i_1 \dots i_{q-1} i_{q+1} \dots i_R}(t)) \right\rangle \\ &+ \sum_{q=1}^{R-1} \sum_{s=1+q}^{R} \left\langle g_{i_q i_s}(t, X(t)) \overset{\circ}{X}_{i_1 \dots i_{q-1} i_{q+1} \dots i_{s-1} i_{s+1} \dots i_R}(t) \right\rangle \\ &+ \sum_{q=1}^{R} \left\langle \overset{\circ}{X}_{i_q}(t) \lambda(t, X(t), \dot{Y}(t)) \right\rangle \mu_{i_1 \dots i_{q-1} i_{q+1} \dots i_R}(t) \\ &+ \left\langle \lambda(t, X(t), \dot{Y}(t)) \right\rangle \mu_{i_1 \dots i_R}(t) - \left\langle \overset{\circ}{X}_{i_1}(t) \dots \overset{\circ}{X}_{i_R}(t) \lambda(t, X(t), \dot{Y}(t)) \right\rangle,\end{aligned}$$

where

$$\overset{\circ}{X}_{i_1 \dots i_{q-1} i_{q+1} \dots i_R}(t) = \overset{\circ}{X}_{i_1}(t) \dots \overset{\circ}{X}_{i_{q-1}}(t) \overset{\circ}{X}_{i_{q+1}}(t) \dots \overset{\circ}{X}_{i_R}(t),$$

$$\overset{\circ}{X}_{i_1 \dots i_{q-1} i_{q+1} \dots i_{s-1} i_{s+1} \dots i_R}(t) = \overset{\circ}{X}_{i_1}(t) \dots \overset{\circ}{X}_{i_{q-1}}(t) \overset{\circ}{X}_{i_{q+1}}(t) \dots \overset{\circ}{X}_{i_{s-1}}(t) \overset{\circ}{X}_{i_{s+1}}(t) \dots \overset{\circ}{X}_{i_R}(t).$$

These equations can be rewritten in a brief form using the multi-index notation such that

$$\begin{aligned}\dot{\mu}_{\bar{i}}(t) = &\sum_{q=1}^{R} \left\langle f_{i_q}(t, X(t))(\overset{\circ}{X}_{\bar{i}(q)}(t) - \mu_{\bar{i}(q)}(t)) \right\rangle + \sum_{q=1}^{R-1} \sum_{s=1+q}^{R} \left\langle g_{i_q i_s}(t, X(t)) \overset{\circ}{X}_{\bar{i}(q,s)}(t) \right\rangle \\ &+ \left\langle \left(\sum_{q=1}^{R} \overset{\circ}{X}_{i_q}(t) \mu_{\bar{i}(q)}(t) + \mu_{\bar{i}}(t) - \overset{\circ}{X}_{\bar{i}}(t) \right) \lambda(t, X(t), \dot{Y}(t)) \right\rangle, \end{aligned} \tag{24.7}$$

where $\bar{i} = (i_1 \dots i_R)$, $\bar{i}(q) = (i_1 \dots i_{q-1} i_{q+1} \dots i_R)$, $\bar{i}(q, s) = (i_1 \dots i_{q-1} i_{q+1} \dots i_{s-1} i_{s+1} \dots i_R)$.

A methodical approach to the high-precision filtering for the signal observation model (Eqs. 24.1–24.2) based on a truncation of moments by cumulant closed techniques [30–32] is considered in [29]. The high accuracy of developed algorithms for the optimal nonlinear filtering problem is due to the use of conditional higher order central moments defined by differential Eq. 24.7, the adaptability of the high-precision filtering is provided by calculating in real time the conditional skewness and excess kurtosis for all components of the state. The most effective computation of expectations $\langle \cdot \rangle$ in differential Eq. 24.7 can be carried out using the method developed in [33] for the statistical approximation of an arbitrary nonlinearity.

The unnormalized conditional probability density $\varphi(t, x|Y_0^t)$ satisfies Duncan–Mortensen–Zakai equation [11, 12, 17, 19]:

$$\frac{\partial\varphi(t, x|Y_0^t)}{\partial t} = \mathcal{A}\varphi(t, x|Y_0^t) + \lambda(t, x, \dot{Y}(t))\varphi(t, x|Y_0^t), \quad \varphi(t_0, x) = \varphi_0(x). \quad (24.8)$$

Equation 24.8 is the linear SPDE in Stratonovich interpretation. The function $\lambda(t, x, z)$ is called an absorption and recovering intensity [21, 34] or a potential function [20].

Let us assume that

$$\lambda^-(t, x, z) = \begin{cases} -\lambda(t, x, z) & \lambda(t, x, z) < 0 \\ 0 & \lambda(t, x, z) \geqslant 0 \end{cases},$$

$$\lambda^+(t, x, z) = \begin{cases} \lambda(t, x, z) & \lambda(t, x, z) > 0 \\ 0 & \lambda(t, x, z) \leqslant 0 \end{cases},$$

i.e., $\lambda(t, x, z) = -\lambda^-(t, x, z) + \lambda^+(t, x, z)$. Using such representation for the function $\lambda(t, x, z)$, we can rewrite Duncan–Mortensen–Zakai equation as the generalized Fokker–Planck–Kolmogorov equation with absorption and recovering functions:

$$\frac{\partial\varphi(t, x|Y_0^t)}{\partial t} = \mathcal{A}\varphi(t, x|Y_0^t) - \lambda^-(t, x, \dot{Y}(t))\varphi(t, x|Y_0^t) + \lambda^+(t, x, \dot{Y}(t))\varphi(t, x|Y_0^t),$$
$$\varphi(t_0, x) = \varphi_0(x).$$

Note that the generalized Fokker–Planck–Kolmogorov equations have been introduced, e.g., in [34, 35], for the switching diffusions or stochastic dynamical systems with variable or random structure.

Then we can define a special random process with terminating and branching paths [21, 34]. Paths of such process are completely determined by SDE (Eq. 24.1), and the observations described by SDE (Eq. 24.2) affect on the terminating and branching rates (or intensities). The probabilities of terminating and branching on the time interval $[t, t + \Delta t]$ at $X(t) = x$ and $\dot{Y}(t) = z$ for small Δt are $\Pr^-(t, \Delta t) = \lambda^-(t, x, z)\Delta t + o(\Delta t)$ and $\Pr^+(t, \Delta t) = \lambda^+(t, x, z)\Delta t + o(\Delta t)$, respectively. Such probabilistic interpretation has been used for solving approximately the filtering and prediction problems [21, 34, 36].

To find the conditional probability density $p(t, x|Y_0^t)$, we should use the normalized representation:

$$p(t, x|Y_0^t) = \frac{\varphi(t, x|Y_0^t)}{\int_{\mathbb{R}^n} \varphi(t, x|Y_0^t)dx}, \quad t \in \mathbb{T}.$$

Rewrite Eq. 24.8 as

$$\frac{\partial \varphi(t, x|Y_0^t)}{\partial t} = \mathcal{L}\varphi(t, x|Y_0^t) + c^{\mathrm{T}}(t, x)q(t)\dot{Y}(t),$$

where

$$\mathcal{L}\varphi(t, x|Y_0^t) = \mathcal{A}\varphi(t, x|Y_0^t) - \frac{1}{2}c^{\mathrm{T}}(t, x)q(t)c(t, x)\varphi(t, x|Y_0^t).$$

Then define functions

$$h_k(t, x) = \sum_{r=1}^{m} q_{kr}(t)c_r(t, x), \quad k = 1, 2, \ldots, m,$$

and a new unnormalized conditional probability density

$$\begin{aligned}\rho(t, x|Y_0^t) &= \exp\left\{-\sum_{k=1}^{m} h_k(t, x)Y_k(t)\right\}\varphi(t, x|Y_0^t)\\ &= \exp\{-h^{\mathrm{T}}(t, x)Y(t)\}\varphi(t, x|Y_0^t).\end{aligned} \tag{24.9}$$

The unnormalized conditional probability density $\rho(t, x|Y_0^t)$ satisfies the robust Duncan–Mortensen–Zakai equation [17, 21]:

$$\begin{aligned}\frac{\partial \rho(t, x|Y_0^t)}{\partial t} &= \mathcal{L}\rho(t, x|Y_0^t) - \sum_{k=1}^{m} Y_k(t)\mathcal{L}_k\rho(t, x|Y_0^t)\\ &+ \sum_{k=1}^{m}\sum_{r=1}^{m} Y_k(t)Y_r(t)\mathcal{L}_{kr}\rho(t, x|Y_0^t) - \sum_{k=1}^{m}\frac{\partial h_k(t, x)}{\partial t}Y_k(t)\rho(t, x|Y_0^t),\end{aligned} \tag{24.10}$$

where $\mathcal{L}_k = [\mathcal{H}_k, \mathcal{L}]$ and $\mathcal{L}_{kr} = \frac{1}{2}[\mathcal{H}_k, \mathcal{L}_r] = \frac{1}{2}[\mathcal{H}_k, [\mathcal{H}_r, \mathcal{L}]]$, $\mathcal{H}_k$ are the multiplication operators with multipliers $h_k(t, x)$, $[\cdot, \cdot]$ denotes Lie bracket.

Expressions for operators $\mathcal{L}_k$ and $\mathcal{L}_{kr}$ have been obtained in [21] for the stationary observation model

$$dY(t) = c(X(t))dt + dV(t)$$

instead of the nonstationary model (Eq. 24.2), where $c(x) : \mathbb{R}^n \to \mathbb{R}^m$ is the m-dimensional function, $V(t)$ is the m-dimensional standard Wiener process.

Let us obtain expressions for operators $\mathcal{L}_k$ and $\mathcal{L}_{kr}$ with respect to nonstationary observation model (Eq. 24.2). Firstly,

$$\begin{aligned}\mathcal{L}_k\rho(t, x|Y_0^t) =&[\mathcal{H}_k, \mathcal{L}]\ \rho(t, x|Y_0^t) = [\mathcal{H}_k, \mathcal{A}]\ \rho(t, x|Y_0^t)\\ =&(\mathcal{H}_k \circ \mathcal{A})\rho(t, x|Y_0^t) - (\mathcal{A} \circ \mathcal{H}_k)\rho(t, x|Y_0^t)\end{aligned}$$

$$
\begin{aligned}
&= - h_k(t,x) \sum_{i=1}^{n} \frac{\partial}{\partial x_i}[f_i(t,x)\rho(t,x|Y_0^t)] \\
&+ \frac{1}{2} h_k(t,x) \sum_{i=1}^{n}\sum_{j=1}^{n} \frac{\partial^2}{\partial x_i \partial x_j}[g_{ij}(t,x)\rho(t,x|Y_0^t)] \\
&+ \sum_{i=1}^{n} \frac{\partial}{\partial x_i}[f_i(t,x)h_k(t,x)\rho(t,x|Y_0^t)] \\
&- \frac{1}{2} \sum_{i=1}^{n}\sum_{j=1}^{n} \frac{\partial^2}{\partial x_i \partial x_j}[g_{ij}(t,x)h_k(t,x)\rho(t,x|Y_0^t)].
\end{aligned}
$$

Further, we will use the following properties. If $u(t,x)$ and $v(t,x)$ are twice differentiable functions with respect to x, then

$$
\frac{\partial\ (uv)}{\partial x_i} = \frac{\partial u}{\partial x_i} v + u \frac{\partial v}{\partial x_i}, \quad \frac{\partial^2 (uv)}{\partial x_i \partial x_j} = \frac{\partial^2 u}{\partial x_i \partial x_j} v + \frac{\partial}{\partial x_j}\left[u \frac{\partial v}{\partial x_i}\right] + \frac{\partial}{\partial x_i}\left[u \frac{\partial v}{\partial x_j}\right] - u \frac{\partial^2 v}{\partial x_i \partial x_j}
$$

for all $i, j = 1, 2, \ldots, n$. The matrix function $g(t,x)$ is symmetric, i.e., $g_{ij}(t,x) = g_{ji}(t,x)$. Consequently,

$$
\begin{aligned}
\mathcal{L}_k \rho(t,x|Y_0^t) &= \sum_{i=1}^{n} f_i(t,x) \frac{\partial h_k(t,x)}{\partial x_i} \rho(t,x|Y_0^t) \\
&- \frac{1}{2} \sum_{i=1}^{n}\sum_{j=1}^{n} \frac{\partial}{\partial x_j}\left[g_{ij}(t,x) \frac{\partial h_k(t,x)}{\partial x_i} \rho(t,xY_0^t)\right] \\
&- \frac{1}{2} \sum_{i=1}^{n}\sum_{j=1}^{n} \frac{\partial}{\partial x_i}\left[g_{ij}(t,x) \frac{\partial h_k(t,x)}{\partial x_j} \rho(t,x|Y_0^t)\right] \\
&+ \frac{1}{2} \sum_{i=1}^{n}\sum_{j=1}^{n} g_{ij}(t,x) \frac{\partial^2 h_k(t,x)}{\partial x_i \partial x_j} \rho(t,x|Y_0^t) \\
&= \sum_{i=1}^{n} f_i(t,x) \frac{\partial h_k(t,x)}{\partial x_i} \rho(t,x|Y_0^t) \\
&- \sum_{i=1}^{n} \frac{\partial}{\partial x_i}\left[\sum_{j=1}^{n} g_{ij}(t,x) \frac{\partial h_k(t,x)}{\partial x_j} \rho(t,x|Y_0^t)\right] \\
&+ \frac{1}{2} \sum_{i=1}^{n}\sum_{j=1}^{n} g_{ij}(t,x) \frac{\partial^2 h_k(t,x)}{\partial x_i \partial x_j} \rho(t,x|Y_0^t)
\end{aligned}
$$

or

$$\mathcal{L}_k\rho(t,x|Y_0^t) = f^k(t,x)\rho(t,x|Y_0^t) - \sum_{i=1}^{n}\frac{\partial}{\partial x_i}[g_i^k(t,x)\rho(t,x|Y_0^t)] + h^k(t,x)\rho(t,x|Y_0^t),$$

where

$$\begin{aligned} f^k(t,x) &= \sum_{i=1}^{n} f_i(t,x)\frac{\partial h_k(t,x)}{\partial x_i} = \nabla^{\mathrm{T}}h_k(t,x)f(t,x),\ g_i^k(t,x) \\ &= \sum_{j=1}^{n} g_{ij}(t,x)\frac{\partial h_k(t,x)}{\partial x_j}, \quad i = 1,2,\ldots,n, \\ h^k(t,x) &= \frac{1}{2}\sum_{i=1}^{n}\sum_{j=1}^{n} g_{ij}(t,x)\frac{\partial^2 h_k(t,x)}{\partial x_i \partial x_j} = \frac{1}{2}\mathrm{tr}[g(t,x)\nabla\nabla^{\mathrm{T}}h_k(t,x)]. \end{aligned}$$

Secondly,

$$\begin{aligned} \mathcal{L}_{kr}\rho(t,x|Y_0^t) &= [\mathcal{H}_k,\mathcal{L}_r]\ \rho(t,x|Y_0^t) \\ &= (\mathcal{H}_k\circ\mathcal{L}_r)\rho(t,x|Y_0^t) - (\mathcal{L}_r\circ\mathcal{H}_k)\rho(t,x|Y_0^t) \\ &= -\frac{1}{2}h_k(t,x)\sum_{i=1}^{n}\frac{\partial}{\partial x_i}[g_i^r(t,x)\rho(t,x|Y_0^t)] \\ &\quad + \frac{1}{2}\sum_{i=1}^{n}\frac{\partial}{\partial x_i}[g_i^r(t,x)h_k(t,x)\rho(t,x|Y_0^t)] \\ &= \frac{1}{2}\sum_{i=1}^{n} g_i^r(t,x)\frac{\partial h_k(t,x)}{\partial x_i}\rho(t,x|Y_0^t) \end{aligned}$$

or

$$\mathcal{L}_{kr}\rho(t,x|Y_0^t) = \frac{1}{2}g^{kr}(t,x)\rho(t,x|Y_0^t),$$

where

$$\begin{aligned} g^{kr}(t,x) &= \sum_{i=1}^{n} g_i^r(t,x)\frac{\partial h_k(t,x)}{\partial x_i} = \nabla^{\mathrm{T}}h_k(t,x)g^r(t,x) \\ &= \nabla^{\mathrm{T}}h_k(t,x)g(t,x)\nabla h_r(t,x) \quad (g^k(t,x) = g(t,x)\nabla h_k(t,x)). \end{aligned}$$

Thus,

$$\frac{\partial\rho(t,x|Y_0^t)}{\partial t} = \mathcal{A}\rho(t,x|Y_0^t) + \sum_{k=1}^{m} Y_k(t)\sum_{i=1}^{n}\frac{\partial}{\partial x_i}[g_i^k(t,x)\rho(t,x|Y_0^t)]$$

$$
\begin{aligned}
&-\sum_{k=1}^{m} Y_k(t)(f^k(t,x)+h^k(t,x))\rho(t,x|Y_0^t)\\
&+\frac{1}{2}\sum_{k=1}^{m}\sum_{r=1}^{m} Y_k(t)Y_r(t)g^{kr}(t,x)\rho(t,x|Y_0^t)\\
&-\frac{1}{2}\sum_{k=1}^{m} h_k(t,x)c_k(t,x)\rho(t,x|Y_0^t)-\sum_{k=1}^{m} Y_k(t)\frac{\partial h_k(t,x)}{\partial t}\rho(t,x|Y_0^t).
\end{aligned}
$$

Let us introduce new notations

$$
\begin{aligned}
\tilde{f}_i(t,x,y) &= f_i(t,x)-\sum_{k=1}^{m} y_k g_i^k(t,x), \quad i=1,2,\ldots,n,\\
v(t,x,y) &= -\sum_{k=1}^{m} y_k(f^k(t,x)+h^k(t,x))+\frac{1}{2}\sum_{k=1}^{m}\sum_{r=1}^{m} y_k y_r g^{kr}(t,x)\\
&\quad -\frac{1}{2}\sum_{k=1}^{m} h_k(t,x)c_k(t,x)-\sum_{k=1}^{m} y_k\frac{\partial h_k(t,x)}{\partial t}
\end{aligned}
$$

or

$$
\tilde{f}(t,x,y) = f(t,x)-g(t,x)\left[\frac{\partial h(t,x)}{\partial x}\right]^{\mathrm{T}} y, \tag{24.11}
$$

$$
\begin{aligned}
v(t,x,y) = &-y^{\mathrm{T}}\frac{\partial h(t,x)}{\partial x}f(t,x)-\frac{1}{2}\mathrm{tr}[g(t,x)\nabla\nabla^{\mathrm{T}}(y^{\mathrm{T}}h(t,x))]\\
&+\frac{1}{2}y^{\mathrm{T}}\frac{\partial h(t,x)}{\partial x}g(t,x)\left[\frac{\partial h(t,x)}{\partial x}\right]^{\mathrm{T}} y\\
&-\frac{1}{2}h^{\mathrm{T}}(t,x)c(t,x)-y^{\mathrm{T}}\frac{\partial h(t,x)}{\partial t}.
\end{aligned} \tag{24.12}
$$

Consequently, Eq. 24.10 can be rewritten in the form:

$$
\frac{\partial \rho(t,x|Y_0^t)}{\partial t} = \widetilde{\mathcal{A}}\rho(t,x|Y_0^t)+v(t,x,Y(t))\rho(t,x|Y_0^t), \tag{24.13}
$$

where

$$
\begin{aligned}
\widetilde{\mathcal{A}}\rho(t,x|Y_0^t) = &-\sum_{i=1}^{n}\frac{\partial}{\partial x_i}[\tilde{f}_i(t,x,Y(t))\rho(t,x|Y_0^t)]\\
&+\frac{1}{2}\sum_{i=1}^{n}\sum_{j=1}^{n}\frac{\partial^2}{\partial x_i\partial x_j}[g_{ij}(t,x)\rho(t,x|Y_0^t)]
\end{aligned}
$$

$$= -\nabla^{\mathrm{T}}(\tilde{f}(t,x,Y(t))\rho(t,x|Y_0^t)) + \frac{1}{2}\mathrm{tr}[\nabla\nabla^{\mathrm{T}}(g(t,x)\rho(t,x|Y_0^t))].$$

The initial condition for Eqs. 24.10 and 24.13 is determined by $\rho(t_0, x) = \varphi_0(x)$, since $\exp\{-h^{\mathrm{T}}(t_0, x)Y_0\} = 1$. Equation 24.13 can also be represented as the generalized Fokker–Planck–Kolmogorov equation with absorption and recovering functions [21, 34]:

$$\frac{\partial \rho(t,x|Y_0^t)}{\partial t} = \widetilde{\mathcal{A}}\rho(t,x|Y_0^t) - \nu^-(t,x,Y(t))\rho(t,x|Y_0^t) + \nu^+(t,x,Y(t))\rho(t,x|Y_0^t),$$

where

$$\nu^-(t,x,y) = \begin{cases} -\nu(t,x,y) & \nu(t,x,y) < 0 \\ 0 & \nu(t,x,y) \geqslant 0 \end{cases},$$

$$\nu^+(t,x,y) = \begin{cases} \nu(t,x,y) & \nu(t,x,y) > 0 \\ 0 & \nu(t,x,y) \leqslant 0 \end{cases},$$

i.e., $\nu(t,x,y) = -\nu^-(t,x,y) + \nu^+(t,x,y)$.

The function $\nu(t,x,y)$ is the absorption and recovering intensity or a potential function similar to Eq. 24.6. This equation describes an evolution of the unnormalized conditional probability density of the special state $\tilde{X} \in \mathbb{R}^n$ defined by the following Itô SDEs:

$$d\tilde{X}(t) = \tilde{f}(t,\tilde{X}(t),Y(t))dt + \sigma(t,\tilde{X}(t))d\tilde{W}(t), \quad \tilde{X}(t_0) = X_0, \qquad (24.14)$$

where $t \in \mathbb{T}$, $\tilde{f}(t,x,y)$ is introduced by Eq. 24.11, $\tilde{W}(t)$ is the s-dimensional standard Wiener process.

Then we can define a special random process with terminating and branching paths determined by SDE (Eq. 24.14) and the observations described by SDE (Eq. 24.2) affect on the terminating and branching rates (or intensities). The probabilities of terminating and branching on the time interval $[t, t+\Delta t]$ at $X(t) = x$ and $Y(t) = y$ for small Δt are $\Pr^-(t,\Delta t) = \nu^-(t,x,y)\Delta t + o(\Delta t)$ and $\Pr^+(t,\Delta t) = \nu^+(t,x,y)\Delta t + o(\Delta t)$, respectively.

24.4 Particle Filters

To solve approximately Duncan–Mortensen–Zakai Eq. 24.8, we can use the particle method or sequential Monte Carlo method [18, 19]. Let $\omega(t)$ denote the weight function defined by the following equation:

$$\omega(t) = \exp\left\{\int_{t_0}^{t} \lambda(\tau, X(\tau), \dot{Y}(\tau))d\tau\right\}$$
$$= \exp\left\{\int_{t_0}^{t} c^{\mathrm{T}}(\tau, X(\tau))q(\tau)dY(\tau) - \frac{1}{2}\int_{t_0}^{t} c^{\mathrm{T}}(\tau, X(\tau))q(\tau)c(\tau, X(\tau))d\tau\right\}.$$

So, the estimate $\hat{X}^{\mathrm{MMSE}}(t)$ is the normalized weighted mean [19]:

$$\hat{X}^{\mathrm{MMSE}}(t) = \frac{\mathrm{E}[\omega(t)X(t)]}{\mathrm{E}\omega(t)}.$$

To find the estimate $\hat{X}^{\mathrm{MMSE}}(t)$, it is necessary to simulate M sample paths $X^j(t)$ of the random process $X(t)$ and corresponding paths $\omega^j(t)$ of the weight function $\omega(t)$ by a numerical method for Itô SDE (Eq. 24.1) (a pair $(X^j(t), \omega^j(t))$ is called a particle), $j = 1, 2, \ldots, M$. For example, using Euler–Maruyama method [37], we have

$$X_{k+1} = X_k + hf(t_k, X_k) + \sqrt{h}\sigma(t_k, X_k)\Delta W_k,$$
$$\omega_{k+1} = \omega_k \mathrm{e}^{c^{\mathrm{T}}(t_k, X_k)q(t_k)(Y(t_{k+1}) - Y(t_k)) - \frac{1}{2}c^{\mathrm{T}}(t_k, X_k)q(t_k)c(t_k, X_k)h}, \quad \omega_0 = 1,$$

where $h = (T - t_0)/N$ is the time discretization step, $t_k = t_0 + kh$, ΔW_k is the s-dimensional random vector with independent components having a standard normal distribution, $k = 0, 1, \ldots, N-1$. Thus,

$$\hat{X}^{\mathrm{MMSE}}(t_k) \approx \hat{X}_k = \frac{1}{\Omega_k}\sum_{j=1}^{M}\omega_k^j X_k^j, \quad \Omega_k = \sum_{j=1}^{M}\omega_k^j. \tag{24.15}$$

The unnormalized conditional probability density $\varphi(t, x|Y_0^t)$ and the conditional probability density $p(t, x|Y_0^t)$ can be represented as

$$\varphi(t_k, x|Y_0^{t_k}) \approx \sum_{j=1}^{M}\omega_k^j\delta(x - X_k^j), \quad p(t_k, x|Y_0^{t_k}) \approx \frac{1}{\Omega_k}\sum_{j=1}^{M}\omega_k^j\delta(x - X_k^j), \tag{24.16}$$

where $\delta(x - x^*)$ is the Dirac delta function concentrated at x^* [18, 19].

Similarly, the particle method can be used to solve approximately the robust Duncan–Mortensen–Zakai Eq. 24.13. Define the weight function as follows:

$$\tilde{\omega}(t) = \exp\left\{\int_{t_0}^{t} \nu(\tau, X(\tau), Y(\tau))d\tau\right\},$$

where $\nu(t, x, y)$ is introduced in Eq. 24.12. Then, we can simulate M sample paths $\tilde{X}^j(t)$ of the random process $\tilde{X}(t)$ and corresponding paths $\tilde{\omega}^j(t)$ of the weight function $\tilde{\omega}(t)$ also using Euler–Maruyama method [37] for Itô SDE (Eq. 24.14) (a pair $(\tilde{X}^j(t), \tilde{\omega}^j(t))$ is also a particle), $i = 1, 2, \ldots, M$:

$$\tilde{X}_{k+1} = \tilde{X}_k + h\tilde{f}(t_k, \tilde{X}_k, Y(t_k)) + \sqrt{h}\sigma(t_k, \tilde{X}_k)\Delta\tilde{W}_k,$$
$$\tilde{\omega}_{k+1} = \tilde{\omega}_k \mathrm{e}^{\nu(t_k, \tilde{X}_k, Y(t_k))h}, \quad \tilde{\omega}_0 = 1.$$

Here, $\Delta\tilde{W}_k$ is the s-dimensional random vector with independent components having a standard normal distribution, $k = 0, 1, \ldots, N-1$.

Hence, the unnormalized conditional probability density $\rho(t, x|Y_0^t)$ can be represented in the form similar to Eq. 24.16:

$$\rho(t_k, x|Y_0^{t_k}) \approx \sum_{j=1}^{M} \tilde{\omega}_k^j \delta(x - \tilde{X}_k^j).$$

Also note that the function $\exp\{-h^{\mathrm{T}}(t, x)Y(t)\}$ in Eq. 24.9 is the additional weight function, which can be used to approximate conditional probability densities $\varphi(t, x|Y_0^t)$ and $p(t, x|Y_0^t)$. Thus, we can conclude that

$$\hat{X}^{\mathrm{MMSE}}(t_k) \approx \hat{X}_k = \frac{1}{\tilde{\Omega}_k^*} \sum_{j=1}^{M} \tilde{\omega}_k^{j*} \tilde{X}_k^j, \quad \tilde{\Omega}_k^* = \sum_{j=1}^{M} \tilde{\omega}_k^{j*}, \tag{24.17}$$

where $\tilde{\omega}_k^{j*} = \tilde{\omega}_k^j \exp\{h^{\mathrm{T}}(t_k, \tilde{X}_k^j)Y(t_k)\}$, and

$$\varphi(t_k, x|Y_0^{t_k}) \approx \sum_{j=1}^{M} \tilde{\omega}_k^{j*} \delta(x - \tilde{X}_k^j), \quad p(t_k, x|Y_0^{t_k}) \approx \frac{1}{\tilde{\Omega}_k^*} \sum_{j=1}^{M} \tilde{\omega}_k^{j*} \delta(x - \tilde{X}_k^j).$$

We can also use different representations for conditional probability densities $p(t, x|Y_0^t)$, $\varphi(t, x|Y_0^t)$, and $\rho(t, x|Y_0^t)$. For example, it is possible to construct the histogram or another density estimates using particles. The kernel estimation [38] can also be applied for the conditional probability density $p(t, x|Y_0^t)$:

$$p(t_k, x|Y_0^{t_k}) \approx \frac{1}{\Omega_k h_x^n} \sum_{j=1}^{M} \omega_k^j K\left(\frac{x - X_k^j}{h_x}\right)$$

or

$$p(t_k, x|Y_0^{t_k}) \approx \frac{1}{\tilde{\Omega}_k^* h_x^n} \sum_{j=1}^{M} \tilde{\omega}_k^{j*} K\left(\frac{x - \tilde{X}_k^j}{h_x}\right),$$

where $K(x)$ is the kernel, i.e., some probability density function, $h_x > 0$ is the smoothing parameter. For instance, $K(x)$ is the probability density for the

n-dimensional normal distribution:

$$K(x) = \frac{1}{(2\pi)^{n/2}} e^{-\frac{1}{2}|x|^2}.$$

To find the estimate $\hat{X}^{\text{MAP}}(t)$ or mode of the conditional distribution, the kernel estimations for the conditional probability density $p(t, x|Y_0^t)$ can be used, but the calculation time increases in this case. Furthermore, this estimate depends on the kernel and the smoothing parameter. As an alternative, it is suggested to use Edgeworth series [10] for the expansion of marginal conditional probability densities. This approach allows to reduce significantly the computation time in contrast to finding the mode by consistent estimating the conditional probability density.

Let us denote marginal conditional probability densities by $p_i(t, x_i|Y_0^t)$:

$$p_i(t, x_i|Y_0^t) = \int_{\mathbb{R}^{n-1}} p(t, x|Y_0^t) dx_1 \dots dx_{i-1} dx_{i+1} \dots dx_n, \quad i = 1, 2, \dots, n.$$

For the normalized random value $\xi_i = (X_i - \bar{X}_i)/\sigma_i$, where $\bar{X}_i = \text{E}X_i$ and $\sigma_i = \sqrt{\text{E}(X_i - \bar{X}_i)^2}$, we can write the marginal probability density as

$$\begin{aligned} p_i^*(\xi) &= \phi(\xi) - \frac{1}{3!}\frac{\mu_{3,i}}{\sigma_i^3}\phi^{(3)}(\xi) + \frac{1}{4!}\left[\frac{\mu_{4,i}}{\sigma_i^4} - 3\right]\phi^{(4)}(\xi) \\ &+ \frac{10}{6!}\frac{\mu_{3,i}^2}{\sigma_i^6}\phi^{(6)}(\xi) - \frac{1}{5!}\left[\frac{\mu_{5,i}}{\sigma_i^5} - 10\frac{\mu_{3,i}}{\sigma_i^3}\right]\phi^{(5)}(\xi) \\ &- \frac{35}{7!}\frac{\mu_{3,i}}{\sigma_i^3}\left[\frac{\mu_{4,i}}{\sigma_i^4} - 3\right]\phi^{(7)}(\xi) - \frac{280}{9!}\frac{\mu_{3,i}^3}{\sigma_i^9}\phi^{(9)}(\xi) + \dots \\ &= \phi(\xi)\Bigg(1 + \frac{1}{3!}\frac{\mu_{3,i}}{\sigma_i^3}H_3(\xi) + \frac{1}{4!}\left[\frac{\mu_{4,i}}{\sigma_i^4} - 3\right]H_4(\xi) + \frac{10}{6!}\frac{\mu_{3,i}^2}{\sigma_i^6}H_6(\xi) \\ &+ \frac{1}{5!}\left[\frac{\mu_{5,i}}{\sigma_i^5} - 10\frac{\mu_{3,i}}{\sigma_i^3}\right]H_5(\xi) + \frac{35}{7!}\frac{\mu_{3,i}}{\sigma_i^3}\left[\frac{\mu_{4,i}}{\sigma_i^4} - 3\right]H_7(\xi) + \frac{280}{9!}\frac{\mu_{3,i}^3}{\sigma_i^9}H_9(\xi) + \dots\Bigg), \end{aligned}$$

where $\phi(\xi)$ is the probability density for the standard normal distribution and $H_k(\xi)$ is the Hermite polynomial of degree k:

$$\phi(\xi) = \frac{1}{\sqrt{2\pi}} e^{-\frac{1}{2}\xi^2}, \quad \phi^{(k)}(\xi) = \frac{d^k\phi(\xi)}{d\xi^k} = (-1)^k\phi(\xi)H_k(\xi),$$

and $\mu_{R,i}$ are the Rth central moments of the random value X_i, i.e., $\mu_{R,i} = \text{E}(X_i - \bar{X}_i)^R$.

According to [10], the mode of the distribution with probability density $p_i^*(\xi)$ can be approximately calculated by

$$\xi_i^* = -\frac{1}{2}\frac{\mu_{3,i}}{\sigma_i^3},$$

consequently,

$$\hat{X}^{\mathrm{MAP}}(t_k) \approx \Xi_k, \quad (\Xi_k)_i = (\hat{X}_k)_i - \frac{1}{2}\frac{\hat{\mu}_{3,i,k}}{\hat{\sigma}_{i,k}^2}, \quad i = 1, 2, \ldots, n, \tag{24.18}$$

where $\hat{\sigma}_{i,k}$ is the estimate of the conditional standard deviation and $\hat{\mu}_{3,i,k}$ is the estimate of the conditional third central moment for the ith component X_i of the state X at time $t = t_k$. These statistics can be obtained similar to $\hat{X}_k$, i.e.,

$$\hat{\sigma}_{i,k} = \sqrt{\frac{1}{\Omega_k}\sum_{j=1}^{M}\omega_k^j((X_k^j)_i - (\hat{X}_k)_i)^2}, \quad \hat{\mu}_{3,i,k} = \frac{1}{\Omega_k}\sum_{j=1}^{M}\omega_k^j((X_k^j)_i - (\hat{X}_k)_i)^3,$$

or

$$\hat{\sigma}_{i,k} = \sqrt{\frac{1}{\tilde{\Omega}_k^*}\sum_{j=1}^{M}\tilde{\omega}_k^{j*}((\tilde{X}_k^j)_i - (\hat{X}_k)_i)^2}, \quad \hat{\mu}_{3,i,k} = \frac{1}{\tilde{\Omega}_k^*}\sum_{j=1}^{M}\tilde{\omega}_k^{j*}((\tilde{X}_k^j)_i - (\hat{X}_k)_i)^3,$$

where Ω_k and $\tilde{\Omega}_k^*$ have been defined in Eqs. 24.15 and 24.17, respectively.

We can also obtain another mode approximation based on necessary conditions for extrema of the marginal probability density approximation using a partial sum of Edgeworth series. Thus,

$$p_i^*(\xi) \approx \phi(\xi) - \frac{1}{3!}\frac{\mu_{3,i}}{\sigma_i^3}\phi^{(3)}(\xi) = \phi(\xi)\left(1 + \frac{1}{3!}\frac{\mu_{3,i}}{\sigma_i^3}H_3(\xi)\right).$$

The mode ξ_i^* is a root of the quartic equation:

$$\xi^4 - 6\xi^2 + \frac{6\sigma_i^3}{\mu_{3,i}}\xi + 3 = 0, \quad \mu_{3,i} \neq 0 \tag{24.19}$$

that is a consequence of the equation

$$\frac{dp_i^*(\xi)}{d\xi} = 0$$

and the recurrence relation for Hermite polynomials

$$\frac{dH_k(\xi)}{d\xi} = \xi H_k(\xi) - H_{k+1}(\xi), \quad H_0(\xi) = 1.$$

All roots of Eq. 24.19 can be found by Descartes–Euler method or Ferrari's method (Descartes–Euler method is more preferable because Eq. 24.19 has a "reduced" form [39]), and the marginal mode approximation $\xi^*_{i,k}$ is the root of Eq. 24.19 with substitutions $\sigma_i = \hat{\sigma}_{i,k}$ and $\mu_{3,i} = \hat{\mu}_{3,i,k}$ if $\hat{\mu}_{3,i,k} \neq 0$, and for this root the probability density $p^*_i(\xi)$ has the largest value. Consequently,

$$\hat{X}^{\mathrm{MAP}}(t_k) \approx \Xi_k, \quad (\Xi_k)_i = (\hat{X}_k)_i + \hat{\sigma}_{i,k}\xi^*_{i,k}, \quad i = 1, 2, \ldots, n, \tag{24.20}$$

and $\xi^*_{i,k} = 0$, $(\Xi_k)_i = (\hat{X}_k)_i$ if $\hat{\mu}_{3,i,k} = 0$.

The algorithms for solving approximately the optimal filtering problem that provide two estimates for the state are given below. These estimates are the unbiased estimate $\hat{X}^{\mathrm{MMSE}}(t)$ with a minimum mean squared error and the maximum a posteriori estimate $\hat{X}^{\mathrm{MAP}}(t)$ (see Eqs. 24.3–24.4 for details). Algorithm 1 is based on the sample paths simulation of the random process $X(t)$, and Algorithm 2 is based on the sample paths simulation of the random process $\tilde{X}(t)$. These algorithms correspond to Duncan–Mortensen–Zakai equation and the robust Duncan–Mortensen–Zakai equation, respectively.

Algorithm 1

1. Specify M, the number of paths for the random process $(X(t), \omega(t))$ to be simulated (sample size), specify h, the time discretization step such that there exists the natural number N for which $h = (T - t_0)/N$. Generate initial states X^j_0 according to a given distribution with the probability density $\varphi_0(x)$ and let $k = 0$, $\omega^j_0 = 1$, $j = 1, 2, \ldots, M$.
2. Estimate the conditional mean of the state and estimate the mode of the conditional distribution at time $t = t_k$ using the sample $(\{X^j_k\}^M_{j=1}, \{\omega^j_k\}^M_{j=1})$ as follows:

$$(\hat{X}^{\mathrm{MMSE}}_k)_i = (\hat{X}_k)_i = \frac{1}{\Omega_k}\sum_{j=1}^{M} \omega^j_k (X^j_k)_i, \quad (\hat{X}^{\mathrm{MAP}}_k)_i = (\hat{X}_k)_i - \frac{1}{2}\frac{\hat{\mu}_{3,i,k}}{\hat{\sigma}^2_{i,k}},$$

where $\hat{\sigma}_{i,k}$ is the estimate of the conditional standard deviation and $\hat{\mu}_{3,i,k}$ is the estimate of the conditional third central moment for the ith component X_i of the state X at time $t = t_k$ $(i = 1, 2, \ldots, n)$, i.e.,

$$\hat{\sigma}_{i,k} = \sqrt{\frac{1}{\Omega_k}\sum_{j=1}^{M} \omega^j_k ((X^j_k)_i - (\hat{X}_k)_i)^2}, \quad \hat{\mu}_{3,i,k} = \frac{1}{\Omega_k}\sum_{j=1}^{M} \omega^j_k ((X^j_k)_i - (\hat{X}_k)_i)^3,$$

and

$$\Omega_k = \sum_{j=1}^{M} \omega_k^j.$$

If $T - t_k = 0$, terminate the estimation process. Otherwise, let $j = 1$.

3. Obtain a realization of the state and corresponding weight at time $t = t_k + h$:
$X_{k+1}^j = X_k^j + hf(t_k, X_k^j) + \sqrt{h}\sigma(t_k, X_k^j)\Delta W_k^j,$
$\omega_{k+1}^j = \omega_k^j e^{c^{\mathrm{T}}(t_k, X_k^j)q(t_k)(Y(t_{k+1}) - Y(t_k)) - \frac{1}{2}c^{\mathrm{T}}(t_k, X_k^j)q(t_k)c(t_k, X_k^j)h},$

where ΔW_k^j is the realization of the s-dimensional random vector with independent components having a standard normal distribution.

4. If $j = M$, let $t_{k+1} = t_k + h$ and $k := k + 1$, go to Step 2. Otherwise, let $j := j + 1$ and go to Step 3.

Algorithm 2

1. Specify M, the number of paths for the random process $(\tilde{X}(t), \tilde{\omega}(t))$ to be simulated (sample size), specify h, the time discretization step such that there exists the natural number N for which $h = (T - t_0)/N$. Generate initial special states $\tilde{X}_0^j$ according to a given distribution with the probability density $\varphi_0(x)$ and let $k = 0$, $\tilde{\omega}_0^j = 1$, $j = 1, 2, \ldots, M$.
2. Estimate the conditional mean of the state and estimate the mode of the conditional distribution at time $t = t_k$ using the sample $(\{\tilde{X}_k^j\}_{j=1}^M, \{\tilde{\omega}_k^j\}_{j=1}^M)$ as follows:

$$(\hat{X}_k^{\mathrm{MMSE}})_i = (\hat{X}_k)_i = \frac{1}{\tilde{\Omega}_k^*} \sum_{j=1}^{M} \tilde{\omega}_k^{j*} \tilde{X}_k^j, \quad (\hat{X}_k^{\mathrm{MAP}})_i = (\hat{X}_k)_i - \frac{1}{2} \frac{\hat{\mu}_{3,i,k}}{\hat{\sigma}_{i,k}^2},$$

where $\hat{\sigma}_{i,k}$ is the estimate of the conditional standard deviation and $\hat{\mu}_{3,i,k}$ is the estimate of the conditional third central moment for the ith component X_i of the state X at time $t = t_k$ $(i = 1, 2, \ldots, n)$, i.e.,

$$\hat{\sigma}_{i,k} = \sqrt{\frac{1}{\tilde{\Omega}_k^*} \sum_{j=1}^{M} \tilde{\omega}_k^{j*} ((\tilde{X}_k^j)_i - (\hat{X}_k)_i)^2}, \quad \hat{\mu}_{3,i,k} = \frac{1}{\tilde{\Omega}_k^*} \sum_{j=1}^{M} \tilde{\omega}_k^{j*} ((\tilde{X}_k^j)_i - (\hat{X}_k)_i)^3,$$

and

$$\tilde{\Omega}_k^* = \sum_{j=1}^{M} \tilde{\omega}_k^{j*}, \quad \tilde{\omega}_k^{j*} = \tilde{\omega}_k^j \exp\{h^{\mathrm{T}}(t_k, \tilde{X}_k^j) Y(t_k)\}.$$

If $T - t_k = 0$, terminate the estimation process. Otherwise, let $j = 1$.

3. Obtain a realization of the special state and corresponding weight at time $t = t_k + h$:

$$\tilde{X}_{k+1}^j = \tilde{X}_k^j + h\tilde{f}(t_k, \tilde{X}_k^j, Y(t_k)) + \sqrt{h}\sigma(t_k, \tilde{X}_k^j)\Delta\tilde{W}_k^j,$$
$$\tilde{\omega}_{k+1}^j = \tilde{\omega}_k^j e^{\nu(t_k, \tilde{X}_k^j, Y(t_k))h},$$

where $\Delta\tilde{W}_k^j$ is the realization of the s-dimensional random vector with independent components having a standard normal distribution.

4. If $j = M$, let $t_{k+1} = t_k + h$ and $k := k+1$, go to Step 2. Otherwise, let $j := j+1$ and go to Step 3.

Note that in algorithms given above, Eq. 24.18 is used for obtaining approximately the mode of the conditional distribution. Equation 24.20 can also be applied to this. Thus, we will have

$$(\hat{X}_k^{\mathrm{MAP}})_i = (\hat{X}_k)_i + \hat{\sigma}_{i,k}\xi_{i,k}^*,$$

where $\xi_{i,k}^*$ is a root of the quartic equation

$$\xi^4 - 6\xi^2 + \frac{6\hat{\sigma}_{i,k}^3}{\hat{\mu}_{3,i,k}}\xi + 3 = 0, \quad \hat{\mu}_{3,i,k} \neq 0,$$

for which the function

$$\phi(\xi)\left(1 + \frac{1}{3!}\frac{\hat{\mu}_{3,i,k}}{\hat{\sigma}_{i,k}^3}H_3(\xi)\right), \quad H_3(\xi) = \xi^3 - 3\xi,$$

has the largest value. If $\hat{\mu}_{3,i,k} = 0$, then $(\hat{X}_k^{\mathrm{MMSE}})_i = (\hat{X}_k^{\mathrm{MAP}})_i$ for the considered mode approximation.

Also note that various methods for solving SDEs numerically, such as Runge–Kutta type methods [37, 40, 41], Rosenbrock type methods [40, 42], Platen's methods [37], Milshtein's methods [43], and Kuznetsov's methods [44, 45] instead of the Euler–Maruyama method should be applied to increase the estimation accuracy.

24.5 Conclusions

The filtering algorithms based on the particle method for nonlinear continuous-time stochastic observation systems for the unbiased estimate with a minimum mean squared error and the maximum a posteriori estimate are suggested in this chapter.

For obtaining the maximum a posteriori estimate, Edgeworth series for the expansion of marginal conditional probability densities is applied. This approach allows to estimate the mode of the conditional distribution approximately, but it significantly reduces the computation time in contrast to finding the mode by consistent estimating the conditional probability density. Particle method is used not only for Duncan–Mortensen–Zakai equation, but also for the robust Duncan–Mortensen–Zakai equation.

Acknowledgements This work is partially supported by the Russian Foundation for Basic Research, project no. 17-08-00530.

References

1. Yavin, Y.: An alternative approach to non-linear filtering: maximizing the probability of hitting a target set. Int. J. Syst. Sci. **13**(3), 289–299 (1982)
2. Bar-Shalom, Y., Li, X.R., Kirubarajan, T.: Estimation with Applications to Tracking and Navigation: Theory, Algorithms and Software. Wiley, New York (2001)
3. Pany, T.: Navigation Signal Processing for GNSS Software Receivers. Artech House, Boston (2010)
4. Stepanov, O.A., Koshaev, D.A., Motorin, A.V.: Identification of gravity anomaly model parameters in airborne gravimetry problems using nonlinear filtering methods. Gyroscopy Navig. **6**(4), 318–323 (2015)
5. Khalaf, W., Chouaib, I., Wainakh, M.: Novel adaptive UKF for tightly-coupled INS/GPS integration with experimental validation on an UAV. Gyroscopy Navig. **8**(4), 259–269 (2017)
6. Shavrin, V.V., Tislenko, V.I., Lebedev, VYu., Filimonov, V.A., Konakov, A.S.: Sigma-point Kalman filter algorithm in the problem of GNSS signal parameters estimation in non-coherent tracking mode in spacecraft autonomous navigation equipment. Gyroscopy Navig. **9**(4), 255–266 (2018)
7. Rudenko, E.A.: Autonomous path estimation for a descent vehicle using recursive Gaussian filters. J. Comput. Syst. Sci. Int. **57**(5), 695–712 (2018)
8. Rybakov, K.A.: Solving the nonlinear problems of estimation for navigation data processing using continuous particle filter. Gyroscopy Navig. **10**(1), 27–34 (2019)
9. Stepanov, O.A., Vasiliev, V.A., Toropov, A.B., Loparev, A.V., Basin, M.V.: Efficiency analysis of a filtering algorithm for discrete-time linear stochastic systems with polynomial measurements. J. Franklin Inst. **356**(10), 5573–5591 (2019)
10. Cramér, H.: Mathematical Methods of Statistics. Princeton University Press, Princeton (1999)
11. Zakai, M.: On the optimal filtering of diffusion processes. Zeitschrift für Wahrscheinlichkeitstheorie und Verwandte Gebiete **11**(3), 230–243 (1969)
12. Hazewinkel, M.: Lectures on linear and nonlinear filtering. In: Schiehlen, W.O., Wedig, W. (eds.) Analysis and Estimation of Stochastic Mechanical Systems. International Centre for Mechanical Sciences (Courses and Lectures), vol. 303, pp. 103–136. Springer, Vienna (1988)
13. Baras, J.S., Blankenship, G.L., Mitter, S.K.: Nonlinear filtering of diffusion processes. In: 8th IFAC Congress, article id 23.1 (1981)
14. Baras, J.S., Blankenship, G.L., Hopkins, W.E.: Existence, uniqueness, and asymptotic behavior of solutions to a class of Zakai equations with unbounded coefficients. IEEE Trans. Autom. Control **28**(2), 203–214 (1983)
15. Yau, S.-T., Yau, S.S.-T.: Real time solution of Duncan-Mortensen-Zakai equation without memory. In: 47th IEEE Conference on Decision and Control, pp. 5086–5091 (2008)

16. Rybakov, K.A.: Robust Duncan-Mortensen-Zakai equation for non-stationary stochastic systems. In: International Multi-Conference on Engineering, Computer and Information Sciences (SIBIRCON), pp. 151–154 (2017)
17. Luo, X., Yau, S.S.-T.: Complete real time solution of the general nonlinear filtering problem without memory. IEEE Trans. Autom. Control **58**(10), 2563–2578 (2013)
18. Crisan, D.: Exact rates of convergence for a branching particle approximation to the solution of the Zakai equation. Ann. Probab. **31**(2), 693–718 (2003)
19. Bain, A., Crisan, D.: Fundamentals of Stochastic Filtering. Springer, New York (2009)
20. Del Moral, P.: Feynman-Kac Formulae: Genealogical and Interacting Particle Systems with Applications. Springer, New York (2004)
21. Rybakov, K.A.: Statistical Methods of Analysis and Filtering for Continuous Stochastic Systems. MAI Publ, Moscow (in Russian) (2017)
22. Øksendal, B.: Stochastic Differential Equations. An Introduction with Applications. Springer, Berlin (2000)
23. Pugachev, V.S., Sinitsyn, I.N.: Stochastic Systems: Theory and Applications. World Scientific, Singapore (2002)
24. Veretennikov, A.J.: On strong solutions and explicit formulas for solutions of stochastic integral equations. Math USSR Sbornik **39**(3), 387–403 (1981)
25. Veretennikov, A.J.: On stochastic equations with degenerate diffusion with respect to some of the variables. Math USSR Izvestiya **22**(1), 173–180 (1984)
26. Jazwinski, A.H.: Stochastic Processes and Filtering Theory. Academic Press, New York (1970)
27. Stratonovich, R.L.: Conditional Markov Processes and Their Application to the Theory of Optimal Control. Elsevier (1968)
28. Kushner, H.J.: On the differential equations satisfied by conditional probability densities of Markov processes, with applications. J. SIAM Ser. A: Control **2**(1), 106–119 (1964)
29. Kosachev, I.M.: Methodology of high-precision nonlinear filtering of random processes in stochastic dynamical systems with a fixed structure. Mil. Acad. Republic of Belarus Proc. **4**, 125–161 (un Russian) (2014)
30. Dashevskii, M.L.: Cumulant method for closing moment equations in analyzing nonlinear systems. Prob. Control Inf. Theory **4**, 317–328 (in Russian) (1975)
31. Kashkarova, A.G., Shin, V.I.: Modified cumulant methods of analyzing stochastic systems. Autom. Remote Control **2**, 69–79 (in Russian) (1986)
32. Socha, L.: Linearization Methods for Stochastic Dynamic Systems. Springer, Berlin (2008)
33. Kosachev, I.M., Eroshenkov, M.G.: Analytical Modeling of Stochastic Systems. Nauka i tekhnika Publ, Minsk (in Russian) (1993)
34. Rybakov, K.A.: Solving approximately an optimal nonlinear filtering problem for stochastic differential systems by statistical modeling. Numer. Anal. Appl. **6**(4), 324–336 (2013)
35. Kazakov, I.E., Artemiev, V.M., Bukhalev, V.A.: Analysis of systems with random structure. Fizmatlit, Moscow (in Russian) (1993)
36. Averina, T.A., Rybakov, K.A.: An approximate solution of a prediction problem for stochastic jump-diffusion systems. Numer. Anal. Appl. **10**(1), 1–10 (2017)
37. Kloeden, P.E., Platen, E., Schurz, H.: Numerical Solution of SDEs Through Computer Experiments. Springer, Berlin (1994)
38. Silverman, B.W.: Density Estimation for Statistics and Data Analysis. Chapman & Hall/CRC, London (1986)
39. Korn, G.A., Korn, T.M.: Mathematical Handbook for Scientists and Engineers. Dover Publ, New York (2000)
40. Artemiev, S.S., Averina, T.A.: Numerical Analysis of Systems of Ordinary and Stochastic Differential Equations. VSP, Utrecht (1997)
41. Roberts, A.J.: Model Emergent Dynamics in Complex Systems. SIAM, Philadelphia (2014)
42. Averina, T.A., Karachanskaya, E.V., Rybakov, K.A.: Statistical modeling of random processes with invariants. In: International Multi-Conference on Engineering, Computer and Information Sciences (SIBIRCON), pp. 34–37 (2017)

43. Milshtein, G.N., Tretyakov, M.V.: Stochastic Numerics for Mathematical Physics. Springer, Berlin (2004)
44. Kuznetsov, D.F.: On numerical modeling of the multidimensional dynamic systems under random perturbations with the 1.5 and 2.0 orders of strong convergence. Autom. Remote Control **79**(7), 1240–1254 (2018)
45. Kuznetsov, D.F.: On numerical modeling of the multidimensional dynamic systems under random perturbations with the 2.5 order of strong convergence. Autom. Remote Control **80**(5), 867–881 (2019)

Chapter 25
Essentials of Fractal Programming

Alexander S. Semenov

Abstract Fractal programming is a programming paradigm based on the concept of "elastic objects", which can transformed (unfolded and folded) dynamically at run time using strategy planning model and production rules. These rules are keeping the object structure self-similar, thus it has fractal property: parts similar to the whole. The paradigm aims to optimize searching of suitable workflow structure of the object at run time. The elastic object models, production rules integrated by iterated algebraic system, and adjustment strategy planning model are introduced.

25.1 Introduction

In response to fluctuating demand of the society and the increased connectivity of people, things, and services, IT systems themselves are becoming highly dynamic. Run time factors on demand are increasingly determining the elasticity of a system. The elasticity is becoming main characteristic of the technologies: cloud computing, blockchain, Internet of things, digital platforms, robotics, biometrics, persuasive technology, and augmented reality.

The term elasticity means "the degree to which a system is able to adapt to workload changes by provisioning and de-provisioning resources in an autonomic manner, such that at each point in time, the available resources match the current demand as closely as possible" [1]. The elastic computing is one of the most important design goals facing software developers.

The traditional static analysis is based on the assumption that a system consists of objects with unchanged structure. If a system is expected to be elastic at run time, its ability to elasticity must be considered, when it is designed. Hence, the elasticity of object must be modeled and be built into it at design time. All this means an evolved

A. S. Semenov (✉)
Moscow Aviation Institute (National Research University), 4, Volokolamskoe Shosse, Moscow 125993, Russian Federation
e-mail: Semenov_Alex@yahoo.com

L. C. Jain et al. (eds.), *Advances in Theory and Practice of Computational Mechanics*, Smart Innovation, Systems and Technologies 173,
https://doi.org/10.1007/978-981-15-2600-8_25

"shift" [?] from the static analysis to analysis with the elastic dynamicity of objects at run time.

In this chapter, a technique that is able to evaluate the effects of elasticity at design time is introduced. This technique is based on the concept of elastic objects and fractal Iterated Algebraic System (IAS). Elastic objects are transformed dynamically at run time by fractal algebra operations. The term fractal highlights the properties of elastic objects: autoscaling (ability for unfolded and folded structure of an object), pattern ability, similarity, and symmetry. The operation of replication (prototyping) makes possible prototype elastic objects in the manner of prototype-based programming. An elastic object structure is optimized by the planning model. This underpins the relevance of the ideas presented in this chapter.

The chapter is organized as follows. In Sect. 25.2, the elastic object model is considered. The elastic object model is based on a system of predefined scaling conditions that automatically handles IT resources from resource pools [3]. Elastic objects can be scalable with respect to its size, meaning that they automatically add objects to the system. Modeling replication is a technique for achieving scalability [4]. Replication is a key for providing high availability and fault tolerance in elastic objects [5].

In Sect. 25.3, models of fractal elastic objects, such as container–component, architectural model based on fractal graphs, and fractal Petri nets, are considered. All these elastic objects models are based on IAS. Operations of IAS are defined and exemplified for each model of elastic object. The basic idea of the fractal [6] scalability in this context took into account. Fractal programming is a technique for achieving the goal of elasticity objects and their computing. Each elastic object changes its structure on demand by the fractal rules. The model of strategy planning of elasticity as a composite part of elastic object is included. Assumption is made that adding elastic objects to a system independently increases a processing potential [5, 7]. Lastly, Sect. 25.4 concludes the chapter.

25.2 The Elastic Object Model

The elasticity is a paradigm that initially known in physics and widely used in economics. In physics, it means the property of some material returning to an initial form or state after the following deformation. For computer systems, this means that with the increased workload of the service take place "deformation", i.e., it negatively impacted the users. Due to this impact, the capacity of the service needs to be increased by IT resources. This may be repeated several times over a certain period of time until the workload returns to its initial value. The service workload and capacity are at the core of the elastic computing.

For coordinating elasticity, a system needs concepts of elastic objects—fundamental building blocks for engineering an end-to-end elasticity. Modern software development techniques evolve around the use of object-orientation approach [8].

Service-oriented approach [9] is an evolution of the object-oriented approach. Hence, designing elastic object is considered as an integral part of a service.

To design an elastic object, it is necessary to analyze it from the point of view of autoscaling (replicating and composition of objects) that depends on the workload of planning strategy, provisioning and binding of IT resources, and synthesis. Summarizing these points of view, the next formula is written as follows:

$$\text{Elasticity} = \text{Autoscaling} + \text{Planning strategy} + \text{Binding} + \text{Synthesis} \quad (25.1)$$

Automatic scaling (autoscaling) is a prerequisite for the elastic object. Method autoscaling of the object is the ability to handle fluctuated workload (without adding IT resources to the object) or by adding IT resources to the object for extending a system's capacity. Scalability of the object needs to be built into it at the design time.

The elasticity is a degree, to which a service autonomously adapts capacity to a workload over time. The elasticity needs to be considered over time for changing workloads. Such an adaptation must be timely minimized [10]. The timeliness entails that adaptation process is autonomous [11]. The workload characterizes the data to be processed by a service's operations on fluctuating users demand during the day, see a typical example Table 25.1.

Let a service has enough capacity at a workload of 2,000 users (see Fig. 25.1). Thus, the service's capacity defines the maximum workload, which the system can handle in terms of number of users. If the workload during some period of time increased up to 3,000 users, then the capacity of the elastic computer system must be transformed dynamically.

There are two main inheritance models in the object-oriented approach: class-based [8] and prototype-based [12]. In prototype-based model, only objects exist. New objects are produced by prototyping and modification of prototypes. Prototyping objects and mechanisms for scalable data replication are the important techniques of

Table 25.1 Timetable of workload observations

S	Time	Workload
1	2	1,000
2	4	2,000
3	6	3,000
4	8	4,000
5	10	5,000
6	12	6,000
7	14	7,000
8	16	8,000
9	18	9,000
10	20	6,000
11	22	4,000
12	24	3,000

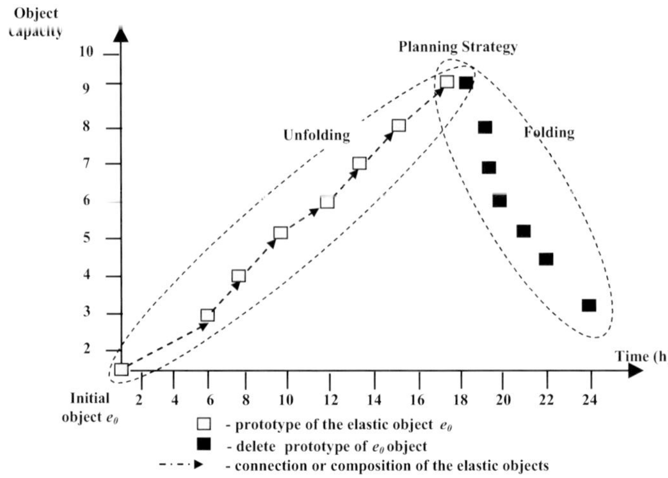

Fig. 25.1 Strategic planning model

achieving structure of services with good performance, high availability, and fault tolerance in a distributed system [13].

When an application requires more resources than are available, it negatively impacted the users. Because of it, elastic computing must be provided for IT resources: hardware and software entities, such as servers, CPU cores, memory, program components, and modules. If a workload was changed, the capacity must be transformed dynamically at run time on demand (see Fig. 25.1) by prototyping the initial object e_0.

Thus, if a workload was decreased, for example, after 9,000 users, the capacity of the system must be decreased (see Fig. 25.1) by deleting the prototyped objects. The elastic computing acquires and releases IT resources, when demand changes that requires support for late binding of IT resources. This technique has been successfully applied in programming language environments but also for operating systems, where modules can be loaded and unloaded at will. An application eliminates the binding between software and hardware through virtualization. The elastic applications would benefit if the underlying infrastructure provided the binding and replication for the elasticity. Operations must be themselves elastic. There are the following binding techniques: resource-to-node, client-to-server, process-to-resource, and geographical binding [14].

The granularity is usually used to characterize how many objects must be prototyped with defined capacity at a certain time. The level of granularity should be sufficient. The granularity trade-off is important at the cost of precision, accuracy, and scope [2].

The cost-efficiency is closely related to the optimal provisioning. The optimal provisioning strikes a balance between over-provisioning and under-provisioning.

Transformations of the elastic object depend on provisioning of IT resources. It must be taken into account by planning strategy.

Figure 25.2 generalizing Eq. 25.1, the model of the elastic object, is introduced. The elastic object is a container which consists of components: attributes, autoscaling operations (prototyping, composition), model of planning strategy, synthesis or modification models, and pool of IT resources for the object (software and hardware).

The container is an object (in terms of object-oriented approach), which aggregates components or containers. A component is an object, which aggregates by the container.

In Fig. 25.3, the elastic computing platform is presented as the elastic object. At any given level of abstraction, meaningful collections of elastic objects collaborate to achieve some higher level behavior. This is exactly the fractal organization of complexity [15].

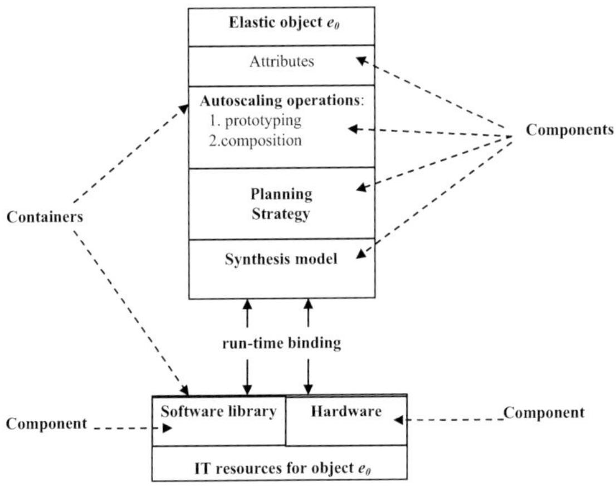

Fig. 25.2 The elastic object model

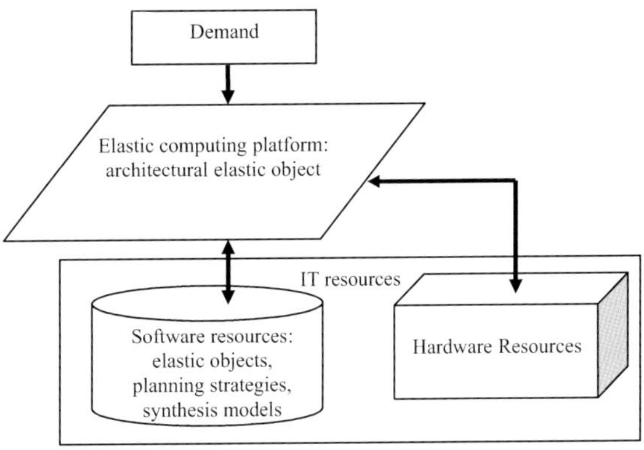

Fig. 25.3 The elastic computing platform is the elastic object

In the next section, the following models of elastic objects are defined: the container–component model, the architectural model based on graphs [16], and Petri nets model. These models are essentials of fractal programming.

25.3 Fractal Programming

Fractal programming is a programming paradigm based on the concept of "elastic objects", which can be transformed (unfolded and folded) dynamically at run time using strategy planning model and production rules. These rules are keeping the object structure self-similar, thus it has the fractal property: parts similar to the whole. The paradigm aims to optimize searching suitable workflow structure of the object at run time. The elastic object models, production rules integrated by IAS, and adjustment strategy planning model are introduced. IAS is used for autoscaling: unfolding and folding elastic objects mainly by operation prototyping. In dependence on the model of elastic object, the operation prototyping is overloaded, and additional operations are included in the fractal algorithm.

The fractal algorithms are based on constructing objects with requirement properties, such as self-similarity, scaling, and fractal dimension [6, 17]. The term fractal is used to highlight the properties of elastic objects: abstraction, ability for unfolding and folding, pattern ability, similarity, symmetry, and scalability. In many works on fractals, self-similarity is used as defining property. In this chapter, the ordered bag of self-similar prototyping objects is used. The ordered bag makes possible to define the position of object in it.

The bag (for example, implemented by container) could include repeatable objects the big number of times. The assumption was made that self-similarity describes the objects, in which the same objects are repeated over and over again on different levels of the scalability, for example, binary tree (the part looks like the whole).

Hereinafter, a model of container–component elastic objects is considered in Sect. 25.3.1, while a model of architectural elastic objects is represented in Sect. 25.3.2. Section 25.3.3 provides a model of fractal Petri nets as elastic objects.

25.3.1 The Model of Container–Component Elastic Objects

In all self-similar constructions, there is a relationship between the scaling factor and the number of parts that the original object is divided into. The relationship is the power law:

$$Nr^d = 1, \tag{25.2}$$

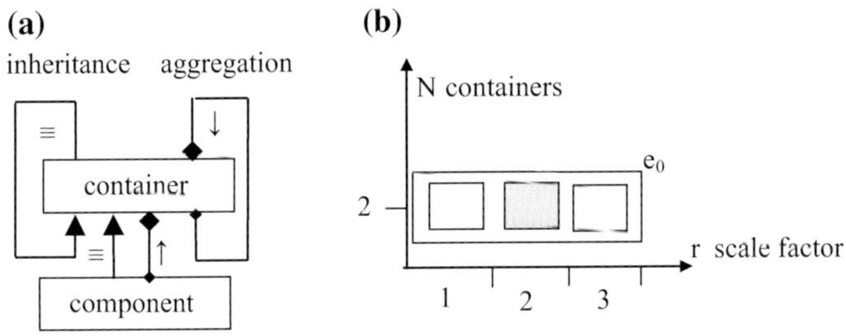

Fig. 25.4 The elastic object as recursive relationships between containers and components: **a** inheritance and aggregation relationships, **b** the elastic object e_0 ($r = 1/3$, $N = 2$, $k = 1$)

where N is the number of self-similar objects, $1/r^d$-, is the reduction scaling factor, d is the fractal dimension.

Let E be the container–component elastic object. The operation prototyping and composition (aggregation of prototyping objects) are carried out repeatedly; the output of iteration is the input for the next one. For components, operations prototyping and composition are undefined. As a whole, the container–component elastic object is an object based on recursive relationships between containers and components.

Definition 1 Elastic object E = < e_0, ≡ ($1/r$, N), ↓ > , where e_0 is an initial container, $\equiv : e_0 \longrightarrow E$ is the operation prototyping object e_0 in an ordered multi bag $E = [e_1, e_2, o_1, o_2, \ldots, o_{\text{r-}N}, \ldots, e_N]$, which consists from N containers (e_N are parts) and $1/r$–N components ($o_{1/r-N}$) (here, the index shows the position of each self-similar object in the bag), $\downarrow : E \longrightarrow e_0$ is the operation aggregation. An ordered multi bag E aggregates by container e_0 (it is whole).

In Fig. 25.4a, the relationships of inheritance and aggregation between containers and components are shown. In Fig. 25.4b, an object e_0 with scale factor $r = 1/3$, two containers $N = 2$, and one component $1/r-N$ is shown.

Algorithm 1 The scaling of the elastic object.

$$E = \{e_0\}, n = 1$$

$$\forall e \in E \text{ repeat } \downarrow (\equiv (1/r, \text{N})e)n \text{ times.}$$

Graphically, container is depicted as a rectangle. Nested containers are depicted as nested rectangles. Component is depicted as gray rectangle. Positions of the containers are ordered by indexes.

In Fig. 25.5, the second iteration $k = 2$ (level 2) of the elastic object e_0 is shown. IAS generalizes the composition of elastic objects and makes possible unfolded and folded its structure [14].

Definition 2 Fractal algebra is a tuple fa = <E, Δ> , where E is the ordered multi bag of elastic objects, Δ is the set of uniquely invertible operations. Δ = { ≡ ($1/r$,

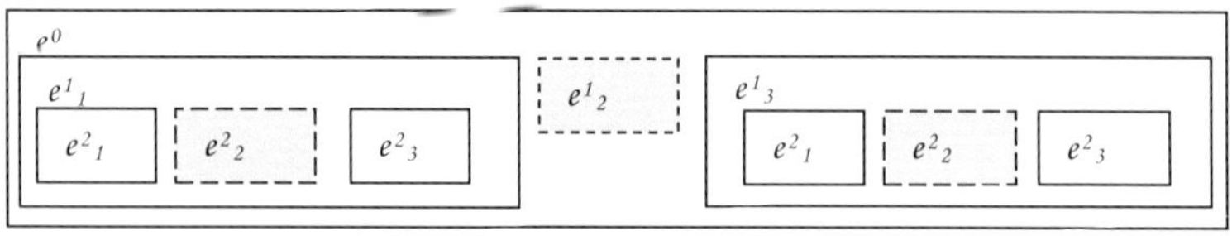

Fig. 25.5 Elastic object: $r = 1/3$, $N = 2$, $k = 2$

N), $\downarrow$}, where $\equiv$ $(1/r;\ N)$ is an operation of prototyping (see Definition 1) $\equiv$ $(1/r, N)^{-1}$ is an invertible operation, $\downarrow$ is the operation of aggregation E containers (see Definition 1), $\downarrow^{-1}$ is the invertible operation.

Let $f : E \longrightarrow E$ be a recursive mapping of an ordered multi-bag E in self-similar an ordered multi bag E by *fa* operations.

Definition 3 Iterated Algebraic System (IAS) $f^n : E \longrightarrow E$ is the recursive mapping of an ordered multi-bag E, where $n = 0,1,\ldots,k$ is the step of mapping, n++ is the unfolding operation, $n = n + 1$, n−− is the folding operation, $n = n - 1$.

Definition 4 Fractal algorithms, in which the *fa* operations presented by rules, have the following notation:

$$F = f^n(E, [R]), \tag{25.3}$$

where F is the ordered multi bag or fractal object (result of mapping), $R = [r_1, r_2, \ldots, r_m]$ is the ordered set of rules. Rules consist of *fa* operations and conditions.

For consideration, the example of the elastic object algorithm is written as follows, $k = 3$ (level):

$$e_0 = f^n(E = \{e_0\}, [n++, \forall e \in E^n, e \downarrow (E^n == \equiv (3, 2)e) | 0 < n =< k]). \tag{25.4}$$

In all the above examples, the elastic object unfolded its structure. For the purpose of folding the elastic object, the uniquely invertible operations were included in fractal algebra. For generalized unfolded and folded elastic object, Eq. 25.5 is used.

$$\begin{aligned} e_0 = f^n(E = \{e_0\} \quad & [n++, \forall e \in E^n,\ \ e \downarrow (E^n == \equiv (3, 2)e) | 0 < n =< k] \\ & [n--, \forall e \in E^n,\ \ e \downarrow^{-1} (E^n == \equiv^{-1} (3, 2)e) | 0 < n >= k]) \end{aligned} \tag{25.5}$$

The container–component model of the elastic object is based on fractal interpretation of constructing the container and component. It makes possible to build a model of adjustment capacity planning.

Adjustment planning strategy is adding or reducing the capacity of IT system in small or large amounts due to consumer's demand.

Let w be the workload and c be the capacity needed to handle workload w. A timetable of observations of w_n is given in Table 25.1. Our objective is to build a

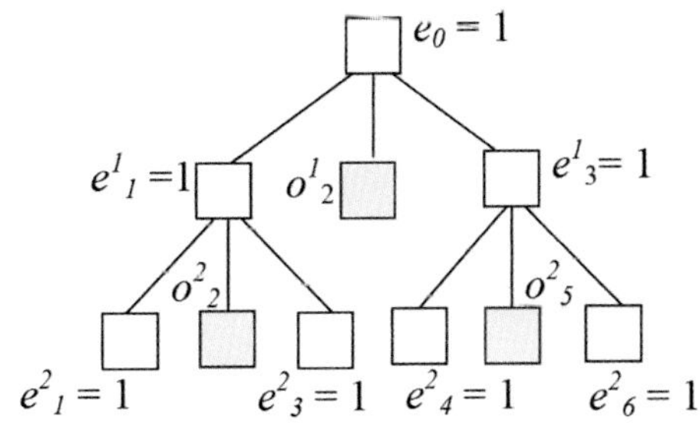

Fig. 25.6 The elastic object as a tree: $r = 1/3$, $N = 2$, $k = 2$

model using fractal algorithm that captures the relationship between changes in the workload and capacity requirement.

Let e_0 be a server implements by elastic object. In general, IT capacity planning involves estimating the storage, computer hardware, software, and connection infrastructure resources required over some future period of time. Let the server has a capacity $c = 1$ unit for 3,000 workload and each prototyping container of the server has the capacity of 1 unit.

In Fig. 25.6, a container–component tree of the elastic object with marked capacity is depicted, $k = 2$. Components are used for replication the shared data, which, in its turn, are applied by container. For workload of 9,000 will need k containers, $k = w_{n=18}/c$:

$$e_0 = f^n(E = \{e_0\}, [n++, \forall e \in E^n, e \downarrow (E^n =\equiv (3,2)e)|0 < n =< w_n/c],$$
$$[n--, \forall e \in E^n, e \downarrow^{-1} (E^n =\equiv^{-1} (3,2)e)|0 < n >= w_n/c]). \tag{25.6}$$

The elastic computing is a run time optimization of elastic objects. One of the potential problems is that the elasticity takes time.

25.3.2 The Model of Architectural Elastic Objects

Fractal graphs by applying elementary operations of fractal algebra to the initial graph g are defined axiomatically [15]. A set of elementary operations forms the basis of fractal algebra.

Definition 5 Fractal algebra $\Delta = \{g \mid \equiv, \leftrightarrow, \div\}$ for graphs over g is called the set of uniquely invertible operations: prototyping ($\equiv^{-1}$), connection ($\leftrightarrow^{-1}$), and subdivision ($\div^{-1}$).

Definition 6 Fractal graph g is a graph generated by the operations of fractal algebra. Subdivision (denoted by $\div$) edges $\{v, v'\}$ of the graph g connecting vertices v,v' by inserting an additional vertex v'' and two edges incident to v'' $\{v,v''\}$, $\{v'',v'\}$ is denoted by $(v \in g) \div (v' \in g)$. As a result, the graph g' and g are homeomorphic.

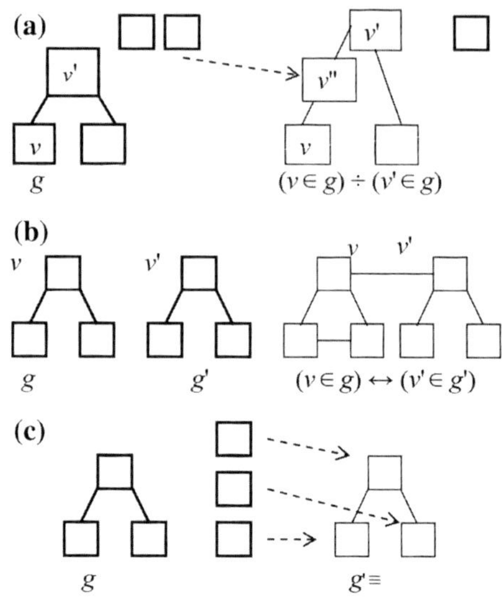

Fig. 25.7 Operations of the fractal algebra Δ for graphs: **a** operation on subdivision, **b** operation connection, **c** operation prototyping

The operation of the subdivision is illustrated in Fig. 25.7a. Let the graph g has two vertices, then $g_1 = (v \in g) \div (v' \in g)$. The result of the inverse operation applied to the graph g_1 is the graph $g = (v \in g_1) \div^{-1} (v' \in g_1)$.

Connection (denoted by $\leftrightarrow$) of the distinguished vertex v of the graph g and its prototype v' in the graph g' by the edge $\{v, v'\}$ is denoted by $(v \in g) \leftrightarrow (v' \in g')$. The connection operation is illustrated in Fig. 25.7b. The result of the connection graph $g_1 = (v \in g) \leftrightarrow (v' \in g')$. The result of the inverse operation applied to the graph g_1 is $\{g, g'\} = (v \in g_1) \leftrightarrow^{-1} (v' \in g_1)$.

Prototyping (denoted by$\equiv$) making a copy g' of a connected graph-sample g. Figure 25.7c illustrates the prototyping operation. Prototype g' of a graph-sample g is created. The result of the inverse operation applied to the graph $g' \equiv^{-1}$ is the graph g.

The class of the graph is determined by rules that implement autoscaling: line (L), mesh (M), hypercube (H), and tree (T) graphs.

The following algorithms are based on IAS (see Eq. 25.3), operation prototyping omitted, as well as, a graph g' is a prototype of a graph g.

Algorithm 2 $L = f^{n}(g_0, [(v \in g) \leftrightarrow (v' \in g')])$ is the class of linear graphs see (Fig. 25.8a).

Algorithm 3 $M = f^{n}(g_0, [(v_i \in g) \leftrightarrow (v_i' \in g'), (v_j \in g) \leftrightarrow (v_j' \in g')])$ is the class of lattices, where v_i, v_j, $i \neq j$ are two pairs of isomorphic vertices of the graph g and the prototype graph g' (see Fig. 25.8b).

Algorithm 4 $H = f^{n}(g_0, [\forall\, v,v' ((v \in g) \leftrightarrow (v' \in g'))])$ is the class of hypercubes, the connection operation is performed for all isomorphic vertices of the graphs g and g' (see Fig. 25.8c).

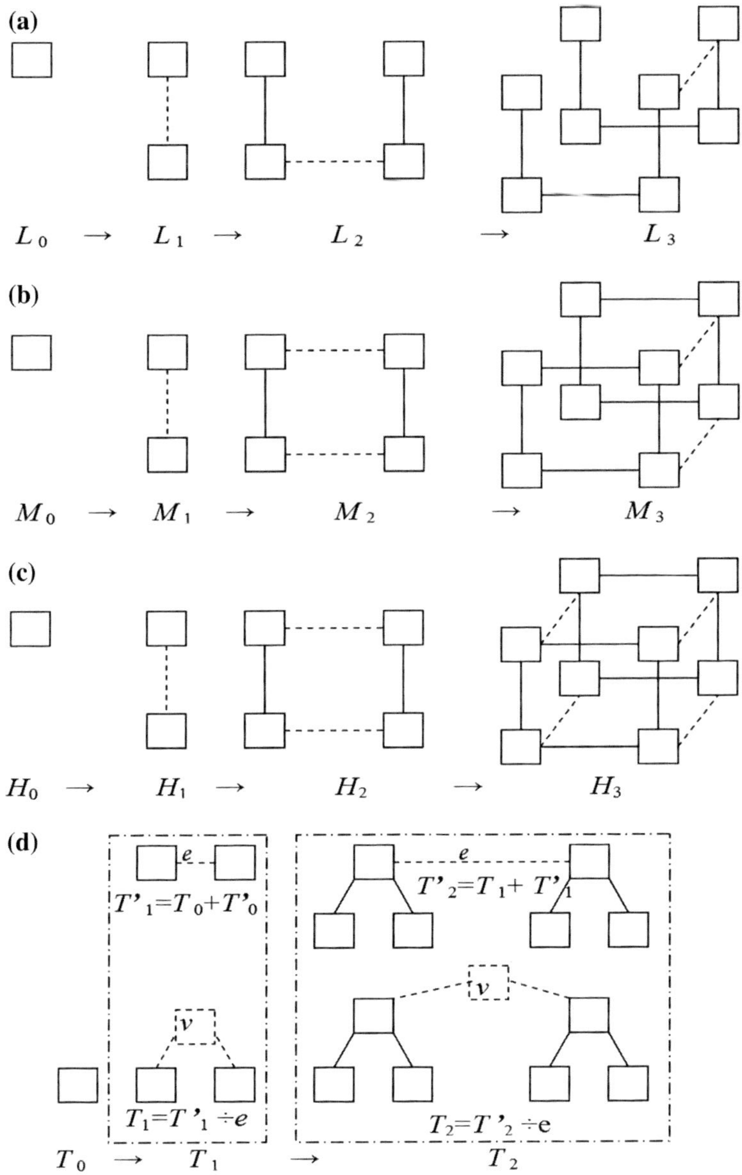

Fig. 25.8 Autoscaling of the fractal graphs: **a** line, **b** mesh, **c** hypercube, **d** tree

Algorithm 5 $T = f^n(g_0, [(v \in g) \leftrightarrow (v' \in g'), (v \in g) \div (v' \in g')])$ is the class of trees, where v, v' are the vertices of the graph g and its prototype (see Fig. 25.8d).

For the presented classes of graphs, the initial graph is the graph g_0 with one vertex.

In Fig. 25.8, the elements of fractal graphs depicted by the operations of connection and subdivision are highlighted by a dotted line.

Uniquely invertible operations are used for destructing subgraphs of initial graph. The elastic object planning strategy can be nested in these algorithms, for example, class line (the operation prototyping omitted):

$$L = f^n(g_0, [n++, (v \in g) \leftrightarrow |(v' \in g')|0 < n =< w_n/c],$$
$$[n--, (v \in g) \leftrightarrow^{-1} (v' \in g')|0 < n >= w_n/c]) \quad (25.7)$$

25.3.3 The Model of Fractal Petri Nets as Elastic Objects

Fractal Petri (FP) nets dynamically synthesized from self-similar Place/Transition subnets on the base of IAS have been introduced in [18, 19] and analyzed in [20]. Here, the example consumer–producer modeled by FP-net is considered as elastic object [19]. FP-net consists of one producer, one consumer, and buffer between them:

1. The buffer (*BUF*) may contain at most three tokens (messages).
2. The producer *PRODUCE* two tokens in each step and *SEND* them into buffer. The production steps of the producer are counted.
3. Consumer *RECIEVE* one token when accessing the buffer and *CONSUME* it.

A demand of the consumer is modeled by N^1, a capacity of the producer is modeled by N^2. FP-nets N^1 and N^2 are the elastic objects.

In Fig. 25.9, FP-net composed of two elastic objects (net N^1 and N^2 with shared resource place—buffer) is shown. For producer and consumer, each of these objects has elasticity: unfolded or folded.

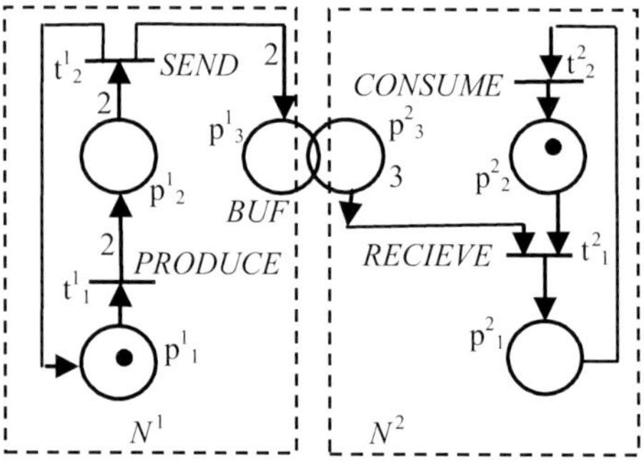

Fig. 25.9 A Producer–consumer FP-net

There are three cases of elasticity in this FP-net: *1:M* (one producer to many consumers), *M:1* (many producers to one consumer), *M:M* (many producers to many consumers).

FP-net for *1:M* (one producer to many consumers) system can be written:

$$f^n(FP, [N^n =\equiv N^2, p_3^1 \in N^1 \# p_3^n \in N^n | n > 1]), \tag{25.8}$$

where # is the operation overlapping places that makes possible to control and planning capacity of the producer.

FP-net for *M:1* (many producers to one consumer) system can be written:

$$f^n(FP, [N^n =\equiv N^1, p_3^n \in N^n \# p_3^2 \in N^2 | n > 1]). \tag{25.9}$$

FP-net for *M:M* (many producers to many consumers) system can be written:

$$\begin{aligned} f^{n,m}(FP, [N^n &= \equiv N^1, p_3^n \in N^n \# p_3^2 \in N^2 | n > 1, N^m \\ &= \equiv N^2, p_3^m \in N^m \# p_3^1 \in N^1 | m > 1]). \end{aligned} \tag{25.10}$$

Uniquely invertible operations are used for destructing prototyping subnets of FP-net. The elastic object planning strategy can be nested in these algorithms, for example, producer N^1 must change capacity depends on consumer workload N^2:

$$\begin{aligned} &f^n(FP, [n++, N^n =\equiv N^1, p_3^1 \in N^2 \# p_3^n \in N^n | 0 < n =< w_n/c], \\ &\quad [n--, N^n =\equiv^{-1} N^1, p_3^1 \in N^2 \#^{-1} p_3^n \in N^n | 0 < n >= w_n/c]). \end{aligned} \tag{25.11}$$

25.4 Conclusions

The main contribution of the chapter is the integration theory of fractal objects by means of IAS, the elastic object models, and object-oriented approach into a new paradigm called as Fractal-Oriented Programming (FOP). FOP aims to optimize searching suitable workflow structure of the object at run time and consolidates different statements to elastic systems. FOP is a way of programming that takes into account the automatic elasticity of objects at run time based on fractal scaling. It does so by adding additional behavior to existing code modifying the code itself by strategy planning model. Moreover, the properties of the result code remain predictable due to the fractal properties of elastic objects. This gives a new type of abstraction, encapsulation, inheritance, modularity, and concurrency of objects, which are indicated in this chapter. FOP forms a basis for fractal-oriented analysis and development of elastic systems.

References

1. Herbst, N.R., Kounev, S., Reussner, R.: Elasticity in cloud computing: what It is, and what it is not. In: 10th International Conference Autonomic Computing, San Jose, CA, pp. 23–27 (2013)
2. Becker, S., Brataas, G., Lehrig, S. (Eds.): Engineering Scalable, Elastic, and Cost-efficient Cloud Computing Applications: the Cloudscale Method. Springer International Publishing AG (2017)
3. Erl, T., Mahmood, Z., Puttini, R.: Cloud Computing: Concepts, Technology & Architecture. Prentice Hall, New Jersey (2013)
4. Tanenbaum, A.S., Steen, M.: Distributed Systems: Principles and Paradigms. Martin van Steen (2016)
5. Coulouris, G., Dollimore, J., Kindberg, T., Blair, G.: Distributed systems concepts and design. Addison Wesley, Boston, Massachusetts (2012)
6. Crownover, R.: Introduction to Fractals and Chaos. Jones and Bartlett Publishers, Inc. (1995)
7. Antonopoulos, N., Gillam, L. (eds.): Cloud Computing Principles, Systems and Applications. Springer, London Limited (2010)
8. Booch, G., Jacobson, I., Rumbaugh, J. (eds.): Object-Oriented Analysis and Design with Applications, 3rd edn. Addison-Wesley, Pearson Education, Inc (2007)
9. Erl, T.: Service-Oriented Architecture: Concepts, Technology, and Design. Prentice Hall PTR, Publisher (2005)
10. Furht, B., Escalante, A. (eds.): Handbook of Cloud Computing. Springer Science-Business Media, LLC (2010)
11. Wilder, B.: Cloud Architecture Patterns. O'Reilly Media (2012)
12. Noble, J., Taivalsaari, A., Moore, I. (eds.): Prototype-Based Programming: Concepts, Languages and Applications, 1st edn. Springer (1999)
13. Fehling, C., Leymann, F., Retter, R., Schupeck,W., Arbitter, P. (eds.): Cloud Computing Patterns: Fundamentals to Design, Build, and Manage Cloud Applications. Springer, Wien (2014)
14. Mahmood, Z., Hill, R. (eds.): Cloud Computing for Enterprise Architectures. Springer, London Limited (2011)
15. Semenov, A.S.: Simulation Self-Organised Evolution Processes: Fractoid-Oriented Approach. Moscow Aviation Institute press, Moscow (in Russia) (2013) Please make this Ref. as [16]
16. Warnecke, H.J.: The Fractal Company: A Revolution in Corporate Culture. Springer, Berlin (1993) Please make this Ref. as [15]
17. Peitgen, H., Jurgens, H., Saupe, D.: Chaos and Fractals. New Frontiers of Science. Springer, New York (2004)
18. Semenov, A.S.: Fractal Petri nets. In: 4th International Conference Control, Decision and Information Technologies, pp. 1174–1179. Barcelona, Spain (2017)
19. Semenov, A.S.: Pattern-type reachability analysis of distributed systems based on fractal Petri nets. In: 5th International Conference Control, Decision and Information Technologies, pp. 346–351. Thessaloniki, Greece (2018) Please make this Ref. as [20]
20. Semenov, A.S.: Pattern recognition technique for synthesis fractal Petri nets. In: 6th International Conference Control, Decision and Information Technologies. Paris, France, pp. 1798–1803 (2019) Please make this Ref. as [19]